S0-DJW-473

Springer
Proceedings in Physics 7

Springer Proceedings in Physics

Managing Editor: H. K. V. Lotsch

Springer Proceedings in Physics is a new series dedicated to the publication of conference proceedings. Each volume is produced on the basis of camera-ready manuscripts prepared by conference contributors. In this way, publication can be achieved very soon after the conference and costs are kept low; the quality of visual presentation is, nevertheless, very high. We believe that such a series is preferable to the method of publishing conference proceedings in journals, where the typesetting requires time and considerable expense, and results in a longer publication period. Springer Proceedings in Physics can be considered as a journal in every other way: it should be cited in publications of research papers as *Springer Proc. Phys.*, followed by the respective volume number, page number and year.

Nonlinear Optics: Materials and Devices

Proceedings of the International School of Materials Science and Technology, Erice, Sicily July 1–14, 1985

Editors: C. Flytzanis and J. L. Oudar

With 185 Figures

Springer-Verlag
Berlin Heidelberg New York Tokyo

Professor Christos Flytzanis
Laboratoire d'Optique Quantique du Centre National de la Recherche Scientifique, Ecole Polytechnique. F-91128 Palaiseau Cedex, France

Dr. Jean Louis Oudar
C.N.E.T., 196, rue de Paris, F-92220 Bagneux, France

ISBN 3-540-16260-7 Springer-Verlag Berlin Heidelberg New York Tokyo
ISBN 0-387-16260-7 Springer-Verlag New York Heidelberg Berlin Tokyo

Offset printing: Weihert-Druck GmbH, 6100 Darmstadt
Bookbinding: J. Schäffer OHG, 6718 Grünstadt
2153/3150-543210

Preface

The field of nonlinear optics has witnessed a tremendous evolution since its beginnings in the early sixties. Its frontiers have been extended in many directions and its techniques have intruded upon many areas of both fundamental and practical interest. The field itself has been enriched with many new phenomena and concepts that have further extended its scope and strengthened its connection with other areas.

As a consequence, it is becoming increasingly unrealistic to expect to cover the different facets and trends of this field in the lectures or proceedings of a summer school, however advanced these may be. However much of the current progress and interest in this field springs to a large extent from the promise and expectation that highly performing all-optical devices that exploit and operate on the principles of nonlinear optics will constitute an important branch of future technology and will provide new alternatives in information processing and transmission. The conception of new devices, in general, requires an intricate and bold combination of facts and methods from most diverse fields, in order to perform functions and operations that fit into an overall technological ensemble. In the case of nonlinear optical devices, in addition one is faced with the double difficulty that such a technological ensemble has not been fully shaped yet and, while waiting for it to come, the isolated nonlinear optical devices that see the light in the laboratories quite often have to be compared with the well-established and highly sophisticated electronic ones. This is due to a large extent to the tremendous impact that the achievements of modern electronics have had in the undergraduate and graduate training of scientists and engineers, and material scientists in particular. In many instances, however, it is becoming clear that the electronic devices can provide only complicated or transient solutions, and in order for the nonlinear optical ones and the light wave technology in general to make its full impact and cease to be considered as a substitute for electronics, the above attitude has to be reversed in many respects.

The International School of Materials Science and Technology course on Nonlinear Optics: Materials and Devices, held in Erice, Sicily, in July 1985, was organized with this in mind. It also aimed to provide yound and experienced engineers and scientists in physics and materials science with an overall view of nonlinear optics that is relevant to the conception of nonlinear optical devices, the function and operation of the latter and in particular the very close interconnection between their performances and the fundamental properties of the underlying nonlinear optical material. A solid general background in nonlinear optics, solid-state physics, electromagnetism and quantum mechanics was necessary for the participants to benefit from the course. The key themes of the school were the description and understanding of the nonlinear interactions of optical beams and short light pulses in reduced space geometries and shapes appropriate to small-size all-optical devices and the response of the nonlinear material in such reduced time and space extensions. They were organized into four groups of lecture series.

A group of three lecture series (Haus, Stegeman, Chemla) was devoted to the nonlinear optics in guided structures and quantum wells. The very intricate problem of nonlinear optical interactions in different restricted geometries was extensively discussed in these lectures and illustrated with the most recent progress in all optical nonlinear device technology related to optical signal processing and transmission and optical computing. Along with the propagation problem, in these lectures the modification of the nonlinear optical properties and dynamics of the material in the reduced space was also discussed and exploited for the conception of new devices.

In a second group of three lecture series (Tang, Oudar, Göbel), the ultrafast dynamical processes that may occur in photoexcited semiconductor crystals and devices was presented. Particular emphasis was put on the production and use of very short optical pulses (in the range of a few picoseconds to a few femtoseconds) to study processes like the rise and fall of the photocarrier response in these materials. These results bring to light some fundamental and still unresolved problems in charge dynamics in condensed matter that are presently receiving great attention.

In a third group of lecture series (Meredith, Huignard, Nayar, Ricard) the very sensitive problem of the nonlinear optical materials was tackled and some new trends in the research for new materials were sketched. The point of view adopted here was to go beyond the well-known classes of inorganic materials, like ferroelectrics and semiconductors, and propose new ones or ways to enhance the optical nonlinearities of the second or third order in small-size materials.

In the last group of lecture series (Wherrett, Grun, Flytzanis) a particular class of optical devices was singled out and discussed in some detail, namely the optical bistable device, which besides its potential use for information processing and optical computing, provides the appropriate ground to study some very interesting aspects of intrinsically nonlinear optical phenomena and instabilities similar to the ones that occur in other areas of physics and chemistry and are intensely studied there.

Although intended as lectures for a summer school, the present notes constitute at the same time reports on the most recent status of the areas that are covered there, and all authors made a particular effort to meet these two goals in a most exemplary way. At the same time, they introduced connections between the different lecture series so that a rather coherent ensemble emerged, which, it is hoped, will be of use for scientists and engineers of different horizons interested in this field.

Palaiseau, France, January 1986 *C. Flytzanis J.L. Oudar*

Contents

Part IV **Optical Bistability and Instabilities in Nonlinear Optical Devices**

Part I

Nonlinear Optics in Guide Structures

Nonlinear Optical Waveguide Devices

H.A. Haus

Department of Electrical Engineering and Computer Science and
Research Laboratory of Electronics, Massachusetts Institute of Technology,
Cambridge, MA 02139, USA

Introduction

Most Optical Waveguide devices operate by controlling the reactive component of the susceptibility $Re(\chi)$, rather than the resistive component $Im(\chi)$. There is a good reason for this, at least at this stage of nonlinear-optical-material development. Any nonlinear loss component has associated with it a linear loss component which produces unavoidable absorption. More seriously, the loss is difficult to control precisely so it varies from device fabrication to device fabrication. Reactive components do not have loss, in principle, the wave need not experience attenuation. Further, fabrication errors in the construction of reactive waveguide devices can be usually compensated by bias adjustment, as we shall see in the course of these lectures.

Once one decides to use reactive components that are modulated by electric fields or by an optical intensity, one must face the fact that medium nonlinearities are weak. For this reason one needs extended regions of propagation of the waves, if the control is applied onto the wave-propagation constant (phase velocity) and the control is to affect the wave. A single mode optical fiber is a special case of an optical waveguide. If employed for nonlinear optics, one uses the $\chi^{(3)}$ coefficient of the fiber, which is small even when compared with the $\chi^{(3)}$ coefficients of optical crystals such as GaAs. Yet the low-loss, single-mode characteristics of quartz fibers (0.18 db/km) are so extraordinary that nonlinear interactions in fibers are most beautifully realized. Soliton propagation in fibers is one example of a nonlinear optical process made observable, and in fact useful [1]. The fiber propagates only one mode with very small associated loss (and a polarization-preserving fiber maintains one polarization of one mode) and realizes physically the mathematical idealization implicit in the one-dimensional nonlinear Schroedinger equation, which predicts solitons [1,2].

All optical waveguide devices with which we shall be concerned operate as phase (velocity) modulators for the reasons given above. If amplitude modulation is desired, then one must provide for a system that transforms phase modulation into amplitude modulation.

There are two main methods of transforming a phase modulation into an amplitude modulation. One is by means of waveguide couplers, the other by means of interferometers. The waveguide couplers are analyzed by means of coupling-of-modes theory. We begin with a study of the theory and illustrate the operation of couplers graphically in terms of the eigensolutions of the coupled modes. We present a simple theory of the $\Delta\beta$ coupler of Kogelnik and Schmidt [3] which permits bias adjustments for the compensation of fabrication errors. Then we look at the waveguide interferometer operation with applied bias fields. Next, we study a modification of the interferometer that converts it into a multiplexer-demultiplexer. Specific devices are shown that have been fabricated in three laboratories.

Next we take up a modification of the interferometer in which the phase modulation is produced optically. Such structures have the potential of very high rate modulation. An all-optical waveguide modulator is described and its potential use as an all-optical XOR gate is brought out. Modifications of the structure can operate as any one of the basic logic gates.

Finally, we describe three possible uses of one, or a few, all-optical logic gates. One is the encoding and decoding of a low-rate signal with a high-rate pseudorandom binary number (pulse) stream. Another use is in a photon number measurement of a wavepacket, without affecting the number of photons in the wavepacket. The third possible use is as a lossfree tap of a random pulse stream that determines the presence or absence of pulses.

1. Coupling of Modes[1]

A change of the dielectric constant can be produced by either an applied electric field (via $\chi^{(2)}$) or by the optical field itself (via $\chi^{(3)}$). A wave propagating in a medium of time-variable dielectric constant acquires phase modulation and thus a changed spectrum; the intensity is redistributed in the frequency domain, without a change of total intensity. Usually we are interested in intensity modulation. A switch is an intensity modulator and so is a multiplexer or demultiplexer. The problem then is to transform a phase modulation into an amplitude modulation. This is most conveniently accomplished, in a waveguide configuration, by coupled waveguides or by a waveguide interferometer. The analysis of coupled waveguides requires the theory of mode coupling.

Consider for example, two optical waveguides that are coupled to each other via their fringing fields (see Fig. 1.1). A wave set up initially in one guide is transferred to the other guide. Because the transfer can be controlled electrically, this mechanism can be used for switching of guided optical radiation. Another coupling is effected between forward and backward waves by periodic perturbations on an optical waveguide. These perturbations can be produced in optical waveguides by integrated-optics fabrication methods and can be used to build the equivalent of a mirror into an optical waveguide without interrupting the waveguide physically. Nonlinear optical phenomena couple waves at different frequencies. This process is also akin to the coupling of modes analysis presented here.

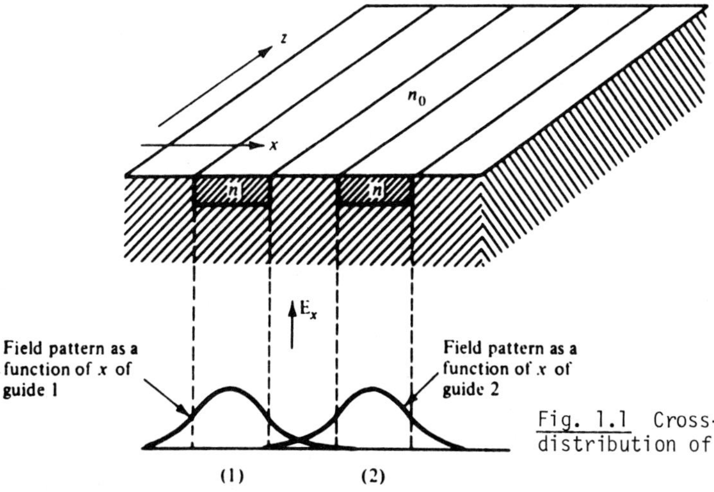

Fig. 1.1 Cross-sectional view of index distribution of two coupled guides

[1]The term "coupling of modes" is deeply ingrained in the literature and can be traced back to J.R. Pierce's original work on the subject [4]. Actually, in coupling of modes in space, waves are coupled, each mode of a waveguide consisting of a forward and backward wave.

Consider two waves a_1 and a_2, or modes 1 and 2 that in the absence of coupling, have propagation constants β_1 and β_2. They obey the equations

$$\frac{da_1}{dz} = i\beta_1 a_1 \tag{1.1}$$

$$\frac{da_2}{dz} = i\beta_2 a_2. \tag{1.2}$$

Suppose next that the two waves are weakly coupled by some means, so that a_1 is affected by a_2 and a_2 is affected by a_1. Then the equations become

$$\frac{da_1}{dz} = i\beta_1 a_1 + \kappa_{12} a_2 \tag{1.3}$$

$$\frac{da_2}{dz} = i\beta_2 a_2 + \kappa_{21} a_1. \tag{1.4}$$

If power is to be conserved, there are restrictions imposed on κ_{12} and κ_{21}. Weak coupling implies that we may evaluate the power in the two waves disregarding the coupling. We normalize a_1 and a_2 so that the power in the modes is $|a_1|^2$ and $|a_2|^2$. Because the waves may carry power in opposite directions, we must distinguish the directions of power flow by a sign. We define $p_{1,2} = 1$ depending upon whether the power flow is in the plus or minus z direction. The net power P is

$$P = p_1|a_1|^2 + p_2|a_2|^2. \tag{1.5}$$

Power conservation requires that the power be independent of distance z.

$$\frac{dP}{dz} = p_1 \frac{d|a_1|^2}{dz} + p_2 \frac{d|a_2|^2}{dz} = 0 \tag{1.6}$$

Multiplication of (1.3) by $a_1{}^*$, (1.4) by $a_2{}^*$, addition of the two equations and their complex conjugate gives: $p_1 a_1{}^* a_2 \kappa_{12} + p_2 a_2{}^* a_1 \kappa_{21} + p_1 a_1 a_2{}^* \kappa_{12}^* + p_2 a_2 a_1{}^* \kappa_{21}^* = 0$. Because a_1 and a_2 can be adjusted arbitrarily, it follows that

$$p_1 \kappa_{12} + p_2 \kappa_{21}^* = 0. \tag{1.7}$$

The determinantal equation for an assumed $\exp(i\beta z)$ dependence is, from (1.3) and (1.4)

$$(\beta - \beta_1)(\beta - \beta_2) + \kappa_{12}\kappa_{21} = 0 \tag{1.8}$$

with the solution

$$\beta = \frac{\beta_1 + \beta_2}{2} \pm \sqrt{\left(\frac{\beta_1 - \beta_2}{2}\right)^2 - \kappa_{12}\kappa_{21}}. \tag{1.9}$$

For waves carrying power in the same direction, $p_1 p_2 = +1$, $\kappa_{12}\kappa_{21} = -|\kappa_{12}|^2$, and β is always real. For $p_1 p_2 = -1$ (i.e., waves carrying power in opposite directions), $\kappa_{12}\kappa_{21} = |\kappa_{12}|^2$ and β is complex for

$$\left|\frac{\beta_1 - \beta_2}{2}\right| < |\kappa_{12}|.$$

Note that appreciable coupling can occur only if $|\beta_1 - \beta_2|$ is of order $|\kappa_{12}|$, which is small compared with $|\beta_1|$ and $|\beta_2|$ (weak-coupling assumption). Thus, $\beta_1 \simeq \beta_2$ and the phase velocities of the two waves must be of the same sign. Nonetheless, the power flow can be in opposite directions $(p_1 p_2 = -1)$ if the group velocities are in opposite directions. This is the case of coupling of forward and backward waves by a periodic structure (not discussed in these lectures).

Suppose that both β_1 and β_2 depend on a parameter V, such as applied electrode voltage, with opposite variations, β_2 increasing with increasing V, β_1 decreasing with increasing V. For each V there is a pair of values β_1 and β_2 which yield the propagation constant β of the coupled modes via (1.9). If both waves have the same direction of power flow as shown in Fig. 1.2a then $\beta = (\beta_1 + \beta_2)/2 \pm |\kappa_{12}|$ at the crossing point of β_1 and β_2. The propagation constants β approach asymptotically the unperturbed propagation constants far from the crossover.

Another case is the one of opposite direction of power flow as shown in Fig. 1.2b. A crossing of two such curves leads to exponentially growing and decaying solutions.

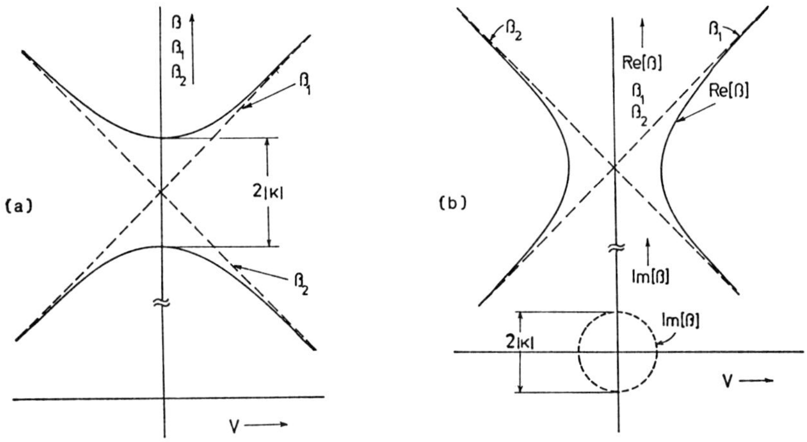

Fig. 1.2 Dispersion diagram for coupling of modes (as function of $\beta_1 - \beta_2$ upon V is assumed)

Consider the case of codirectional, power flows, $p_1 = p_2 = +1$. Suppose that a wave $a_1(0)$ is launched at $z = 0$. Then the solutions of (1.3) and (1.4) are:

$$a_1(z) = a_1(0)\left[\cos \beta_0 z - i\,\frac{\beta_1 - \beta_2}{2\beta_0}\sin \beta_0 z\right] \cdot e^{i[(\beta_1+\beta_2)/2]z} \tag{1.10}$$

$$a_2(z) = \frac{\kappa_{21}}{\beta_0}\, a_1(0) \sin \beta_0 z \cdot e^{i[(\beta_1+\beta_2)/2]z} \tag{1.11}$$

where

$$\beta_0 = \sqrt{\left(\frac{\beta_1 - \beta_2}{2}\right)^2 + |\kappa_{12}|^2}. \tag{1.12}$$

The solutions (1.10) and (1.11) underlie the operation of optical couplers and switches.

Let us consider the case $\beta_2 = \beta_1$ in greater detail, the two waveguides are "synchronous". It will be convenient to suppress the subscripts 12 and 21 on the κ's. For this purpose we note that one may pick phase references for a_1 and a_2 such that κ_{12} is pure imaginary, $\kappa_{12} = -i\kappa$. Then, with $p_1 = p_2 = 1$ we have $\kappa_{21} = -i\kappa$. We shall see later in section 4 that this phase reference is the "natural" one i.e. corresponds to a common phase plane for a_1 and a_2. We then see from (1.10) and (1.11) that

$$a_1(z) = a_1(0) \cos \kappa z \tag{1.13}$$

$$a_2(z) = -ia_1(0) \sin \kappa z. \tag{1.14}$$

This behavior can also be interpreted geometrically. To arrive at the geometric interpretation we note that when the two waveguides are synchronous, the composite system has symmetric and antisymmetric solutions as shown in Fig. 1.3, that travel at different phase velocities. This can be seen from (1.9), (1.3) and (1.4). One has, for $\beta_1 = \beta_2$

$$\frac{a_1}{a_2} = \mp 1 \tag{1.15}$$

for the solutions

$$\beta_0 = \pm\kappa. \tag{1.16}$$

The symmetric solution has a smaller propagation constant than the antisymmetric solution. If only waveguide (1) is excited at $z = 0$, then the solution is made up of equal contributions of the symmetric and antisymmetric solutions (see Fig. 1.4a). After a relative phase shift of $2\beta_0 \ell = 2\kappa\ell = \pi$, the solutions that subtracted in guide (2) at $z = 0$ now add in the guide, whereas they subtract in guide (1). The excitation has been transferred to guide (2) as shown in Fig. 1.4b.

Another case of interest is the case of coupling over a length $2\beta_0 \ell = 2\kappa\ell = \pi/2$. In this case the solutions are shifted by 90°. The excitations add to equal amplitudes and a 90° phase shift as shown in Fig. 1.4c. The coupler acts like a "half-silvered mirror". Thus, depending upon the length of the interaction region, one attains either complete transfer, or 3 db transfer (and, of course, any transfer in between).

Suppose now that the propagation constants are made different. This can be done by the application of an electric field, utilizing $\chi^{(2)}$, as shown in Fig. 1.5. The applied control-field components along y are opposite in the two guides, and thus β of the TE-wave polarized along x is shifted in opposite directions via $\chi^{(2)}_{xxy}$, $\beta_1 - \beta_2 \neq 0$. The eigenvectors are now from (1.3).

$$\frac{a_2}{a_1} = -\frac{\beta - \beta_1}{\kappa} = -\frac{\beta_2 - \beta_1}{2\kappa} \mp \sqrt{\frac{(\beta_2 - \beta_1)^2}{(2\kappa)^2} + 1} \tag{1.19}$$

It is of interest to study the solution (1.19) in the limit when β_1 and β_2 are very different (very weak interaction), when $|\beta_1 - \beta_2| >> |\kappa|$. In this limit $|a_2| << |a_1|$ and β approaches β_1, when $\beta_2 > \beta_1$ for the solution with the upper sign in (1.19), the solution henceforth denoted by +. The solution approaches that of the uncoupled case with guide (1) excited. For the same choice of sign, but with $\beta_2 < \beta_1$, $|a_2| >> |a_1|$, and β approaches β_2. The solution approaches the uncoupled case with guide (2) excited. These cases, and those with the lower sign in (1.19) are summarized in Table 1.1 for the general case $\beta_1 \neq \beta_2$.

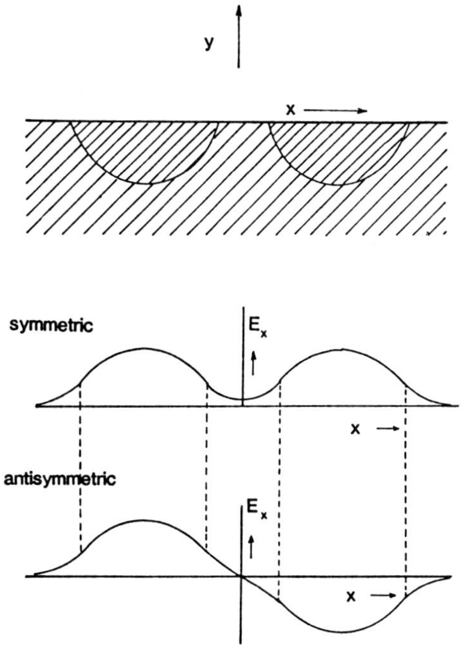

Fig. 1.3 The symmetric and antisymmetric field distributions.

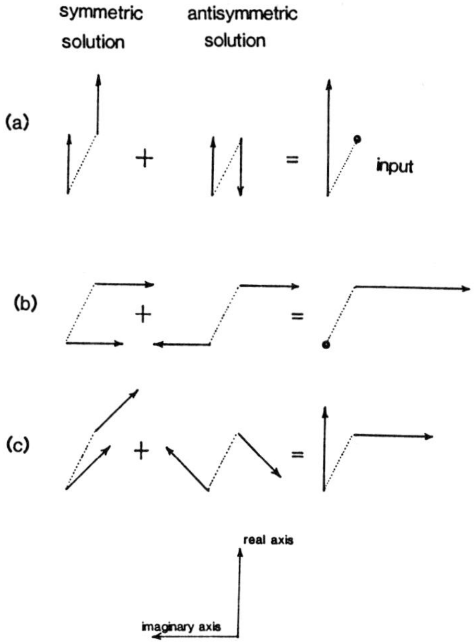

Fig. 1.4 Projective representation of the two eigenvectors in their complex planes. The planes are displaced for better viewing.

Table 1.1 The amplitudes of the eigenvectors

	$\beta_1 < \beta_2$	$\beta_1 > \beta_2$
+ solution	$\lvert a_2\rvert > \lvert a_1\rvert$ "antisymmetric" 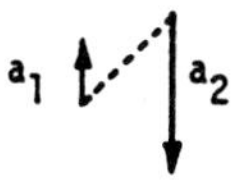 $\lvert a_1\rvert = x\lvert a_2\rvert$	$\lvert a_2\rvert > \lvert a_2\rvert$ "antisymmetric" 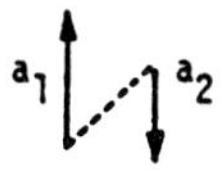 $\lvert a_2\rvert = x\lvert a_1\rvert$
- solution	$\lvert a_2\rvert < \lvert a_1\rvert$ "symmetric" 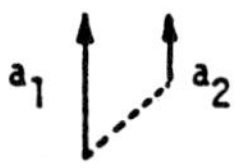 $\lvert a_2\rvert = x\lvert a_1\rvert$	$\lvert a_2\rvert > \lvert a_1\rvert$ "symmetric"  $\lvert a_1\rvert = x\lvert a_2\rvert$
	$x \equiv \sqrt{\left\lvert\frac{\beta_1 - \beta_2}{2\kappa}\right\rvert^2 + 1} - \left\lvert\frac{\beta_1 - \beta_2}{2\kappa}\right\rvert$	

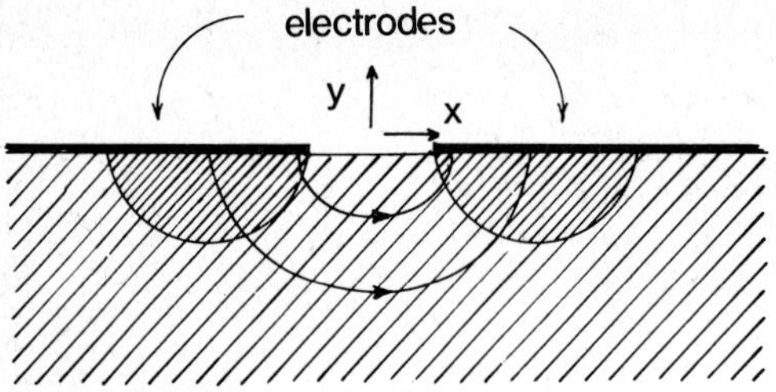

Fig. 1.5 Electric field produced by electrodes on top of guides.

If now an excitation is imposed at $z = 0$ in guide (2), then the input conditions appear as in (1.10). For a phase shift $\beta_0 z = \pi$, the power returns into guide (1). Note that the condition for full transfer at synchronism, $\beta_1 = \beta_2$, gives

$$\kappa\ell = \pi/2. \tag{1.20}$$

The condition $\beta_0\ell = \pi$ gives

$$\sqrt{\left(\frac{\beta_1 - \beta_2}{2}\right)^2 \ell^2 + \kappa^2\ell^2} = \pi$$

or

$$\frac{\beta_1 - \beta_2}{2}\ell = \sqrt{\pi^2 - \kappa^2\ell^2} = \sqrt{\pi^2 - \frac{\pi^2}{4}} = \sqrt{3/4}\,\pi. \tag{1.21}$$

Thus, a detuning of the right amount can return the power into guide (1). The system acts as a switch [5]. For zero-applied voltage, there is complete transfer to guide (2); for an appropriate applied voltage, there is no transfer.

2. Fabrication

Most commonly, waveguides are constructed in $LiNbO_3$ (which has a very large electrooptic effect) by Ti indiffusion. Where a waveguide is to be formed, a thin layer of Ti is deposited by photolithographic means on the $LiNbO_3$ crystal surface (Fig. 2.1a).

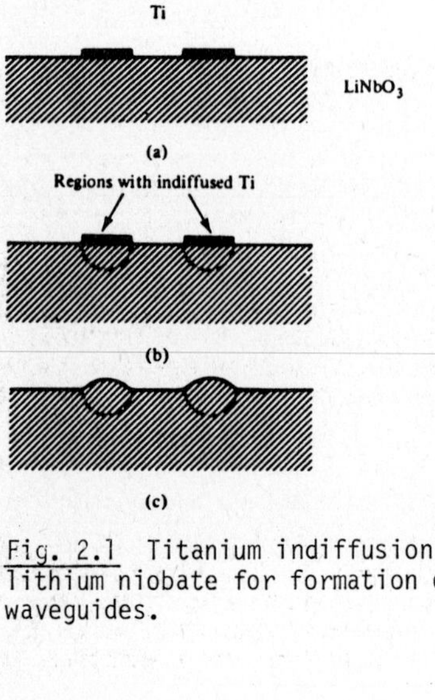

Fig. 2.1 Titanium indiffusion in lithium niobate for formation of waveguides.

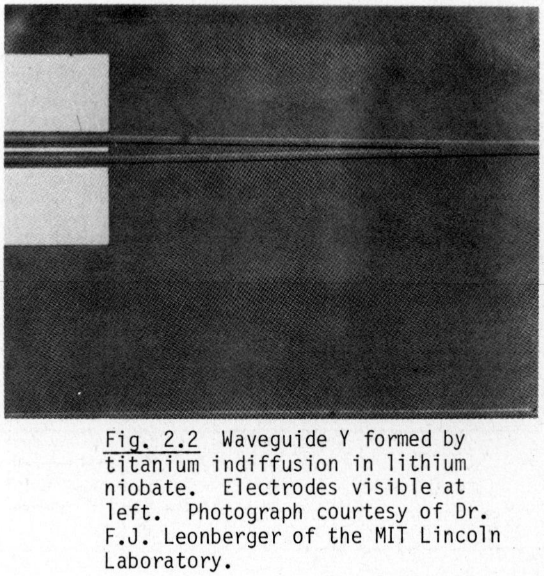

Fig. 2.2 Waveguide Y formed by titanium indiffusion in lithium niobate. Electrodes visible at left. Photograph courtesy of Dr. F.J. Leonberger of the MIT Lincoln Laboratory.

The structure is heated up, which causes the Ti to diffuse into the $LiNbO_3$, producing cylindrical volumes with Ti interspersed in the $LiNbO_3$ structure (Fig. 2.1b). These regions have an index higher than the surrounding pure $LiNbO_3$ and form guides completely analogous to the fiber and slab guides discussed in the preceding section. The indiffusion causes local expansions which cause the surface to bulge. These can be observed under a microscope (Fig. 2.2). Finally, electrodes are deposited on top of the crystal surface.

3. The $\Delta\beta$ Coupler

A practical switch must achieve full transfer and extinction ratios better than 20 dB. Full transfer occurs only when $\kappa\ell = \pi/2$. Because the microstructure tolerances are difficult to control and κ is a strong function of guide spacing (see Section 4), it is desirable to build into the structure flexibility, so that adjustments can be made after fabrication. The $\Delta\beta$ coupler proposed by Kogelnik and Schmidt [3] is such a structure. It consists of two couplers back to back, with electrodes driven "push-pull" as shown in Fig. 3.1. Even when $|\kappa|\ell \neq \pi/2$, but slightly larger, full transfer can be effected by an applied voltage of appropriate amplitude.

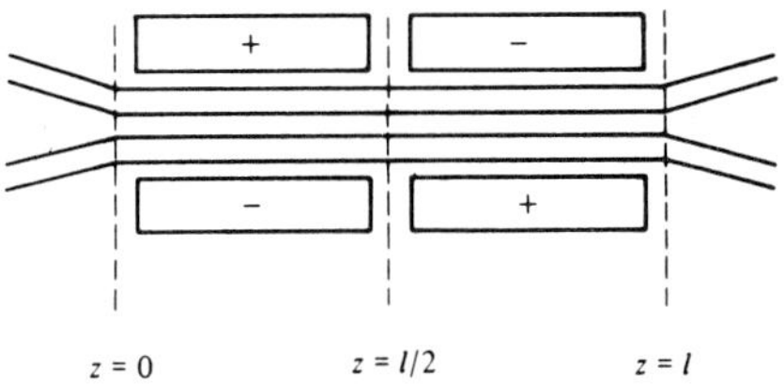

Fig. 3.1 Schematic of coupler

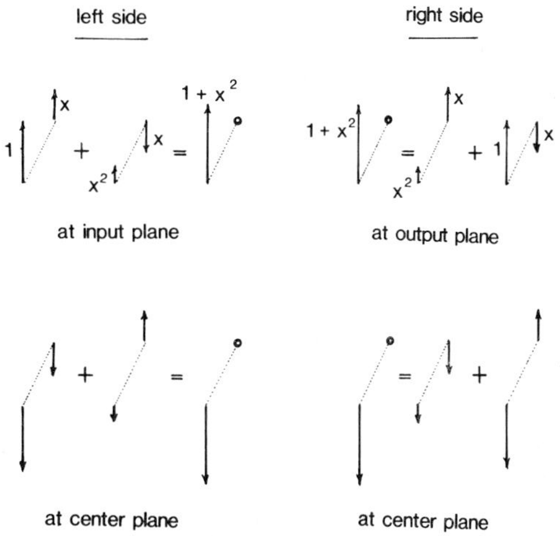

Fig. 3.2 $\Delta\beta$-coupler; case of no transfer

In order to understand the operation of the $\Delta\beta$ switch, we can make use of the same kind of diagram as Fig. 1.4, except that we need now two such diagrams, back to back, one for the first section and one for the second section. Figures 3.2 and 3.3 summarize the eigenvector properties in the two sections. Consider first the case of no transfer. Assume $\beta_1 < \beta_2$ on the left, $\beta_1 > \beta_2$ on the right. The two diagrams with zero excitation in guide (2) at both input and output must meet in the center with the same amplitudes and relative phases. From (1.19) we gather that the ratio of amplitudes in guide (2) to guide (1) is equal to the quantity x defined in Table 1.1 for the symmetric solution on the left-hand side, the antisymmetric solution on the right-hand side. The symmetric and antisymmetric solutions add on each side as shown in Fig. 3.2. No transfer occurs, if the two excitations meet in the center after a net phase shift of $m\pi$ for each of the solutions in each of the half-sections:

$$\beta_o\ell = m\pi. \tag{3.1}$$

Next, consider the case of transfer from guide (1) to guide (2). Here the reversal of phase-velocity-shifts and of guide-roles leads to behaviors on the two sides that are phase-reversed mirror images of each other around the symmetry plane of the two guides. The relative phases of the excitations of modes a_1 and a_2 at the center are the same on each side, as is evident from the construction of Fig. 3.3.

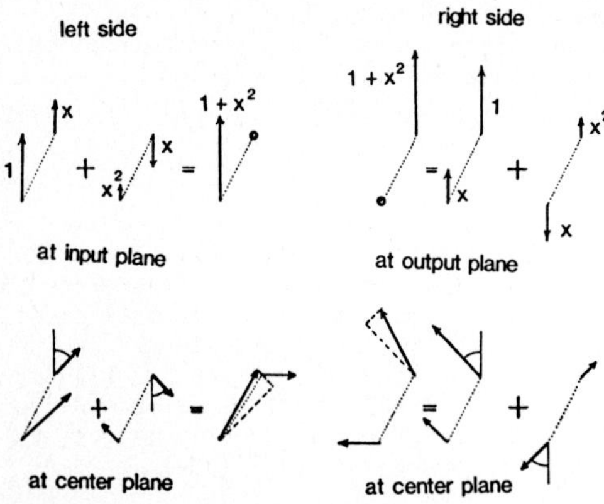

Fig. 3.3 $\Delta\beta$-coupler; case of full transfer. Magnitudes of vectors at center have to be made equal. This picks x, and hence the detuning.

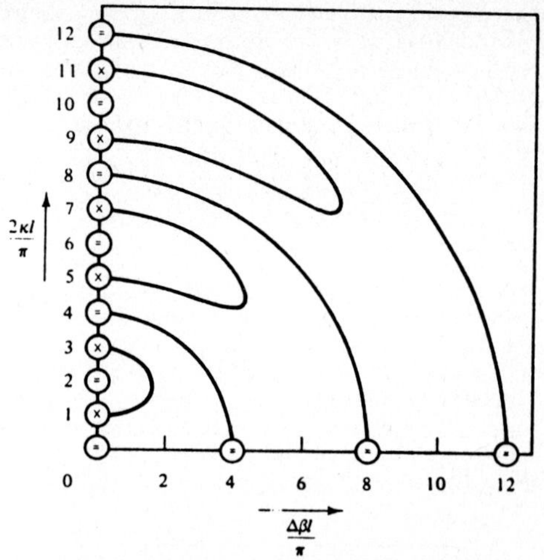

Fig. 3.4 Switching diagram for a switched coupler with two sections of alternating $\Delta\beta$. The ⊗ sign marks the cross-state conditions and the ⊜ sign marks the parallel-state conditions. (From H. Kogelnik and R. V. Schmidt, IEEE J. Quant. Electron. QE-12, 396-401 July 1976 copyright 1976 IEEE)

If the solution be a valid one, then the magnitudes of the net excitation vectors must match at the center. This will be the case if, and only if, the amplitude of the excitation in guide (1) is equal to the amplitude of excitation in guide (2). From (1.10) we find the condition

$$\cos^2 \frac{\beta_o \ell}{2} + \left(\frac{\beta_2 - \beta_1}{2\beta_o}\right)^2 \sin^2 \frac{\beta_o \ell}{2} = \frac{\kappa^2}{\beta_o^{\,2}} \sin^2 \frac{\beta_o \ell}{2}$$

or

$$\cot \frac{\beta_o \ell}{2} = \sqrt{\frac{\kappa^2 - \left(\frac{\beta_2 - \beta_1}{2}\right)^2}{\kappa^2 + \left(\frac{\beta_2 - \beta_1}{2}\right)^2}} \,. \tag{3.2}$$

The conditions for no transfer, (3.2), and full transfer, (3.3), have been plotted by Kogelnik and Schmidt and are shown in Fig. 3.4.

4. The Coupling Coefficient

The preceding analysis of coupling of modes in space assumed knowledge of the coupling coefficient $\kappa_{12}(= \pm\kappa_{21}^*)$. Here we show one way of evaluating it. Denote the normalized field pattern of waveguide 1 by $\bar{e}_1(x, y)$, that of waveguide 2 by $\bar{e}_2(x, y)$ (see Fig. 4.1). The total field in both waveguides is then, by assumption, the superposition of the two field patterns (compare Fig. 1.1):

$$\bar{E}(x, y, z) = a_1(z)\, \bar{e}_1(x, y) + a_2(z)\, \bar{e}_2(x, y). \tag{4.1}$$

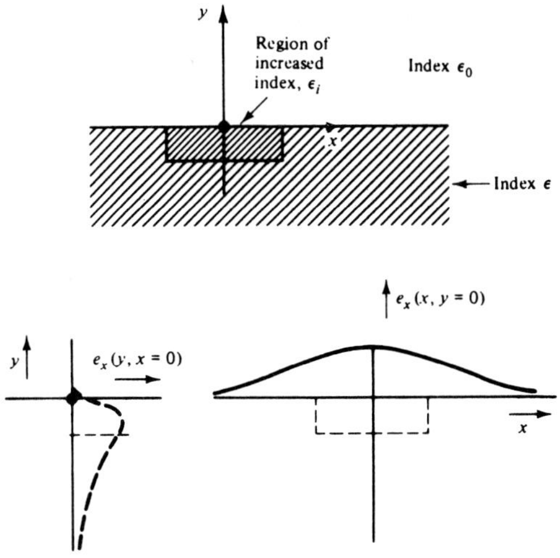

Fig. 4.1 Typical field pattern of TE mode

The field $a_1\bar{e}_1$ is, by definition, the field in the absence of waveguide 2; in other words, the dielectric constant increase that produces the waveguide is "thought away". Figure 4.1 shows a sketch of a typical field pattern. The power transferred from waveguide 1 to waveguide 2 is caused by the polarization current $j\omega P_{21}$ produced in waveguide 2 by the field of waveguide 1:

$$-i\omega\bar{P}_{21} = -i\omega(\varepsilon_i - \varepsilon)\, a_1\bar{e}_1(x, y). \tag{4.2}$$

Note that only $(\varepsilon_i - \varepsilon)$ appears, because the polarization current is equal to $-i\omega\varepsilon a_1\bar{e}_1$ in the absence of guide 2 and must be subtracted. The power transferred is

$$\begin{aligned} &\frac{1}{4}\left[\int_{\substack{\text{cross section}\\ \text{of guide 2}}} \bar{E}_2^* \cdot (i\omega\bar{P}_{21})\,da + \text{c.c.}\right] \\ &= \frac{1}{4}\left[i\omega a_1 a_2^* \int_{\substack{\text{cross section}\\ \text{of guide 2}}} da(\varepsilon_i - \varepsilon)\, \bar{e}_1 \cdot \bar{e}_2^* + \text{c.c.}\right] \end{aligned} \tag{4.3}$$

From coupling of modes we know that the power transfer is

$$\frac{d|a_2|^2}{dz} = \kappa_{21} a_1 a_2^* + \kappa_{21}^* a_1^* a_2. \tag{4.4}$$

Comparison of (4.3) and (4.4) gives

$$\kappa_{21} = \frac{i\omega}{4}\int_{\substack{\text{cross section}\\ \text{of guide 2}}} da(\varepsilon_i - \varepsilon)\, \bar{e}_1 \cdot \bar{e}_2^*. \tag{4.5}$$

The same approach gives

$$\kappa_{12} = \frac{i\omega}{4}\int_{\text{cross section}} da(\varepsilon_i - \varepsilon)\, \bar{e}_2 \cdot \bar{e}_1^* \tag{4.6}$$

In general, $\kappa_{12} \neq -\kappa_{21}^*$ when symmetry is not maintained. In this case, the power is not correctly given as the sum of the powers in the individual modes.

5. The Waveguide Y and the Mach Zehnder Interferometer

The waveguide coupler is one structure that transforms phase change into amplitude change. The Mach Zehnder interferometer in waveguide realization provides another means. Its construction is shown in Fig. 5.1. A single-mode waveguide separates into two waveguides in the first waveguide Y. After propagation over equal distances in the two separate waveguides, the excitation is recombined in a single-mode output guide. Phase changes can be produced in the two arms which lead to amplitude changes in the output guide.

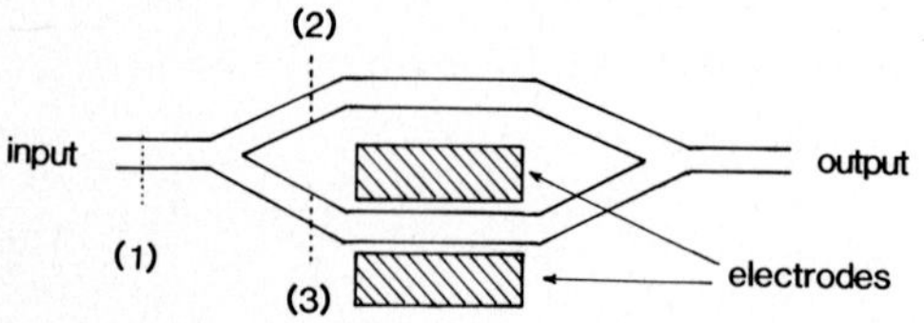

Fig. 5.1 Schematic of Mach-Zehnder waveguide interferometer

The waveguide Y acts as a half-silvered mirror, separating the input signal into excitations of equal amplitudes in the outgoing waveguide pair. This can be understood as follows. An input excitation, symmetric with respect to the symmetry plane, can be gradually transformed into a double-humped symmetric excitation, if the index distribution is "slowly" tapered to transform the mode pattern "adiabatically". If the transition is made sufficiently gradual there need not be any loss associated with the transformation. Losses of Y-junctions have been calculated. Typically, if the angle between the two guides is kept at less than 1 degree, the losses can be kept below 2 percent [6].

Denote the x-component of the TE wave in waveguide (1) by $E_x(1)$, in waveguide (2) by $E_x(2)$. At the output, the incoming waveguide pair has an excitation

$$E_x(1) + e^{i\phi} E_x(2) = e^{i\phi/2}[e^{i\phi/2} E_x(1) - e^{-i\phi/2} E_x(2)]$$

$$= e^{i\phi/2} \{\cos \frac{\phi}{2} [E_x(1) + E_x(2)] - i \sin \frac{\phi}{2} [E_x(1) - E_x(2)]\} \quad (5.1)$$

The excitation consists of two parts, a symmetric part with the coefficient $\cos \frac{\phi}{2}$ and an antisymmetric part with the coefficient $\sin(\phi/2)$. The first can be transformed into the fundamental mode of the single-mode output guide by a slowly tapered transition. The second component becomes the first higher order antisymmetric mode of the output guide, which "leaks" and thus escapes from the structure after a sufficient distance of propagation. Thus the output power is proportional to the input power times $\cos^2(\phi/2)$; a phase modulation has been transformed into an amplitude modulation.

It will be of interest to study the general scattering matrix S of the waveguide Y. If the amplitudes of the incident waves are denoted by a_i (i = 1, 2, 3, see Fig. 5.1) and the outgoing waves by b_i, then

$$b = Sa. \quad (5.2)$$

In fact, we already know some of the elements of the scattering matrix. By proper choice of phase references, for $a_2 = a_3 = 0$

$$b_2 = b_3 = \frac{1}{\sqrt{2}} a_1 \quad (5.3)$$

and

$$b_1 = 0. \tag{5.4}$$

It follows that:

$$S_{11} = 0 \tag{5.5}$$

and

$$S_{21} = S_{31} = \frac{1}{\sqrt{2}} . \tag{5.6}$$

By reciprocity $S_{ij} = S_{ji}$, the scattering matrix is symmetric. Further, if a wave is fed into guide (2), no output is expected in guide (3), $S_{32} = 0$. Finally, one may argue that $S_{22} = S_{33} = 0$, because waves incident upon the very weak spatial index discontinuities experience no reflection. Thus

$$S = \frac{1}{\sqrt{2}} \begin{pmatrix} 0 & 1 & 1 \\ 1 & 0 & 0 \\ 1 & 0 & 0 \end{pmatrix} . \tag{5.7}$$

One interesting consequence is immediately that a wave incident from guide (2) produces

$$b_1 = \frac{1}{\sqrt{2}} a_2 \qquad b_2 = b_3 = 0$$

and thus half of the power is lost. This is an important consideration in the construction of optical logic gates based on the waveguide interferometer principle. The 3 dB loss can be avoided by a different construction than the one envisaged here. We shall return to this point in Section 8.

The preceding discussion shows that the Mach-Zehnder waveguide interferometer has a response of output power, P_{out}, to input power, P_{in}

$$P_{out} = \cos^2 \frac{\phi}{2} P_{in} \tag{5.8}$$

and if ϕ is made time-dependent, the output is an amplitude-modulated version of the input.

Three coupled synchronous waveguides of propagation constant β_o interacting over an appropriate distance can function as a waveguide Y (see Fig. 5.2). To demonstrate this consider the equations of the coupled modes:

$$\frac{da_1}{dz} = i\beta_o a_1 - i\kappa a_2 \tag{5.9}$$

$$\frac{da_2}{dz} = i\beta_o a_2 - i\kappa a_1 - i\kappa a_3 \tag{5.10}$$

$$\frac{da_3}{dz} = i\beta_o a_3 - i\kappa a_2 . \tag{5.11}$$

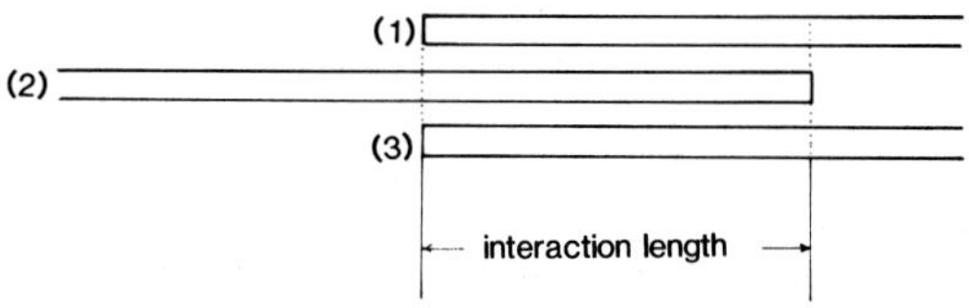

Fig. 5.2 Three-guide coupler functioning as Y.

Note β_o does not have the meaning of (1.12). The solutions of the coupling equations are:

$$\beta = \beta_o \pm \sqrt{2\kappa^2}; \qquad \text{and} \qquad \beta = \beta_o. \tag{5.12}$$

The eigenvectors are shown in Fig. 5.3. An excitation in the center waveguide is made up of the two symmetric solutions. After a propagation distance ℓ with

$$\sqrt{2}\,|\kappa|\ell = \pi/2, \tag{5.13}$$

the excitations in the center guides subtract, and the excitations in the outside guides add. The excitation has been transferred to the outside guides.

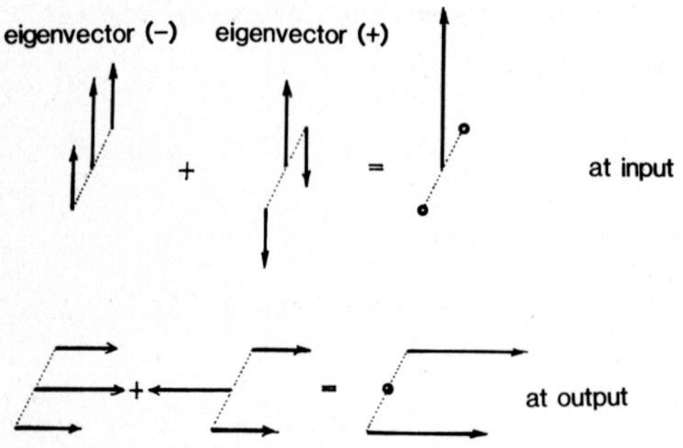

Fig. 5.3 The modes in a three-guide coupler.

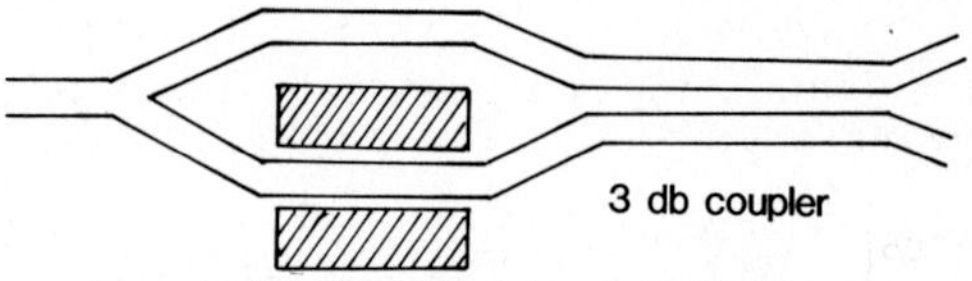

Fig. 5.4 Modified Mach-Zehnder waveguide interferometer.

The Mach Zehnder interferometer can be modified so as to act as a single-pole-double-throw switch that couples the input to either of two output guides. For this purpose, the output Y is replaced by a 3 dB waveguide coupler (Fig. 5.4). We have seen in Fig. 1.4c that an input into one guide of a 3 dB coupler produces outputs of equal amplitudes and phase-shifted by 90°. Because the system is time-reversible, this solution can be "run backwards": inputs of equal amplitudes and phase-shifted by 90° produce an output in one single guide. A -90° phase-shift produces an output in the other guide. The outputs of the two interferometer arms can be phase-shifted with respect to each other by + or -90° and thus can produce outputs in either one of the two output guides of a 3 dB coupler following the interferometer.

For an arbitrary phase-shift of ϕ one may write down the general response using the solutions of (1.3) and (1.4) for the synchronous case, $\beta_1 = \beta_2$. The output power in guide (1) is

$$P_{out}(1) = \frac{1}{2}[1 - \sin\phi]\, P_{in} \tag{5.14}$$

where P_{in} is the total input power in both guides. The output power in guide (2) is

$$P_{out}(2) = \frac{1}{2}[1 + \sin\phi]\, P_{in}. \tag{5.15}$$

6. Multiplexers and Demultiplexers

Switches that switch the input signal into either of two output guides are demultiplexers. Run "backwards", with input and output interchanged, they are multiplexers.

We have investigated two kinds of switches. One kind was based on coupled waveguides, the other on the interferometer. The power transfer for a waveguide coupler is from (1.11):

$$P_{out}(2) = \frac{|\kappa|^2}{\beta_0^2} \sin^2 \beta_0 \ell \; P_{in}. \qquad (6.1)$$

The transfer as a function of time for a sinusoidal applied voltage is of particular interest, because it is relatively easy to apply a high-frequency sinusoidal voltage, causing $\beta_1 - \beta_2$ to change sinusoidally with time. In order to swing back and forth between full transfer to the other guide, and no transfer, $\beta_0 \ell$ has to change between $\pi/2$ and π according to (1.12). The time-dependence of this power transfer is shown in Fig. 6.1. The complement of the power transfer is the time-dependence of the power in guide (1). It is clear from Fig. 6.1 that the time-dependences of the powers in the two guides are different. This may not be of importance in the demultiplexing of short pulses, but for pulses that are long enough to cause "cross-talk" during the switching the asymmetry is undesirable.

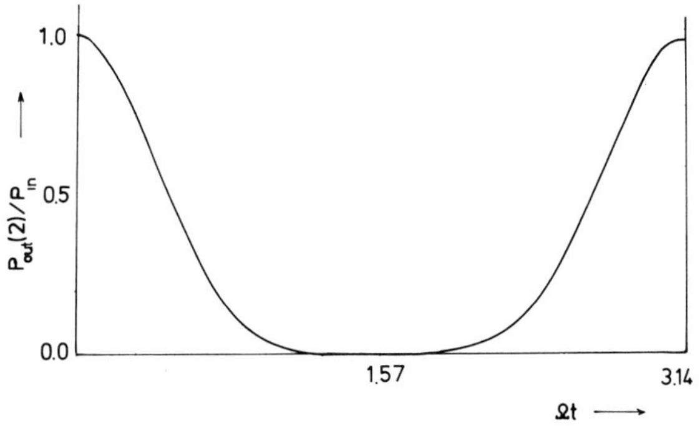

Fig. 6.1 Transfer function for coupler

The Mach Zehnder version of the demultiplexer has the transfer function (5.8) with $\phi = (\pi/2) \sin \Omega t$. The power transfer to the two guides is the same, except of course for the relative time shift.

7. Various Realizations of Switches and Modulators

M. Izutsu and T. Sueta of Osaka University have built a broadband push-pull interferometer-intensity modulator [7] as shown schematically in Fig. 7.1. The modulating

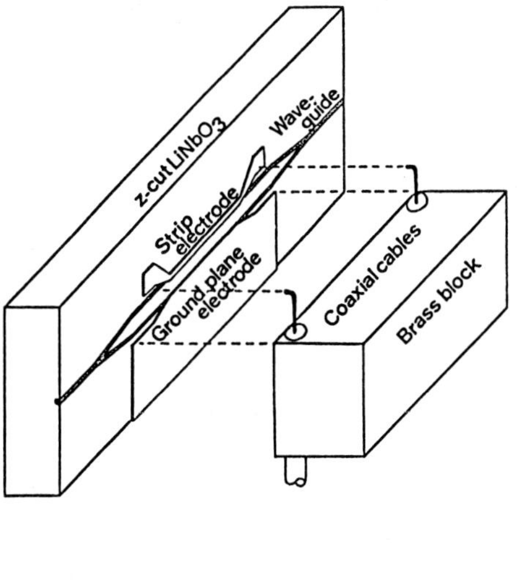

Fig. 7.1 A broadband push-pull interferometer intensity modulator (after M. Izutsu and T. Sueta)

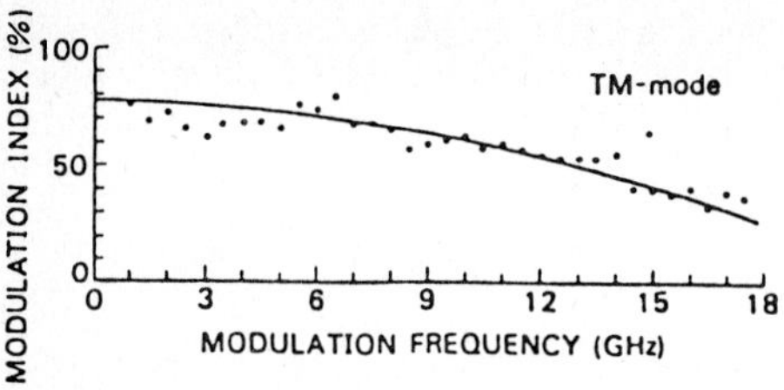

Fig. 7.2 High-frequency modulator characteristic for 100 mw input power (after M. Izutsu and T. Sueta)

fields, normal to the conductor surface on the microwave strip lines, are oppositely directed so that the phase-shift produced in the two optical waveguides are equal and opposite. Fig. 7.2 shows the broadband modulation characteristic of the device with a half-power bandwidth of the order of 18 GHz.

As pointed out earlier, the interferometer can be made into a multiplexer-demultiplexer if the output Y is replaced by a 3 dB coupler. A version of a 3 dB waveguide coupler is the hybrid coupler, which has the advantage of shorter length. It consists of a junction of two symmetric guides with two asymmetric guides as shown in Fig. 7.3. The scattering matrix of this four-port can be made into that of a 3 dB coupler by proper dimensioning of the asymmetric guides. With this compact version of the 3 dB coupler, the Japanese workers constructed a demultiplexer, whose schematic is shown in Fig. 7.4. The low-frequency response at 10 kHz is shown in Fig. 7.5, and the response to a 1 GHz sinusoidal drive is shown in Fig. 7.6.

Our work at MIT was concerned with interferometers driven by resonant microwave structures [8]. Modulation only over a narrow band of frequencies is possible, however, with the advantage of lower drive powers. This offers advantages in multiplexer-demultiplexer applications designed for fixed modulation rates. Fig. 7.7

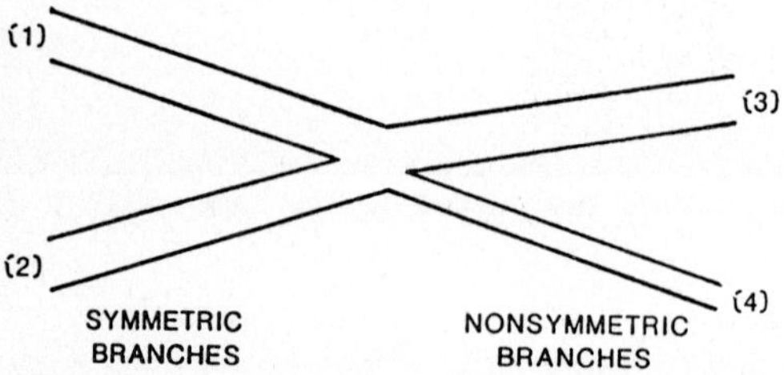

Fig. 7.3 Waveguide coupler equivalent to 3 dB coupler

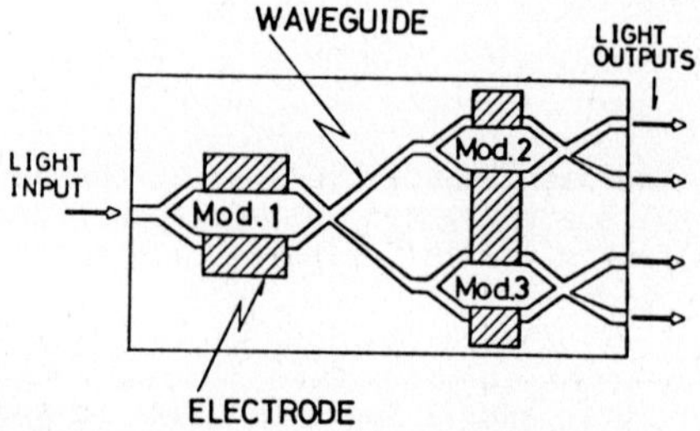

Fig. 7.4 Schematic of multiplexer-demultiplexer (after M. Izutsu and T. Sueta)

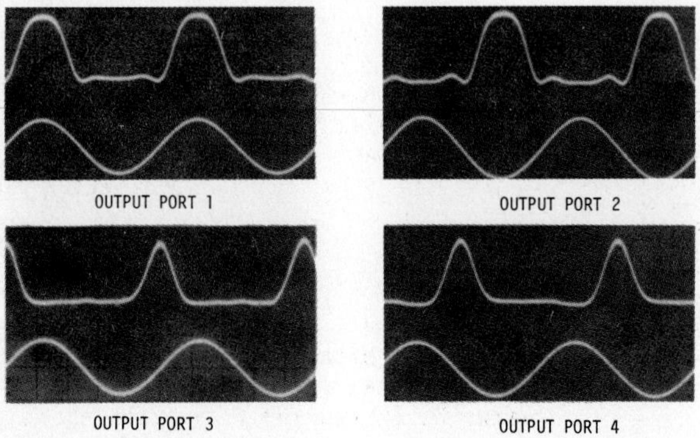

Fig. 7.5 Low-frequency response of demultiplexer (10 kHz drive)

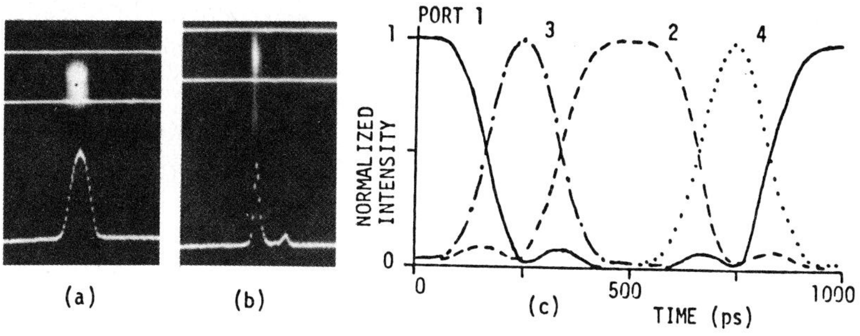

Fig. 7.6 High-frequency response of demultiplexer (1 GHz drive)

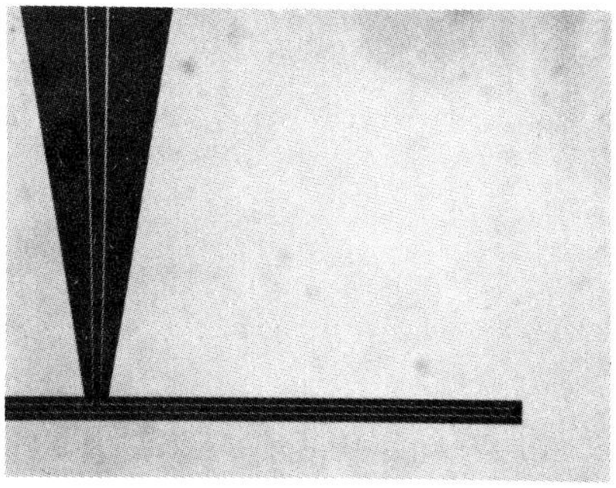

Fig. 7.7 Resonant electrode structure of MIT interferometer modulator. Structure is symmetric, only part of it shows.

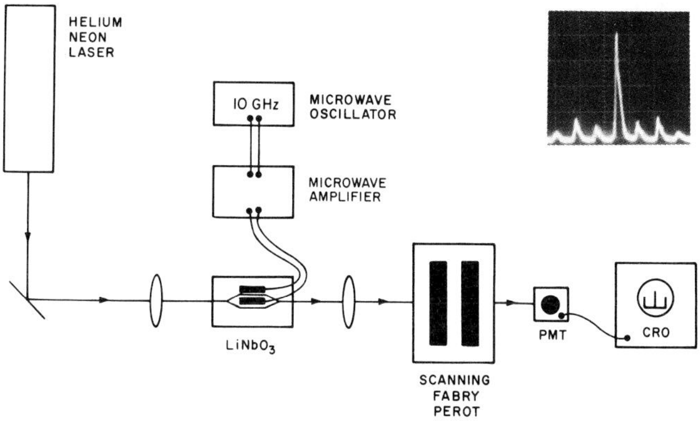

Fig. 7.8 Experimental arrangement for measurement of interferometer response. Inset is observed spectrum.

shows the electrode structure consisting of a center-fed shorted strip line. A dummy electrode provides symmetry for the optical environments of the two interferometer arms. The structure is 3.3 mm long and resonates at 10 GHz because of the high value of the microwave dielectric constant of $LiNbO_3$. The performance of the interferometer was measured via the spectrum of the output light observed through a scanning Fabry-Perot interferometer with a cw laser input illumination (Fig. 7.8).

Dr. R. Alferness of AT&T Bell Laboratories has based most of his modulator work on the waveguide coupler. Fig. 7.9 shows the schematic of a traveling-wave directional coupler switch with the responses as insets. Note how the two waveguides are brought together, and apart, to define the interaction region as coincident with the electrode length. Figure 7.10 is a photograph of the electrodes. In Fig. 7.11 the

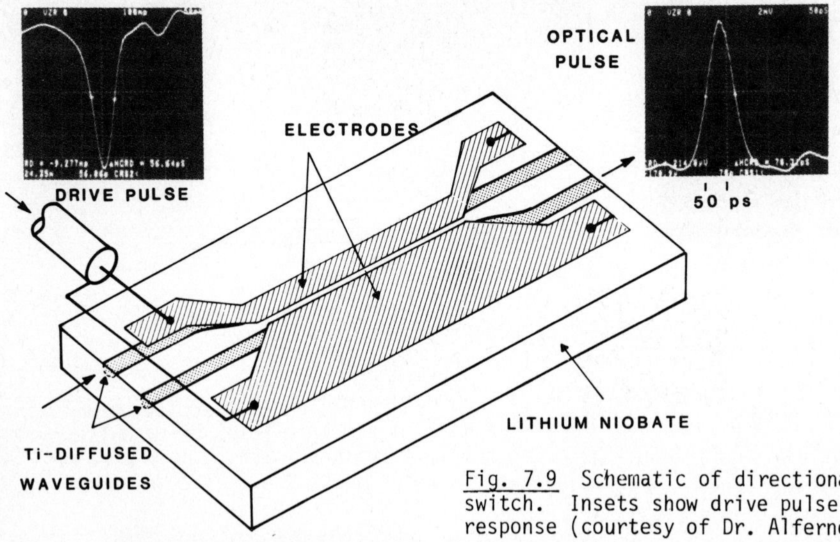

Fig. 7.9 Schematic of directional-coupler-switch. Insets show drive pulse and optical response (courtesy of Dr. Alferness)

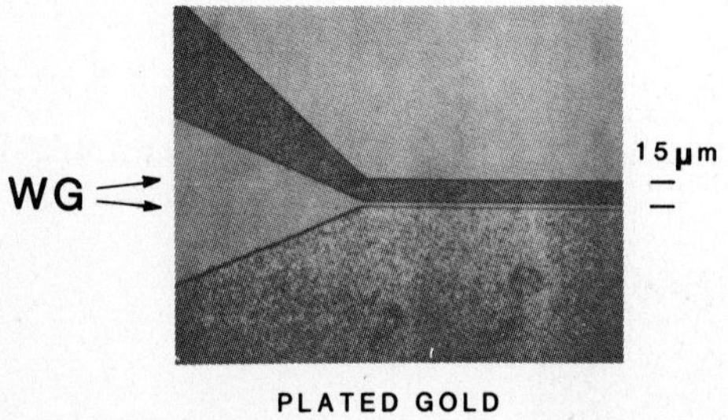

Fig. 7.10 Photograph of travelling wave electrode structure (courtesy of Dr. Alferness)

Fig. 7.11 The modulator response: left, the response to dc voltage; right, the frequency response (courtesy of Dr. Alferness)

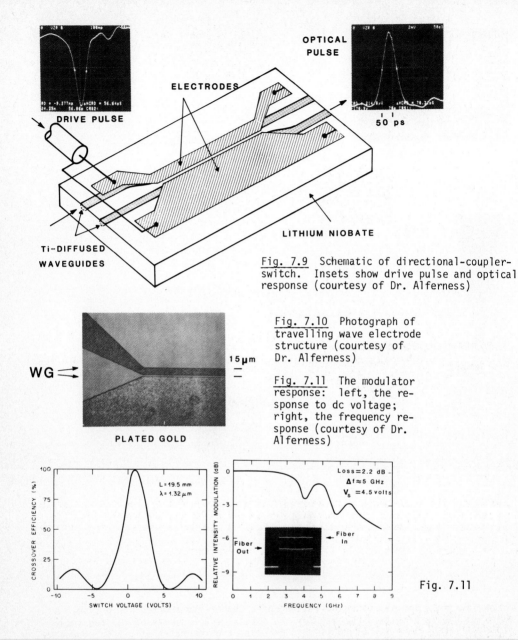

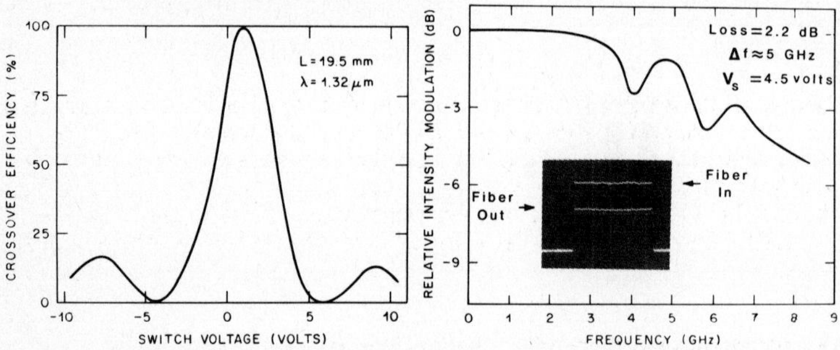

Fig. 7.11

characteristics of the modulator are shown. On the left is the measured optical response as a function of applied voltage. On the right is the frequency response. The total throughput loss in the "on" state is 2.2 dB.

The coupler can be used as a multiplexer-demultiplexer. In Fig. 7.12 are Dr. Alferness' results for a cascade of such couplers. The multiplexer combines three pulses. The pulses are switched out, one in the upper guide, two in the lower guide. The time-scale is 100 ps per division.

A very interesting application of the coupler-modulator is in a "forced" mode-locking arrangement as a modulator of a semiconductor laser. Fig. 7.13 shows the

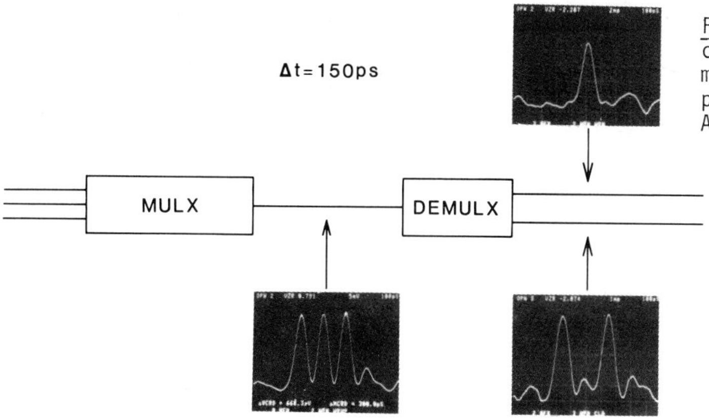

Fig. 7.12 Response of couplers in cascade as multiplexers and demultiplexers (courtesy of Dr. Alferness)

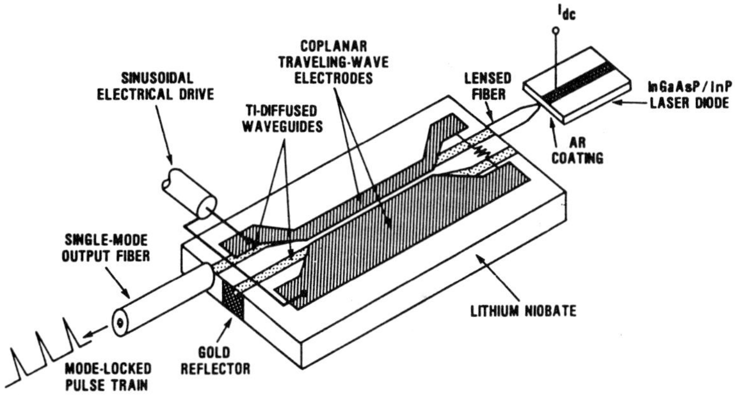

Fig. 7.13 Layout of Ti: $LiNbO_3$ coupler and InGaAsP laser diode in modelocking arrangement (courtesy of Dr. Alferness)

schematic. The modulator-electrode transmission line is fed by a sinusoidal drive and terminated in a matched load. The coupler presents to the diode alternately the gold-reflector termination, and an optical match at the frequency corresponding to the inverse optical transit time from the diode to the reflector and back, or a multiple thereof. This modulator produces actively modelocked pulses that emerge in one of the coupler guides due to the nonideality of the operation of the coupler.

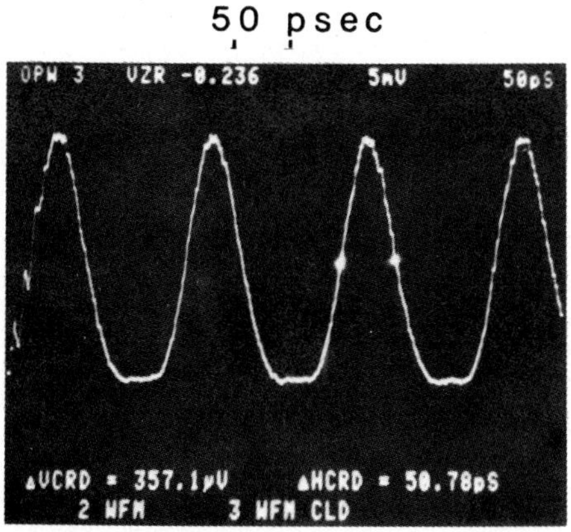

Fig. 7.14 Modelocked pulses at 7.2 GHz repetition rate (courtesy of Dr. Alferness)

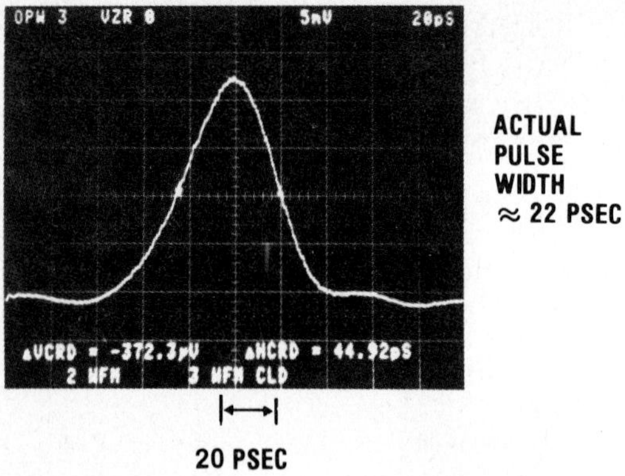

Fig. 7.15 Intensity autocorrelation of modelocked pulse obtained at 1.8 GHz repetition rate (courtesy of Dr. Alferness)

The pulse trains observed experimentally by Dr. Alferness are shown in Fig. 7.14 at a rate as high as 7.2 GHz. The pulse-intensity autocorrelation obtained by Second Harmonic Generation is shown in Fig. 7.15.

8. The All-Optical Interferometer

The Mach-Zehnder waveguide interferometer can perform as a switch, multiplexer or demultiplexer with index modulation by an applied electric field. Such applications can provide fast (20 Gbit) modulation and switching if the switching is regular (the applied signal is narrow-band). Such applications take advantage of the large optical bandwidth inherent in optical waveguides, but utilize driving electrodes that are either narrowband, or low-pass, up to a cutoff frequency determined by the (skin-effect) losses of the electrode structure. The optical bandwidth of the waveguide is much greater than 20 GHz (a precise definition of available bandwidth depends on the application and the details of the structure). Modulation of optical signals by optical signals could achieve, in principle, rates much higher than 20 Gbit. This was the reasoning that led our group at MIT to construct the all-optical waveguide modulator that could function as an XOR gate. The schematic is shown in Fig. 8.1. A continuous stream of pulses is fed into a Mach Zehnder waveguide interferometer that has, in addition, two optical waveguide inputs of different polarization from that of the central input. With a fixed applied voltage on the electrode, so that a relative phase shift of 180 degrees is achieved, no output appears in the output guide. If a pulse is entered into one of the control guides, the 180 degree shift can be compensated and the pulse emerges.

With different biases and different uses of input and output ports, all basic logic gates can be realized (see Fig. 8.2). The speed of the gate is determined by the speed of the optical nonlinearity, the required optical power can be reduced by increasing the length of the interferometer, at a cost of delay of the response, not the rate of the pulse throughput. A logic gate based on the waveguide coupler was proposed by Jensen et al. [9]. They gave no experimental results. The difficulty

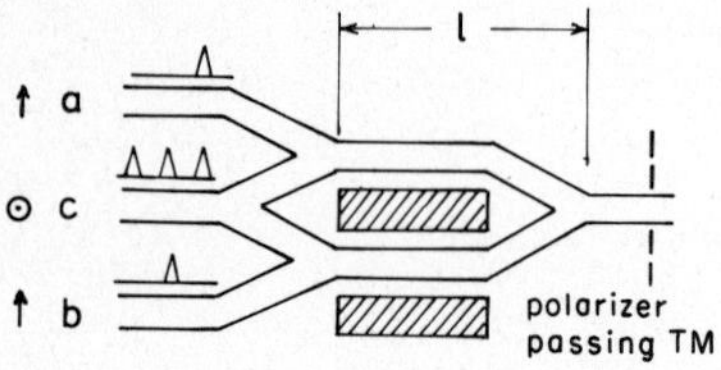

Fig. 8.1 Schematic of Mach-Zehnder interferometer as all-optical XOR gate

FUNCTION	INPUTS	PHASE SHIFT	COMMENTS
inverter	A	0	pulse stream in C
$\overline{XOR}$	A,B	0	pulse stream in C
AND	A,C	π	no pulse stream
XOR	A,B	π	pulse stream in C

Fig. 8.2 Gate operation as function of phase bias and choice of input ports

with the waveguide coupler is that it is excited asymmetrically. The photorefractive effect in $LiNbO_3$ can rapidly destroy the symmetry of the structure and prevent proper operation. A gate in GaAs built on the principle of Jensen was demonstrated by Li Kam Wa et al. [10], using a Multiple Quantum Well structure. No rate of switching was established in the mentioned reference.

We describe first the all-optical interferometer modulation, its fabrication and the experimental results. Because the all-optical logic gate, in any realization, will be much larger in size than the electronic counterpart, one can expect uses of the all-optical gate in special applications and not as a component of an all-purpose processor. Section 9 discusses the possibility of its use in an encoding-decoding system. Section 10 shows the potential use of a version of the interferometer in a "Quantum Nondemolition" (QND) photon number measurement which operates at the limit of quantum detection theory. Section 11 picks up some issues of the quantum theory of a simultaneous measurement of two noncommuting observables, in the present case photon number and phase. It is shown that the proposed use of the interferometer may form part of such a measurement apparatus, again satisfying the associated uncertainties in an optimal way. In Section 12 we show that the same apparatus could be used to tap a communication channel without net extraction of power. A fiber realization of such an interferometer could detect power levels characteristic of a semiconductor laser with a detection bandwidth of 1 MHz.

Sections 10 and 11 are based on work by Dr. Y. Yamamoto of NTT (Nippon Telegraph and Telephone) and on joint work done while the author spent 2 months at the NTT Laboratory in January and February 1985.

An all-optical waveguide interferometer was fabricated in X-cut, Y-propagating $LiNbO_3$ [11]. Lithium Niobate was chosen for convenience because it was the only material in which interferometers had been built successfully. At the time of the fabrication, the value of $\chi^{(3)}$ of $LiNbO_3$ was not known, but was estimated by Miller's rule; the estimated control power was of the order of 1.5 kW peak in an interaction length of 2 cm. One could not hope to achieve such a high input power and thus it was decided, from the outset, to demonstrate the operation as a waveguide modulator.

In the description of the operation of the gate, square pulses were assumed. In reality, the pulses from a modelocked laser are not square, and therefore a phase shift of π cannot be obtained uniformly across the controlled pulse. In an X-cut Y-propagating waveguide configuration the natural birefringence of $LiNbO_3$ imparts different group velocities to the TE and TM mode. The controlled pulses "slide through" the control pulses in the interaction region, thus simulating a longer duration of the control pulses, giving a more uniform phase shift as seen by the controlled pulses. In Ti:$LiNbO_3$ waveguides the index discontinuity is small (weak guidance), and therefore the group velocities of the modes have essentially the unguided bulk value. Therefore, $v_c = v_{TM} = c/n_o$ and $v_{a,b} = v_{TE} = c/n_e$ where, for $\lambda = 0.84\ \mu m$, $n_e = 2.17$ and $n_o = 2.25$ are the extraordinary and ordinary indices, respectively. For an interaction length $L = 2$ cm, the relative "slip" is about 5 ps. If the interaction length were much longer than 2 cm, the pulses would no longer by synchronized and would fail to interact.

Figure 8.3 shows the predicted pulse shape at the output and compares it to the input Gaussian pulse. The peak power of the control pulse was chosen to maximize the output energy. One may observe that the distortion of the pulse shape due to the nonuniform phase shift over the duration of the pulse is kept very small. For $LiNbO_3$, where the pulses travel at the velocities shown above, about 94 percent of the energy is transmitted. In an isotropic material, the two pulses would travel at the same velocity and about 80 percent would be transmitted. The "zeros" obtained with this structure are real "nulls" and are not dependent on the pulse shape, but only on the quality of the interferometer.

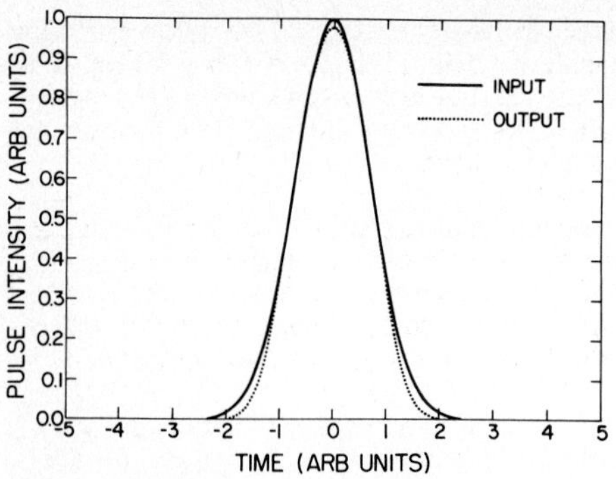

Fig. 8.3 Predicted output pulse shape and comparison with Gaussian input pulse.

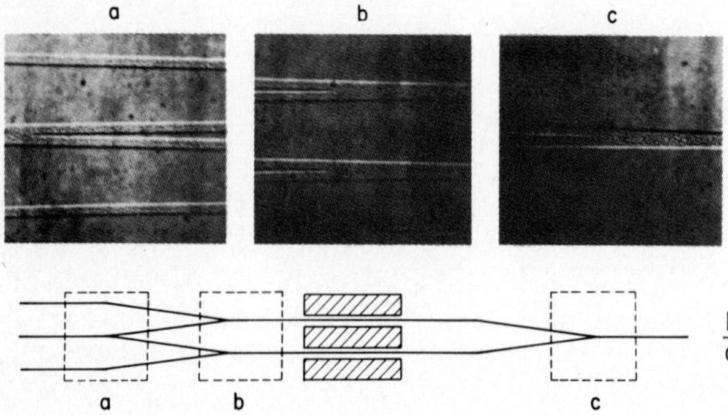

Fig. 8.4 Micrograph of portions of interferometer

Figure 8.4 is a microphotograph of portions of the 3 cm long interferometer chip. Preliminary experiments were performed with a synchronously pumped rhodamine 6G laser, and were found to produce intolerably high levels of optical damage in our Ti indiffused waveguides. To avoid this well-known damage effect [12], a laser was developed for operation at longer wavelengths. Using oxazine 750 [13] dissolved in propylene carbonate and ethylene glycol, and synchronously pumped with a modelocked Kr^+ laser, 5 ps pulses were obtained tunable over the range of 720-900 nm. At the longest wavelengths, where optical damage was less, peak powers from the dye laser were also reduced. A compromise between maximizing the nonlinear gating signal and minimizing waveguide damage was found at 840 nm. At this point the peak power, for approximately transform-limited 5 ps pulses, was about 200 W.

The experimental setup is shown in Fig. 8.5. The train of pulses from the dye laser goes through an acoustooptic modulator, which reduces the average power by a factor of about 300 to reduce the optical damage. The same laser provides both the control and controlled pulses through beam splitters and polarizers.

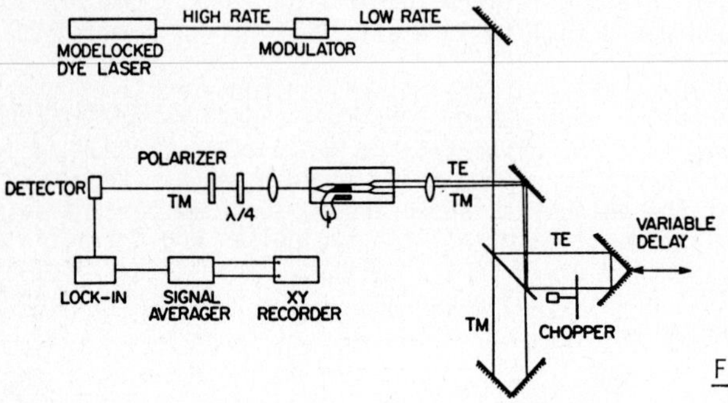

Fig. 8.5 Experimental setup

A translation stage provides a variable delay between the pulses in each channel. The control pulse is modulated at a low frequency. The output is detected and amplified in a lock-in amplifier. The signal is then processed by a multichannel averaging system which is incremented by the variable delay. The output is recorded on a plotter or photographed directly from the screen of the multichannel averager.

To enhance the sensitivity, a dc voltage was applied so that the interferometer was biased halfway between the points of maximum and minimum transmission which corresponds to a phase of $(\pi/2) \pm n\pi$. Two different traces were taken, Fig. 8.6, and stored in the multichannel averager. In one case, the controlled pulse increases the throughput of the control pulse, and in the other case it decreases it.

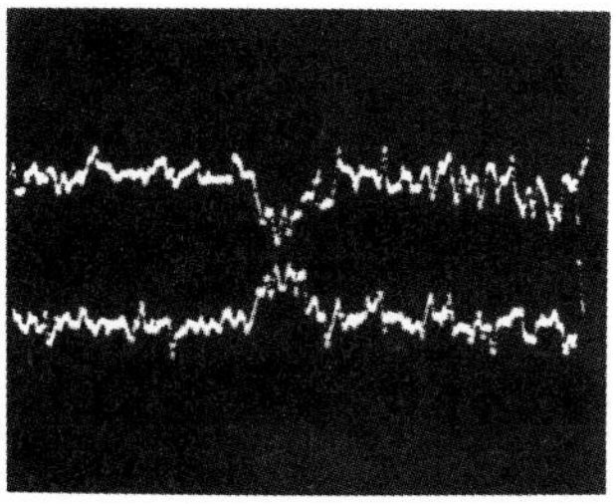

Fig. 8.6 Modulation as function of delay for two bias conditions

From the experiment it was possible to estimate the value of n_2 (proportional to the matrix element $\chi^{(3)}_{xxzz}$)

$$n_2 \simeq 3 \times 10^{-9} \left(\frac{\mathrm{Mw}}{\mathrm{cm}^2}\right)^{-1} .$$

The experiment indicates that the observed index nonlinearity is essentially instantaneous. The measured response is described well by calculations which take into account pulse shape and the velocity difference between control and controlled pulses. For the relatively short interaction lengths, material and waveguide dispersion could be ignored.

Finally, let us look at the 3 dB loss of a waveguide Y. The optical XOR gate feeds the control signal into the Mach-Zehnder waveguide interferometer arm, and in doing so loses 50 percent of the power, as explained in connection with the S-matrix of the Y. A practical logic gate must exhibit gain, the output power must be larger than the control-power. The 3 dB loss of the Y makes it more difficult to obtain gain from the device. It would be desirable to eliminate the 3 dB loss. One way of eliminating the loss is to use Y's composed of the 3-waveguide coupler. The "control" signals can be fed into the outer waveguides in a mode (polarization) different from that of the "controlled" signal. If the propagation constants of the modes are adjusted so as to give proper transfer of the controlled signal, and no net transfer of the control signals, the 3 dB loss of the control signals is avoided.

9. Encoding and Decoding with Optical XOR Gate

The size of the XOR gate, the high energy per pulse required, and the pipeline mode of its operation restrict the use of such a structure to special purpose applications. Some special applications can be envisioned in which one, or only a few, high-speed optical logic gate(s) could perform useful functions.

A linear-shift register and an XOR gate provide a simple way to generate random sequences [14] (Fig. 9.1). The outputs from two specific stages of the shift register are fed into an XOR gate to form the input of the shift register. The output of any stage is then a binary sequence. By choosing the feedback connections properly, the output is a sequence of maximal length. This is the maximum length of a series

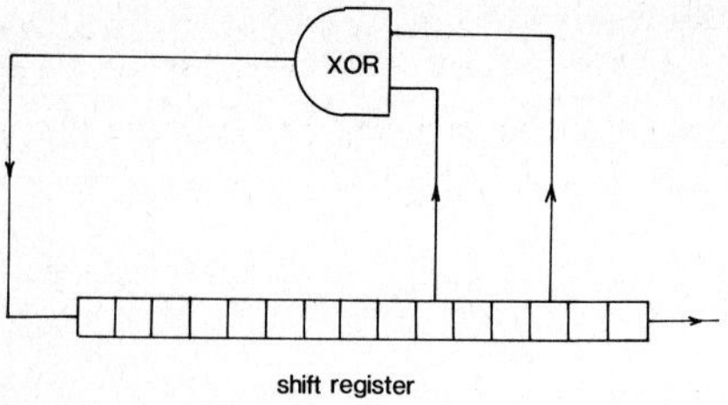

Fig. 9.1 Shift register and XOR gate as pseudorandom number generator

of ones and zeros that can be formed before the sequence is repeated. Electronic systems of this type have been used extensively in radar to generate pseudorandom sequences. An optical system, using a delay line in lieu of the shift register and incorporating a waveguide XOR gate could perform similarly. The high-rate pseudorandom sequence of pulses could be "mixed" with a message at a low rate. The resulting high-rate XOR-ed signal could not be detected by a conventional detection system because the arrival rate of the pulses is too high for conventional detectors. A receiver in possession of the code could XOR the pulse stream to a low rate and then detect it. For this purpose synchronization of the pseudorandom number generator at the receiver end is necessary. That can be accomplished through detection of the phase of the high-frequency transmitter "clock" that controls the transmitter pseudorandom number generator. Through successive correlation tests on a prearranged synchronization signal, the receiver can bring his system in synchronism with the transmitted signal. At this point coded transmission can begin.

10. Interferometric Quantum Nondemolition Measurement

A quantum nondemolition (QND) measurement is the measurement of an observable of a quantum state that yields a value of the observable, which is found unchanged upon repetition of the measurement on the state (now of course modified by the preceding measurement). Thus the observable of the state is not "demolished" by the measurement.

Every quantum mechanical observable has a complementary variable. Measurement of the observable, in general, affects the complementary variable. A precise measurement of the observable with zero uncertainty leads to an infinite mean square deviation of the complementary variable.

In this section we shall show how the nonlinear interferometer can be used to perform a QND measurement of the photon number [15]. The schematic of the interferometer adapted to the QND measurement is shown in Fig. 10.1. The probe at frequency ω_p enters through the interferometer Y, the signal at frequency ω_s enters and leaves through the two couplers. Because the probe and signal frequencies are different, the coupler can be designed to be synchronous at the signal frequency, but asynchronous (and "inoperative") at the probe frequency. A nonlinear medium extends over a length ℓ. Symmetry is preserved through dummy couplers and the same nonlinear medium placed symmetrically in the other arm.

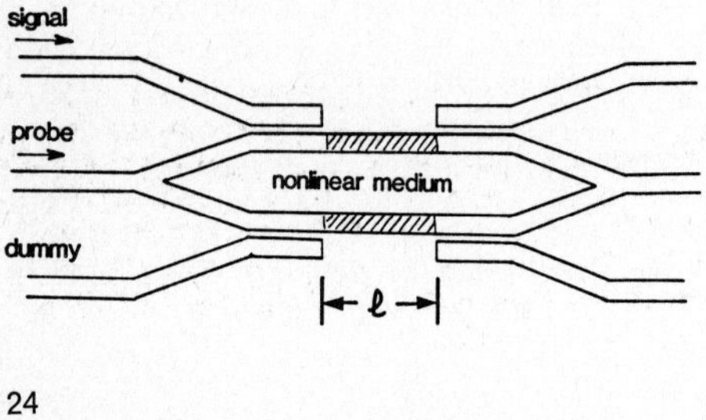

Fig. 10.1 Interferometer adapted to QND measurement

Suppose that the polarization (mode) of the signal is in the direction i, that of the probe in the direction j. The signal propagating through the nonlinear medium with the coefficient $\chi^{(3)}_{iijj}$ produces a change of index as seen by the probe, Δn that is evaluated from

$$\varepsilon_o(n + \Delta n)^2 = \varepsilon_o(n^2 + 2n\Delta n) = \varepsilon_o(1 + \chi^{(1)}_{jj} + \chi^{(3)}_{iijj}(\omega_s, -\omega_s, \omega_p; \omega_p)\, E_i^{(s)} E_i^{(s)*}) \tag{10.1}$$

and thus

$$\Delta n = \frac{1}{2n(\omega_p)} \chi^{(3)}_{iijj}\, E_i^{(s)} E_i^{(s)*}. \tag{10.2}$$

This change of index over a length ℓ causes a phase shift of the probe wave, Φ_p, that is given by

$$\Phi_p = -\frac{\omega_p}{c}\Delta n\ell = -\frac{\omega_p \ell}{2cn(\omega_p)} \chi^{(3)}_{iijj}\, E_i^{(s)} E_i^{(s)*}. \tag{10.3}$$

To express $|E_i^{(s)}|^2$ in photon number we use a volume of normalization V that is made up of the waveguide cross-section and the length of the wavepacket. Then the signal photon number N_s is given by

$$\hbar\omega_s N_s = \frac{\varepsilon}{2} |E_i^{(s)}|^2\, V. \tag{10.4}$$

Thus (10.3) gives a relation between the signal photon number and the phase shift of the probe

$$\Phi_p = -\frac{\hbar\omega_p\omega_s}{\varepsilon V\, cn(\omega_p)} \ell\chi^{(3)}_{iijj}(\omega_s, -\omega_s, \omega_p; \omega_p)N_s \equiv -FN_s \quad \text{where} \tag{10.5}$$

$$F \equiv \frac{\hbar\omega_p\omega_s\ell}{\varepsilon V\, cn(\omega_p)} \chi^{(3)}_{iijj}(\omega_s, -\omega_s, \omega_p; \omega_p). \tag{10.6}$$

The uncertainty in the measurement of the signal photon number is predicated on the accuracy of determination of Φ_p. The phase of the probe signal is subject to the uncertainty relation

$$\overline{\Delta\Phi_p^2} \geq \frac{1}{4\overline{\Delta N_p^2}} \tag{10.7}$$

and thus N_s cannot be determined to within a range better than

$$\overline{\Delta N_s^2}\Big|_{meas} = \frac{1}{F^2}\overline{\Delta\Phi_p^2} = \frac{1}{F^2}\frac{1}{4\overline{\Delta N_p^2}} \tag{10.8}$$

where we assume that the equality sign holds in (10.7). The measurement of ΔN_s via a probe causes unpredictable phase changes of the signal, because the field fluctuations of the probe induce phase shifts of the signal via a formula analogous to (10.5)

$$\Phi_s = -\frac{\hbar\omega_p\omega_s\ell}{\varepsilon V\, cn(\omega_s)} \chi^{(3)}_{jjii}(\omega_p, -\omega_p, \omega_s; \omega_s)N_p = -GN_p \tag{10.9}$$

where

$$G \equiv \frac{\hbar\omega_p\omega_s \ell}{\varepsilon V\ cn(\omega_s)} \chi^{(3)}_{jjii}(\omega_p, -\omega_p, \omega_s; \omega_s). \tag{10.10}$$

Thus, the phase fluctuations induced by the measurement are

$$\overline{\Delta\Phi_s^{\,2}}\Big|_{meas} = G^2\ \overline{\Delta N_p^{\,2}}. \tag{10.11}$$

Substitution of $\overline{\Delta N_p^{\,2}}$ from (10.8) gives

$$\overline{\Delta N_s^{\,2}}\Big|_{meas}\ \overline{\Delta\Phi_s^{\,2}}\Big|_{meas} = \frac{1}{4}\frac{G^2}{F^2} = \frac{1}{4} \tag{10.12}$$

since from Kleinmann's symmetry relations [16] $G = F$. Thus we have found that the product of the uncertainty in the determination of the photon number and the mean square phase-fluctuations induced by the measurement obey the uncertainty relation of number and phase.

Let us put in some numbers to determine the accuracy with which the photon number of a signal can be determined. Suppose we want to measure the number fluctuations of a semiconductor laser to an accuracy better than the Poisson fluctuations emerging from a very stable ideal laser.

$$\overline{\Delta N_s^{\,2}}\Big|_{meas} < \frac{1}{\bar{N}_s}. \tag{10.13}$$

Now from (10.6)

$$F^2\ \overline{\Delta N_s^{\,2}}\Big|_{meas} = \overline{\Delta\Phi_p^{\,2}}\Big|_{ind}. \tag{10.14}$$

$\overline{\Delta\Phi_p^{\,2}}|_{ind}$ must be larger than, or at least equal to, the fluctuations of the probe phase,

$$\overline{\Delta\Phi_p^{\,2}} = 1/4\bar{N}_p. \tag{10.15}$$

Combining (10.13), (10.14), and (10.15) we obtain

$$F^2\ N_s N_p > \frac{1}{4}. \tag{10.16}$$

If we reintroduce the definition of F into (10.16) the relation reads

$$\frac{\hbar\omega_p\omega_s \ell}{\varepsilon V\ cn(\omega_p)} \chi^{(3)}_{iijj} \sqrt{N_s N_p} > \frac{1}{2}. \tag{10.17}$$

Conversion back into field quantities gives

$$\frac{\sqrt{\omega_p\omega_s}\ \ell}{cn} \chi^{(3)}_{iijj}\ \overline{|E_i^{(s)2}\ E_j^{(p)2}|} > 1. \tag{10.18}$$

One may write the above in terms of an effective n_2 coefficient for which numbers are known in the literature, of course now implying very special orientations of $E_i^{(s)}$ and $E_j^{(p)}$ with respect to the crystal axes.

$$\chi_{iijj}^{(3)} \overline{E_i^{(s)2} E_j^{(p)2}} = 2nn_2 \sqrt{I_s I_p}$$

where I_p and I_s are the probe and signal intensities. When the above is introduced into (10.18) we have

$$\frac{\sqrt{\omega_p \omega_s}}{c} \ell n_2 \sqrt{I_p I_s} > 1. \qquad (10.19)$$

This is the desired relation. The left-hand side represents a phase angle produced by the geometric mean of the signal and probe waves. This phase angle must be of the order of unity or greater. We use an n_2 value for GaAs estimated from a third-harmonic-generation $\chi^{(3)}$ measurement at 1.06 μm by Burns and Bloembergen [17]

$$n_2 = 4.6 \times 10^{-13} \frac{cm^2}{W} .$$

In Quartz, n_2 is known to be

$$n_2 = 2.7 \times 10^{-16} \frac{cm^2}{W} .$$

Suppose we look first at an interferometer in GaAs with waveguide cross-sections of $A = 5\ \mu^2$ and an interaction length of 2 cm. If we want to measure the signal from a semiconductor laser with 10 mw output, the required probe power is

$$P_p = AI_p = \frac{c^2 A^2}{n_2^2 \omega_p \omega_s \ell^2} \frac{1}{P_s} .$$

Assume

$$\frac{\sqrt{\omega_p \omega_s}}{c} = \frac{2\pi}{\langle\lambda\rangle} \qquad \text{with}$$

$$\langle\lambda\rangle = 1.5\ \mu.$$

Then

$$P_p \geq \frac{1}{4} \left(\frac{\langle\lambda\rangle}{2\pi} \frac{A}{n_2 \ell}\right)^2 \frac{1}{P_s} = 42\ \text{w}.$$

This is a large power and makes a device in GaAs impractical. The situation with a fiber interferometer is different. Here one may accept lengths of the order of km with negligible loss. Suppose we assume a probe power of 5 w. How long must be the interferometer? We find from (10.19)

$$\ell \geq \frac{Ac}{2n_2 \omega_p \omega_s \sqrt{P_s P_p}} = 99\ \text{m}.$$

This makes for a reasonble length. The delay is

$$T \geq \frac{\ell n}{c} = 0.52 \times 10^{-6}\ \text{sec}.$$

The interferometer has, accordingly, a delay of the order of 1 μs.

11. QND as "Simultaneous" Measurement of Two Nonconjugate Variables

We have seen in the preceding section that the photon number can be measured with an accuracy determined by the amount of fluctuations one is willing to produce in the phase. Of course, one could measure the phase of the wave by homodyning right after the photon number has been determined. We want to show that the cascading of the two measurements doubles the fluctuations (when arranged optimally) just as required by the theory of a simultaneous measurement of two conjugate variables [18].

Thus, suppose that the signal enters with optimal phase-photon number fluctuations, so that

$$\overline{\Delta N_s^{\,2} \Delta \Phi_s^{\,2}} = \frac{1}{4} . \tag{11.1}$$

The probe is similarly optimized as already assumed in the derivation of the preceding section. After the measurement of the photon number, the phase fluctuations introduced by the measurement are given by (10.11).

The total mean square deviation of the signal phase is, therefore,

$$\overline{\Delta \Phi_s^{\,2}}\Big|_{total} = \overline{\Delta \Phi_s^{\,2}}\Big| + \overline{\Delta \Phi_s^{\,2}}\Big|_{meas} = \frac{1}{4\overline{\Delta N_s^{\,2}}} + \frac{1}{4\overline{\Delta N_s^{\,2}}\Big|_{meas}} . \tag{11.2}$$

The total fluctuations of the photon number is

$$\overline{\Delta N_s^{\,2}}\Big|_{total} = \overline{\Delta N_s^{\,2}} + \overline{\Delta N_s^{\,2}}\Big|_{meas} . \tag{11.3}$$

The product of (11.2) and (11.3) gives

$$\overline{\Delta \Phi_s^{\,2}}\Big|_{total} \overline{\Delta N_s^{\,2}}\Big|_{total} = \frac{1}{4} + \frac{1}{4} + \frac{\overline{\Delta N_s^{\,2}}}{4\overline{\Delta N_s^{\,2}}\Big|_{meas}} + \frac{\overline{\Delta N_s^{\,2}}\Big|_{meas}}{4\overline{\Delta N_s^{\,2}}} . \tag{11.4}$$

This product is optimized to unity when

$$\overline{\Delta N_s^{\,2}}\Big|_{meas} = \overline{\Delta N_s^{\,2}} .$$

This corresponds to doubling of the mean square fluctuations. This result has been obtained by Kelly and Arthurs [18] from an analysis of a particular measurement apparatus.

12. Probing of a Channel with PCM

The QND measurement by the interferometer described in the preceding two sections has one most interesting property: it measures the photon number without changing it, i.e. with no absorption of power. Thus, a measurement of this kind could be performed on a stream of pulses along a fiber without extracting energy from the fiber. One can imagine a local area network that can be "tapped" with taps of zero insertion loss. The delay of 1 μs of the interferometer does not limit the bandwidth of detection, it only causes a delay in the detected signal. Thus, high-rate pulse streams could be "probed" by such taps. The action upon the information obtained from the tapping would have to be delayed by 1 μs.

Yuen and Shapiro have proposed similar taps, except that in their scheme the use of squeezed states was essential [19]. Since sources of squeezed states have not been realized as yet, the present scheme has the advantage of being realizable today.

13. Conclusions

We have described the theory of single-mode waveguide couplers and interferometers that transform phase modulation into amplitude modulation. All devices, which we described, operated with a definite mode pattern whose propagation constant was (weakly) perturbed, so that perturbation theory was applicable. The advantage of such devices is, of course, the simplicity and reliability of their coupling to incoming and outgoing waveguides and fibers. Devices that operate via $\chi^{(2)}$, modulated by a dc or microwave field, are in an advanced state of development and their use in communication- and measurement-systems systems has already begun. Devices operating via $\chi^{(3)}$, using optical modulation fields, are much less certain of success, and represent a greater challenge. The Multiple Quantum Well (MQW) systems (D. Chemla, elsewhere in this issue) offer a greatly enhanced nonlinearity. Through proton bombardment their response time can be made as short as 100 ps. For this reason they deserve intensive research and development efforts for nonlinear optical waveguide applications. Because the loss component of $\chi^{(3)}$ dominates in MQW systems, the device realizations may turn out to be different from those discussed in these lectures. Yet, the interferometer principle may still prove useful, since zero output can be achieved as one of the "system states" by proper balance of the interferometer.

The nonlinearities achievable with organic crystals also appear promising, and their incorporation in nonlinear waveguides awaits the development of technologies for their crystal growth and waveguide fabrication.

There is no doubt that optical devices will never acquire the versatility and integrability of electronic semiconductor devices. It is important, therefore, to be aware of the fact that a few optical gates can perform important functions that are out of reach of electronic systems. The high speed encoder-decoder was given as one of the examples. This function could be performed by an optical gate that uses either reactive or resistive nonlinearities. In contrast, the Quantum Nondemolition experiment described here can be realized only with low loss, ideally zero-loss optical components. Because the proposed measurement employs single-mode waveguides or fibers, it can obey the uncertainty relation with an equality sign. The development of all-optical devices is only at its beginning. The potential applications are sufficiently interesting and unique to warrant the research that is currently devoted to them.

Acknowledgments

Professors M. Izutsu and T. Sueta of Osaka University and Dr. R. C. Alferness of AT&T Bell Laboratories kindly made available to the author illustrations of their devices. The author wrote the notes during his four months' stay as a visiting professor at the Technische Hochschule Wien. The application of the QND measurement to a lossless tap was suggested by Dr. G. Reider.

This work was supported, in part, by the Joint Services Electronics Program under Contract DAAG29-83-K-0003 and in part by the National Science Foundation Grant ECS8310718.

References

1. A. Hasegawa and F. Tappert: "Transmission of Stationary Nonlinear Optical Pulses in Dispersive Dielectric Fibers," Appl. Phys. Lett. 23, 142 (1973)
2. L.F. Mollenauer, R.H. Stolen, and M. N. Islam: "Experimental Demonstration of Soliton Propagation in Long Fibers: Loss Compensated by Raman Gain," Opt. Lett. 10, 229 (1985)

3. H. Kogelnik and R.V. Schmidt: "Switched Directional Couplers with Alternating $\Delta\beta$," IEEE J. Quant. Electron. QE-12, 396 (1976)

4. J.R. Pierce: "Coupling of Modes of Propagation," J. Appl. Phys. 25, 179 (1954)

5. H.F. Taylor: "Optical Switching and Modulation in Parallel Dielectric Waveguides," J. Appl. Phys. 44, 3257 (1973)

6. M. Kuznetsov: "Radiation Loss in Dielectric Waveguide Y-Branch Structures," IEEE/OSA J. Lightwave Techn. LT-3 (1985)

7. H. Haga, M. Izutsu, and T. Sueta: "An Integrated Optic 4-Channel Fast Time Demultiplexer," Paper 30BZ-3, IOOC'83 Conference, Tokyo, June 1983; also, T. Sueta and M. Izutsu, "High-Speed Guided-Wave Optical Modulators," J. Opt. Commun. 3, 52 (1982)

8. L.A. Molter-Orr, H.A. Haus, and F.J. Leonberger: "20 GHz Optical Waveguide Sampler," IEEE J. Quant. Electron. QE-19, 1877 (1983)

9. S.M. Jensen, "The Nonlinear Coherent Coupler," IEEE J. Quant. Electron. QE-18, 1580 (1982)

10. P. Li Kam Wa, J.E. Stitch, N.J. Mason, J.S. Roberts, and P.N. Robson: "All-Optical Multiple Quantum Well Waveguide Switch," Electron. Lett. 21, 26 (1985)

11. A. Lattes, H.A. Haus, F.J. Leonberger, and E.P. Ippen: "An Ultrafast All-Optical Gate," IEEE J. Quant. Electron. QE-19, 1718 (1983)

12. A.M. Glass: "The Photorefractive Effect," Opt. Eng. 17, 470 (1978)

13. G.D. Aumiller: "Broadly Tunable Near IR CW Dye Laser Using Propylene Carbonate as Solvent," Opt. Commun. 14, 115 (1982)

14. M. Skolnik: Radar Handbook (McGraw Hill, New York, Sec. 20.18-20.21, 1970)

15. N. Imoto, H.A. Haus, Y. Yamamoto: "Quantum Nondemolition Measurement of Photon Number via Optical Kerr Effect," submitted to Phys. Rev. A (1985)

16. D.A. Kleinmann: "Nonlinear Dielectric Polarization in Optical Media," Phys. Rev. 126, 1977 (1962)

17. W.K. Burns and N. Bloembergen: "Third-Harmonic Generation in Absorbing Media of Cubic or Isotropic Symmetry," Phys. Rev. B 4, 3437 (1971)

18. E. Arthurs and J.L. Kelley, Jr.: "On the Simultaneous Measurement of a Pair of Conjugate Observables," Bell System Tech. J. XLIV, 725 (1965)

19. J.H. Shapiro: "Optical Waveguide Tap with Infinitesimal Insertion Loss," Opt. Lett. 5, 351 (1980)

Nonlinear Guided Waves

G.I. Stegeman [1], *C.T. Seaton* [1], *W.M. Hetherington III* [2], *A.D. Boardman* [3], *and P. Egan* [3]

[1] Optical Sciences Center and Arizona Research Laboratories, University of Arizona, Tucson, AZ 85721, USA
[2] Optical Sciences Center and Department of Chemistry, University of Arizona, Tucson, AZ 85721, USA
[3] Department of Physics, University of Salford, Salford, M5 4WT, U.K.

1. Introduction

Since its inception in the 1960s, nonlinear optics has led to a rich variety of wave-mixing interactions that have applications to basic materials research [1-2], to the generation of new frequencies [1], and most recently to all-optical signal processing [3]. In general, nonlinear optical interactions occur whenever the optical fields associated with one or more laser beams propagating in a material are large enough to produce polarization fields proportional to the product of two or more of the incident fields. These nonlinear polarization fields radiate electric fields at the nonlinear frequency. For some interactions, the generated fields grow linearly with propagation distance under optimum conditions of phase-matching. Typically the efficiency of any nonlinear optical interaction depends on (1) the product of the power densities of the input and output waves, raised to some power, and (2) the interaction distance raised to some power greater than or equal to unity. Since power density is power per unit area, the efficiency of any nonlinear interaction can be enhanced by reducing the cross-sectional area of the interacting beams. For plane waves this can be achieved by focusing with a lens. There is a tradeoff, of course, because the high power density can be maintained only over the depth of focus of the lens, which limits the effective interaction length.

Electromagnetic waves can be guided by the interface between two semi-infinite media, or by single or multiple films bounded by two semi-infinite media. The key feature is that the fields decay exponentially away from the boundaries into both semi-infinite bounding media with 1/e distances of typically a fraction of the wavelength of the radiation being guided. Therefore the effective beam dimension along the direction normal to the surface can be of the order of the wavelength of light, which corresponds to the minimum beam cross-sectional area and hence maximum power density for a given input power level. These attractive features were recognized at an early stage and have, by now, been applied to an impressive number of nonlinear guided wave phenomena.

Guided waves usually result from a coupling between an electromagnetic field and some resonance. The resonances can be geometric, for example as occurs in a thin-film integrated-optics waveguide [4], where constructive interferences in the film result in waveguide modes. The resonances can also be related to material properties. For example, the plasma resonance associated with the electron gas in a metal is coupled to an electromagnetic field via the interaction between the field and the charges, and leads to surface plasmons [5]. Such material resonances are usually accompanied by large losses, and hence the propagation distances attained are limited in these cases. For this reason, most work in guided wave nonlinear optics for device applications has been limited to integrated optics with freely propagating guided waves. In this chapter we will concentrate solely on nonlinear interactions involving freely propagating waves.

The nonlinear guided wave phenomena studied to date have consisted of (1) guided wave analogs of plane wave nonlinear interactions, and (2) a limited number of mixing processes that are unique to guided wave geometries. Certainly the most developed area involves second-order interactions [6]. Second harmonic generation has been demonstrated for waves guided by dielectric films and channels. Other phenomena such as sum and difference frequency generation, and parametric amplification and oscillation have also been reported with integrated optics waveguides. Finally, the mixing of oppositely propagating guided waves to produce a second harmonic radiated normal to the guiding surface, a phenomenon unique to waveguiding geometries, has also been observed.

New developments in nonlinear optics in the last few years have centered on phenomena depending on $\chi^{(3)}$, the third-order susceptibility. This case deals with the mixing of up to three separate incident optical fields, for example at the frequencies ω_a, ω_b, and ω_c. Therefore, the nonlinear polarization and hence the radiated fields can have the frequency components $\omega_a \pm \omega_b \pm \omega_c$, which leads to a large range of phenomena. In contrast to efficient second-order phenomena that are difficult to phase-match, there are third-order processes such as degenerate four-wave mixing and optical bistability that are automatically phase-matched [3]. There are interactions in which tuning the difference frequency between two of the laser fields through characteristic molecular vibrational frequencies leads to resonant enhancements in the signal and hence can be used for spectroscopy [2]. Here we discuss guided-wave versions of these phenomena [7]. In addition, there are third-order interactions that are unique to integrated optics and have no plane-wave analogs. In particular, there is the nonlinear coherent coupler, power-dependent coupling into waveguides via distributed couplers such as prism and gratings with possible applications to switching, and a new class of guided waves whose properties change dramatically with guided wave power. In this chapter we discuss nonlinear phenomena in waveguides that utilize the third-order susceptibility.

2. Guided Waves

We start our discussion of nonlinear guided waves with a brief review of the properties of guided waves [4,8,9]. In this section we assume that the fields are normal modes and that their properties do not depend on the power of the guided wave.

Guided waves are electromagnetic fields that satisfy both Maxwell's equations in every medium into which the fields penetrate, and continuity of the tangential boundary conditions at every interface. For the most general case which includes a nonlinear polarization field, the wave equation for fields and polarization sources at the frequency ω takes the form

$$\nabla^2 \mathbf{E}(\mathbf{r},t) + \omega^2 \frac{n_\gamma(z)^2}{c^2} \mathbf{E}(\mathbf{r},t) = -\mu_0 \omega^2 \mathbf{P}^{NL}(\mathbf{r},t) , \qquad (1)$$

where $n_\gamma(z)$ is the refractive index of the γ'th medium. For the simplest case of an isotropic waveguide media with z normal to the guiding surface(s) and propagation along the x-axis, the normal mode solutions to Eq. (1) with $\mathbf{P}^{NL}(\mathbf{r},t) = 0$ separate into TE (s-polarized) waves with field components E_y, H_x, and H_z, and TM (p-polarized) waves with field components H_y, E_x, and E_z. For s-polarized (TE) waves, the electric field of frequency ω is given by

$$E_y(\mathbf{r},t) = \sum_m E_y^{(m)}(\mathbf{r},t) + \int E_y^{(\nu)}(\mathbf{r},t)\, d\nu, \qquad (2)$$

where the summation is taken over the guided wave modes, and the integral is taken over the radiation fields $E_y^{(\nu)}(\mathbf{r},t)$. Guided wave fields are characterized by exponential field decay into both semi-infinite bounding media, in contrast to radiation fields that exhibit oscillatory behavior in at least one bounding medium and extend to infinity in that medium (in the absence of loss). For the m'th order guided wave mode,

$$E_y^{(m)}(\mathbf{r},t) = \frac{1}{2} E_y^{(m)}(z)a^{(m)}(x)e^{i(\omega t - \beta^{(m)}kx)} + c.c. \qquad (3)$$

The detailed depth-dependence of the guided wave fields is contained in the $E_y^{(m)}$ term, which is normalized so the $|a^{(m)}(x)|^2$ is the guided wave power in watts per meter of wavefront (along the y-axis). Since $\omega = kc$, the term $\beta^{(m)}$ plays the role of a refractive index for propagation along the x-axis and is called "the effective index": Its value is obtained from an eigenvalue equation or dispersion relation obtained by satisfying the boundary conditions across each interface. The radiation fields have the same basic structure as Eq. (3) with the mode index m replaced by ν, and $a^{(m)}(x)$ replaced by $b^{(\nu)}(x)$. Since these solutions are normal modes, they satisfy orthogonality relations:

$$\frac{\beta}{2\mu_0 c}\int_{-\infty}^{\infty} dz\, E_y^{(m)}(z)E_y^{*(r)}(z) = \delta_{m,r}\,. \qquad (4)$$

These orthogonality relations will prove very useful later for evaluating the amplitudes of nonlinearly generated fields.

It is useful to consider the simple geometry of a film of thickness "h" and refractive index n_f bounded by a cladding medium ($0 \geq z$, n_c) and a substrate ($z \geq h$, n_s), as illustrated in Fig. 1. A finite number of discrete modes occur, and the field distributions associated with the first few TE_m waves shown in Fig. 2 exhibit oscillatory behavior in the film, and decay exponentially with distance into the cladding and substrate. Therefore nonlinear interactions can take place in any one of the three media. One of the unique features of thin film guided waves is that the $\beta^{(m)}$ for each mode depend on film thickness, as illustrated in Fig. 3.

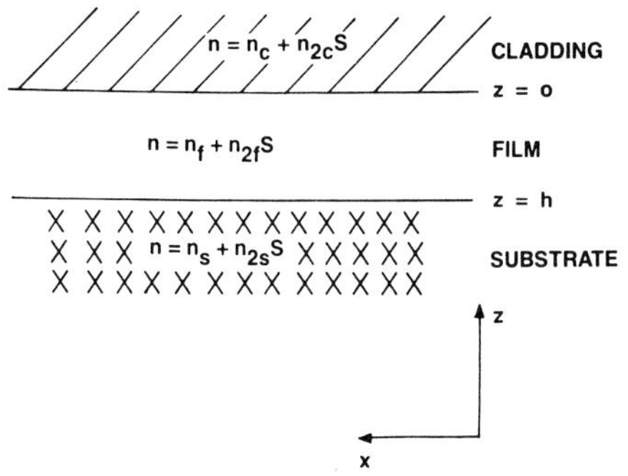

Fig. 1. A thin film dielectric waveguide of thickness "h".

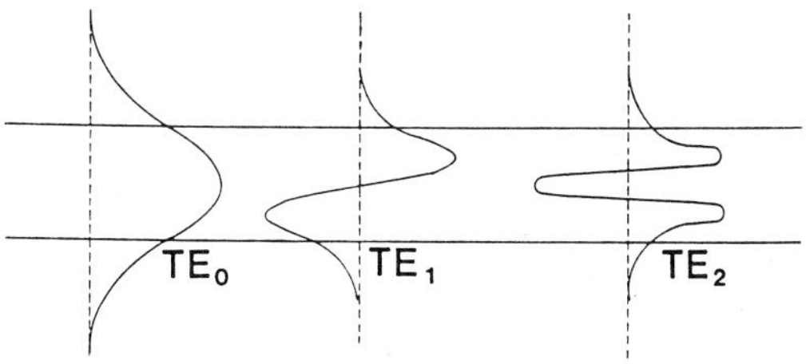

Fig. 2. Typical electric field distributions for TE_m modes.

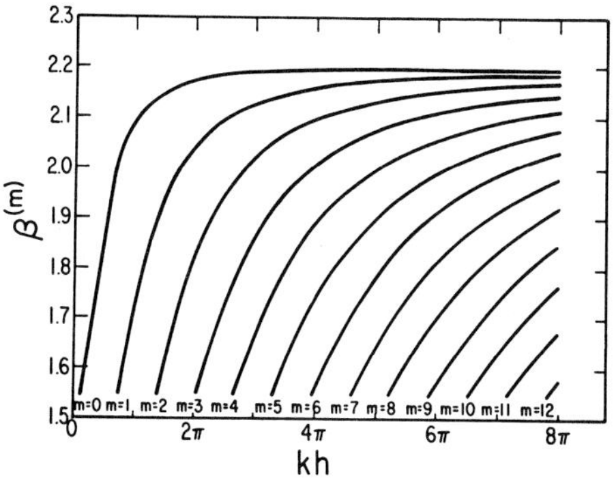

Fig. 3. Dispersion in effective index β with product of film thickness and wavevector.

3. Nonlinear Polarization Fields

The mixing of multiple optical beams leads to nonlinear polarization fields that are usually expanded in products of the mixing fields [1]. A nonlinear polarization field of frequency ω_s can be generated in each of the media in which the guided wave fields exist. It is usually written as

$$\mathbf{P}^{NL}(\mathbf{r},t) = \frac{1}{2}\mathbf{P}^{NL}(z,\omega_s)e^{i(\omega_s t - \beta_p \mathbf{k}_s\cdot\mathbf{r})} + c.c. \tag{5a}$$

If we now assume up to three input guided-wave fields of frequency ω_a, ω_b, and ω_c (of which two or three may be degenerate)

$$\mathbf{P}^{NL}(z,\omega_s) = \varepsilon_0\chi^{(2)}(-\omega_s;\omega_a,\pm\omega_b):\mathbf{E}^{(m,a)}(z)\mathbf{E}^{(m',b)}(z)a^{(m,a)}(x)a^{(m',b)}(x)$$

$$+ \varepsilon_0\chi^{(3)}(-\omega_s;\omega_a,\pm\omega_b,\pm\omega_c):\mathbf{E}^{(m,a)}(z)\mathbf{E}^{(m',b)}(z)\mathbf{E}^{(m'',c)}(z)\; a^{(m,a)}(x)a^{(m',b)}(x)a^{(m'',c)}(x), \tag{5b}$$

where $\omega_a = k_a c$, $\omega_b = k_b c$, etc., $\chi^{(2)}$ and $\chi^{(3)}$ are the second- and third-order susceptibilities, and a minus sign for a frequency corresponds to taking the complex conjugate of the appropriate field. Note that $\beta_p\mathbf{k}_s = \beta^{(m,a)}\mathbf{k}_a \pm \beta^{(m',b)}\mathbf{k}_b \pm \beta^{(m'',c)}\mathbf{k}_c$ is the wavevector associated with the nonlinear polarization source field, and it is <u>not necessarily</u> equal to $\beta^{(n,s)}\mathbf{k}_s$ which is the value appropriate to a propagating field of frequency ω_s. The case $\beta_p k_s = \beta^{(n,s)}k_s$ corresponds to phase-matching, as will be discussed later. For the second-order processes there are two possible input waves with frequencies ω_a and ω_b, which produce polarization and signal fields at $\omega_s = \omega_a \pm \omega_b$. For third-order processes, the nonlinear polarizations can occur at the frequencies $\omega_s = \omega_a \pm \omega_b \pm \omega_c$. Most of these interactions have now been demonstrated in nonlinear guided-wave experiments.

In addition to the nonlinear polarization that occurs inside a material due to the product of electric fields, nonlinear source terms can also occur near surfaces where the field gradients are large [10]. It can easily be shown that this nonlinear polarization term is zero for TE polarized waves. However, for TM modes, there are contributions because there exists an electric field component parallel to the propagation wavevector, and because the E_z TM field component is discontinuous across the boundary. These contributions to the nonlinear polarization are typically less than the usual ($\chi^{(2)}$) terms by a factor of 10 to 10^4, depending on the material and mode. One would therefore expect these terms to be important only for cases in which $\chi^{(2)}$ is uniquely zero because of symmetry reasons. To date, these terms have only played a role in nonlinear surface plasmon interactions [10].

4. Coupled Mode Theory and Nonlinear Wave Generation

It is important to note that the existence of the nonlinear polarization field does not ensure the generation of strong signal fields that must satisfy the driven wave equation, Eq. (1), with $\mathbf{P}^{NL} \neq 0$. With the exception of phenomena based on an intensity-dependent refractive index, the generation of the nonlinearly produced signal waves at frequency ω_s can be treated in the slowly varying amplitude approximation using well-known guided-wave coupled-mode theory [8,9]. For example, for an s-polarized nonlinear polarization source field that generates guided-wave fields of mode order "n", the general form of the solution field [Eq. (2)] is substituted into Eq. (1). As already explicitly assumed in Eq. (3), the amplitudes of the waves are allowed to vary weakly with propagation distance x, that is $(d^2)/(dx^2)\, a^{(n',s)}(x) \ll \beta^{(n',s)}k_s\, d/(dx)\, a^{(n',s)}(x)$, which leads to

$$\sum_{n'} 2i\beta^{(n',s)}k_s \frac{d}{dx}a^{(n',s)}(x)\; E_y^{(n',s)}(z)\; e^{-ik_s\beta^{(n',s)}x}$$

$$+ \int d\nu'\; 2i\beta^{(\nu',s)}k_s \frac{d}{dx}\, b^{(\nu',s)}(x)\; E_y^{(\nu',s)}(z)\; e^{-ik_s\beta^{(\nu',s)}x}$$

$$= -\omega_s^2 \mu_0 P_y^{NL}(z,\omega_s) e^{-ik_s \beta_p x} \,. \tag{6}$$

We now make use of the orthogonality relations, [Eqs. (4)], by multiplying both sides of Eq. (6) by $E^*{}_y^{(n,s)}(z)$ and integrating over z. Note that by choosing a guided-wave field distribution as the multiplier, we are explicitly evaluating the guided-wave field amplitude coefficients. This results in

$$\frac{d}{dx} a^{(n,s)}(x) = i \frac{\omega_s}{4} \int_{-\infty}^{\infty} dz\, \mathbf{P}^{NL}(z,\omega_s) \cdot \mathbf{E}^{*(n,s)}(z)\, e^{-i(\beta_p - \beta^{(n,s)}) k_s x} \,, \tag{7}$$

where we have generalized the result to also include TM guided-wave fields.

The same formalism can be used to calculate the generation of radiation fields due to nonlinear polarization sources in the vicinity of the waveguide. Multiplying Eq. (6) by $\mathbf{E}^{*(\nu,s)}(z)$, integrating over z, and invoking the orthogonality relations leads to the equivalent of Eq. (7) with $a^{(n,s)}(x)$ replaced by $b^{(\nu,s)}(x)$ and the superscript n replaced by its value for radiation fields, namely ν.

5. Second Harmonic Generation (SHG)

SHG is the only nonlinear interaction that has been studied extensively in optical waveguides. Second harmonic signals can be generated by a single guided wave, by the mixing of two co-directional guided waves of different polarization, or by the interaction of two oppositely propagating guided waves. The discussion in each case will be separated into two parts, theory and experiment.

(a) SHG by Co-directional Guided Waves

The simplest case is a single fundamental beam producing a SHG wave via the material nonlinearity. Hence there are two waves present, the fundamental (ω_a = ω, k_a = k) and the second harmonic (ω_s = 2ω). In terms of the d_{ijk} tensor (which is usually used instead of the χ_{ijk} tensor), the nonlinear polarization fields are

$$P_i^{NL}(z,2\omega) = \varepsilon_0 d_{ijk} E_j^{(m,\omega)}(z) E_k^{(m,\omega)}(z) a^{2(m,\omega)} \,, \tag{8a}$$

$$P_i^{NL}(z,\omega) = 2\varepsilon_0 d_{ijk} E^*{}_j^{(m,\omega)}(z) E_k^{(n,2\omega)}(z) a^{(n,2\omega)} a^{*(m,\omega)} \,. \tag{8b}$$

Substituting into Eq. (7) gives

$$\frac{d}{dx} a^{(n,2\omega)}(x)$$

$$= \frac{i\omega\varepsilon_0}{2} \int_{-\infty}^{\infty} dz\, d_{ijk} E^*{}_i^{(n,2\omega)}(z) E_j^{(m,\omega)}(z) E_k^{(m,\omega)}(z)\; a^{(m,\omega)2}(x) e^{-2i\Delta kx} \,, \tag{9a}$$

$$\frac{d}{dx} a^{(m,\omega)}(x) = \frac{i\omega\varepsilon_0}{2} \int_{-\infty}^{\infty} dz\, d_{ijk} E^*{}_i^{(m,\omega)}(z) E_j^{(n,2\omega)}(z) E^*{}_k^{(m,\omega)}(z)$$

$$\times\, a^{(n,2\omega)}(x) a^{*(m,\omega)}(x)\, e^{2i\Delta kx} \,, \tag{9b}$$

which are the two equations that must be solved with $\Delta k = (\beta^{(m,\omega)} - \beta^{(n,2\omega)})k$. At frequencies far from any material resonances, the three indices ijk can be interchanged so that the integrals in Eqs. (9), called the overlap integral, yield the same value, to within a phase factor.

It is very useful to first examine the case in which the fundamental beam is undepleted and the conversion to the harmonic power after an interaction distance L is small. In that case

$$|a^{(n,2\omega)}(L)|^2 = (kL)^2 \, |K|^2 \, \frac{\sin^2(\Delta kL)}{(\Delta kL)^2} \, |a^{(m,\omega)}(0)|^4 \, . \tag{10a}$$

$$K = \frac{c\varepsilon_0}{2} \int_{-\infty}^{\infty} dz \; d_{ijk} \; E_j^{(m,\omega)}(z) E_k^{(m,\omega)}(z) E^{*}_i{}^{(n,2\omega)}(z) \, . \tag{10b}$$

The $\sin^2\phi/\phi^2$ term describes the effect of phase-mismatch, that is, $\Delta k \neq 0$. Efficient conversion can be obtained only for $\Delta kL < \pi/4$. The optimum clearly occurs for $\Delta k = 0$, the phase-matched case.

In the limit $\Delta kL = 0$, Eqs. (9) can be solved in closed form. The results are

$$a^{(m,\omega)}(x) = a^{(m,\omega)}(0) \operatorname{sech} \frac{x}{\ell_{SH}} \, , \tag{11a}$$

$$a^{(n,2\omega)}(x) = i \, \frac{a^{(m,\omega)2}(0)}{|a^{(m,\omega)}(0)|} \tanh \frac{x}{\ell_{SH}} \, , \tag{11b}$$

$$\frac{1}{\ell_{SH}} = k \, K |a^{(m,\omega)}(0)| \, . \tag{11c}$$

Therefore the conversion efficiency is given simply by $\eta = \tanh^2(L)/(\ell_{SH})$.

Although it initially appears that SHG in waveguides should be simpler to achieve than for the plane wave case, in practice this is not true. For example, the existence of multiple modes of the same polarization at any frequency should offer many more possibilities for phase-matching, that is the superscripts n, m and m'(for two orthogonally polarized input waves) can take multiple values. However, there is dispersion in β with film thickness so that $\beta^{(n,2\omega)} > \beta^{(n,\omega)}$. This dispersion reinforces the usual dispersion in material refractive index with wavelength so that SHG with n = m (= m') is not possible with modes of the same polarization. It is, however, possible to produce SHG with n > m,m'. This leads to a greatly reduced cross-section because the overlap integral involves interference effects when products of fields are integrated over the depth coordinate. This is clear from Fig. 2. Since $\beta_{TE}{}^{(m,\omega)} > \beta_{TM}{}^{(m,\omega)}$, the situation improves when the fundamental wave is a TE mode and the harmonic is a TM wave. Nevertheless, it has still proven very difficult to find combinations in which there are no interference effects in the overlap integral.

A number of different approaches for facilitating phase-matching have been proposed and implemented. Temperature has been used to tune the refractive indices [11]. Surface gratings have been used to add additional wavevectors to the interaction [12]. One very elegant solution to this interference effect has been demonstrated by Ito and Inaba [13]. They overcoated the nonlinear waveguide with a second film chosen so that no interference effects occurred in the field overlap inside the nonlinear film, as shown schematically in Fig. 4. In fact, this experiment was the only one in which relatively efficient SHG was generated in a slab waveguide geometry.

The second problem with SHG in thin film waveguides is the very strong dispersion with film thickness that occurs, see Fig. 3. Any variation in film thickness in the film fabrication process leads to changes in $\beta^{(r)}$ where r = n, m, m'. Since the slopes of the dispersion curves are different for different mode numbers, the phase-matching tolerances can vary with propagation distance. The larger the total dispersion in β with film thickness, the more sensitive are

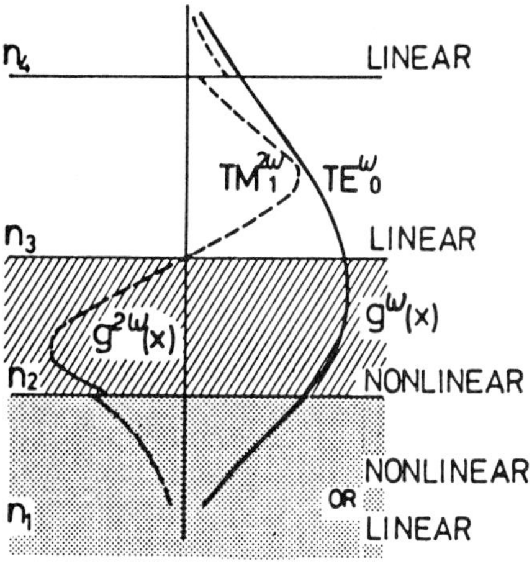

Fig. 4. Guided-wave field overlap in a two-film waveguide for harmonic generation with $TE_0 + TE_0 \rightarrow TM_1$ [13].

the phase-matching conditions to changes in the film thickness. In almost all slab waveguide experiments to date, the film thickness was tapered and the beams were coupled in at different points along the taper until phase-matching was achieved [14].

These problems, along with the small values of nonlinear coefficients available for most dielectric materials that can be easily fabricated in thin film form, stopped development of SHG in nonlinear slab waveguides for a number of years. Interest has been revived recently by the development of new, highly nonlinear, organic materials with high damage thresholds. The nonlinear coefficients $\chi^{(2)}$ are one to two orders of magnitude larger than those for materials such as $LiNbO_3$, thus opening new possibilities for efficient harmonic generation at low laser powers. Preliminary experiments have been reported on para-chloro-phenylurea and MNA with very encouraging results [15,16]. However, much more work is still required for finding better materials and film fabrication techniques.

The current state-of-the-art waveguide SHG was obtained with in-diffused $LiNbO_3$ waveguides. Typically the waveguides are fabricated by diffusing Ti ions into the surface of single crystal $LiNbO_3$, thus increasing the refractive index in the surface region to create a waveguiding region [17]. The net refractive index change in this case is small, less than 0.01 so that the waveguides are quite deep (a few micrometers), and the conditions on maintaining phase-matching are relaxed relative to slab waveguides. Furthermore, the $LiNbO_3$ birefringence can be used to facilitate phase-matching, even between orthogonally polarized modes of the same order. Because of the nature of the in-diffusion process, it has proven possible to fabricate high-quality channel waveguides by depositing the titanium for the diffusion process through a mask. Propagation distances as long as 50 mm have been obtained. Phase-matching has been achieved by tuning the temperature or wavelength, or by applying electric fields to make use of the electro-optic effect in $LiNbO_3$. Such waveguides currently provide the best results reported to date for SHG in waveguides [18,19].

SHG in channel $LiNbO_3$ waveguides has been enhanced by making a resonator out of the channel waveguide [20]. High-reflectivity (0.96 at the fundamental wavelength) mirrors were coated onto the end faces of the channels to increase the fundamental power in a low-loss (0.15 dB/cm) waveguide. Although the structure was not optimized, 1.5 mW of HeNe radiation at 1.15 µm was doubled with an efficiency of 10^{-3} by tuning the temperature until phase-matching was obtained. For a matched resonator, an order of magnitude improvement in the conversion efficiency has been predicted. This requires that the fundamental be trapped in the waveguide until it is either absorbed or converted to the harmonic field.

(b) SHG By Contradirectional Waves

This particular phenomenon is unique to guided-wave geometries because it relies crucially on the strong confinement of the guided-wave fields. For TE incident waves, the nonlinear polarization field has components at $\omega_s = 0$ and $\beta_p = 2\beta^{(m,\omega)}$, and at $\omega_s = 2\omega_a = 2\omega$ and $\beta_p = 0$. The radiative nonlinear polarization field created by the mixing via the d_{yyy} tensor component of guided waves propagating along the ±x axis, $a_+(x)$ and $a_-(x)$ respectively, is given by

$$P_y^{NL}(z,2\omega) = 2\epsilon_0 d_{yyy} E_y^{2(m,\omega)}(z)\, a_+^{(m,\omega)}(x) a_-^{(m,\omega)}(x) \,. \quad (12)$$

Note that other components of the **d** tensor, which have either x or y as the first index, will also lead to the same phenomenon. The source field has no spatial periodicity parallel to the surface, and hence the harmonic fields are radiated in directions normal to the waveguide surfaces. When this polarization is substituted into the radiative version of Eq. (7), $\beta_p = \beta^{(\nu)} = 0$, and integration over the depth coordinate gives the amplitudes of the fields radiated along the ±z directions.

The second harmonic signal has been observed for pulses of 1.06-μm that are coupled by a radiation prism into opposite ends of a Ti:in-diffused $LiNbO_3$ waveguide. The experimental and theoretical results were in good agreement [21].

This interaction is of special interest because it can be used for all-optical signal processing or as a picosecond transient digitizer. This application is summarized in Fig. 5. For two incident optical waveforms of the form $U_+(t - x/v)$ and $U_-(t + x/v)$, the field radiated normal to the surface is proportional directly to the instantaneous overlap of the two waveforms. The resulting waveform is the real-time convolution of the incoming waveforms, as shown in Fig. 5d. This outgoing signal profile can be captured by placing a CCD array above and parallel to the waveguide surface. The convolution aspect of this interaction has been demonstrated experimentally [22]. If one of the input beams is effectively a δ-function in time, the radiated signal is exactly the temporal envelope of the second signal. Hence this device can operate as a picosecond transient digitizer for analyzing picosecond laser pulses.

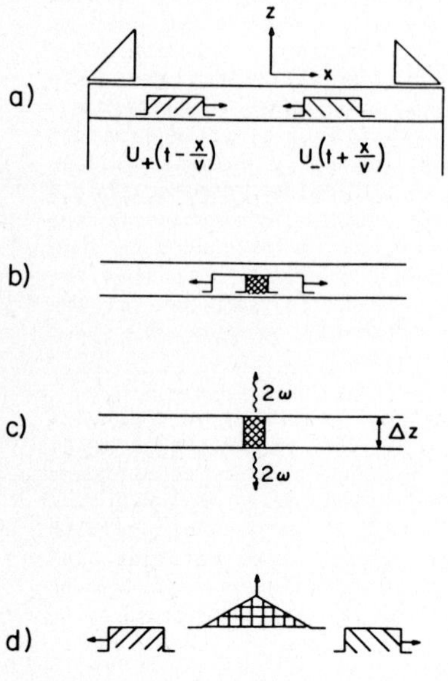

Fig. 5. Schematic diagram of signal convolution by the nonlinear mixing of two counter-propagating waves in an optical waveguide.

6. Parametric Processes

There are a number of other nonlinear processes that utilize the second-order susceptibility. Those implemented to date in waveguide formats include sum and difference frequency generation, parametric amplification, and parametric oscillation. The only waveguides of sufficient quality for producing parametric amplification and oscillation have been Ti:in-diffused $LiNbO_3$ waveguides.

(a) Sum and Difference Frequency Generation

This case is essentially a generalization of the preceding SHG discussion. There are two separate incident guided waves at the frequencies ω_a and ω_b ($\omega_a \geq \omega_b$) that mix via the second-order susceptibility to produce nonlinear polarization fields at the frequencies $\omega_s = \omega_a \pm \omega_b$. The pertinent nonlinear polarization fields are

$$P_i^{NL}(z,\omega_a+\omega_b) = 2\varepsilon_0 d_{ijk} E_j^{(m,a)}(z) E_k^{(m',b)}(z) a^{(m,a)}(x) a^{(m',b)}(x) , \tag{13a}$$

$$P_i^{NL}(z,\omega_a-\omega_b) = 2\varepsilon_0 d_{ijk} E_j^{(m,a)}(z) E^*{}_k^{(m',b)}(z) a^{(m,a)}(x) a^{*(m',b)}(x) , \tag{13b}$$

$$P_i^{NL}(z,\omega_a) = 2\varepsilon_0 d_{ijk} E_j^{(n,s)}(z) E^*{}_k^{(m',b)}(z) a^{(n,s)}(x) a^{*(m',b)}(x) , \tag{13c}$$

$$P_i^{NL}(z,\omega_b) = 2\varepsilon_0 d_{ijk} E_j^{(n,s)}(z) E^*{}_k^{(m,a)}(z) a^{(n,s)}(x) a^{*(m,a)}(x) . \tag{13d}$$

It is of course possible to derive a perfectly general formalism for this case, similar to that just discussed for SHG. Since the reported conversion efficiencies in waveguides have been small, only the weak generation case is developed here. The difference frequency case has been studied in the context of parametric amplification and will be addressed in the next section. For sum frequency generation with phase-matching over a distance L

$$|a^{(n,s)}(L)|^2 = (\omega_a + \omega_b)^2 L^2 \, |K|^2 \, |a^{(m,a)}(0)|^2 \, |a^{(m',b)}(0)|^2 , \tag{14a}$$

$$K = \frac{\varepsilon_0}{2} \int_{-\infty}^{\infty} dz \; d_{ijk} \; E_j^{(m,a)}(z) E_k^{(m',b)}(z) E^*{}_i^{(n,s)}(z) . \tag{14b}$$

The same comments about phase-matching, the overlap integral, etc. made for the SHG case are also valid here.

Two experiments have been reported to date, both in $LiNbO_3$ waveguides. Sum frequency generation has been reported by Uesugi et al. [23] in in-diffused $LiNbO_3$ channel waveguides and by Reutov and Tarashchenko [24] in thin $LiNbO_3$ platelets. They (Uesugi) used either a HeNe (1.19 μm) or a Nd:YAG laser in conjunction with an optical parametric oscillator operating in the near infrared to produce tunable sum-frequency radiation from 0.532 μm to 0.545 μm. Their tuning characteristics are shown in Fig. 6.

(b) Parametric Amplification and Oscillation

Difference frequency generation, usually in the form of parametric amplification or oscillation, is a very useful technique for producing tunable radiation in the infrared. In this case, a whole new nomenclature unique to parametric processes has been developed. The signal, "pump," and "idler" waves have the frequencies ω_s (ω_b), ω_p (ω_c) and ω_i (ω_a), respectively.

In parametric amplification, a weak input signal at ω_s is amplified by the phase-matched conversion of the pump wave into both a signal and an idler wave, that is, $\omega_p = \omega_s + \omega_i$ and $\beta^{(m,p)}k_p = \beta^{(n,s)}k_s + \beta^{(m',i)}k_i$. For parametric oscillation, only the pump beam is incident, and the signal and idler build up from noise in the resonator cavity.

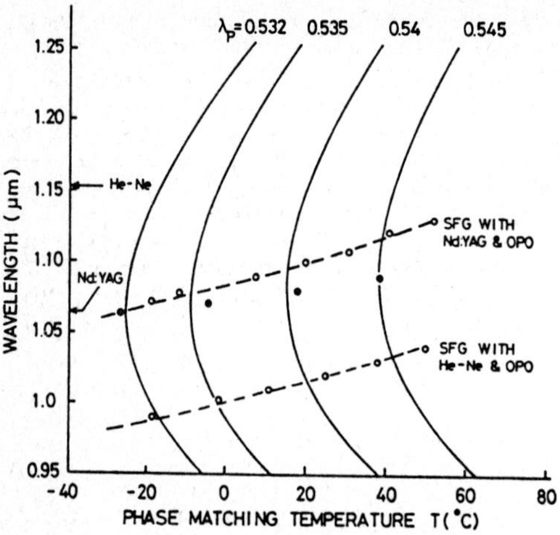

Fig. 6. Tuning characteristics for second harmonic generation (SHG) and sum frequency generation (SFG) in Ti:in-diffused $LiNbO_3$ channel waveguides. Closed and open circles are experimental data for SHG and SFG respectively. The solid lines correspond to tuning curves. The broken lines are theoretical curves for SFG [23].

The appropriate nonlinear polarization fields are given by Eqs. (13c) and (13d). Substituting into Eq. (7), assuming phase-matching and no significant depletion of the pump beam,

$$a^{(m',b)}(x) = a^{(m',b)}(0) \cosh\frac{x}{\ell_{OPA}} , \tag{15a}$$

$$a^{(m,a)}(x) = i\sqrt{\frac{\beta^{(m',b)}\omega_b}{\beta^{(m,a)}\omega_a}}\; a^{*(m',b)}(0)\; \frac{a^{(n,s)}(0)}{|a^{(n,s)}(0)|}\; \sinh\frac{x}{\ell_{OPA}} , \tag{15b}$$

$$\frac{1}{\ell_{OPA}} = \sqrt{k_a k_b}\; |K|\; |a^{(n,s)}(0)| . \tag{15c}$$

$$K = \frac{c\varepsilon_0}{2}\int_{-\infty}^{\infty} dz\; d_{ijk}E^{*}_{i}{}^{(m,a)}(z)E_j{}^{(n,s)}(z)E_k{}^{(m',b)}(z) . \tag{15d}$$

In the limit $x \gg \ell_{OPA}$, the hyperbolic functions degenerate into exponentials with gain coefficient $1/\ell_{OPA}$.

Optical parametric amplification and the corresponding difference frequency generation have been studied experimentally in $LiNbO_3$ waveguides by two groups [11,20]. Uesugi [11] reported the generation of a difference frequency signal when an idler wave from a cw Nd:YAG ($\lambda \simeq 1.318$ μm) laser and a dye laser ($\lambda \simeq 0.58$ μm), which served as the pump, were mixed in a Ti:in-diffused $LiNbO_3$ channel waveguide. Using the mode combination $TE_0 + TM_0 \rightarrow TM_0$ with pump and idler powers of 70 mW and 1.7 mW respectively, a generation efficiency of 0.014% was obtained at $\lambda \simeq 1.035$ μm. By tuning the temperature and adjusting the dye laser wavelength appropriately, the phase-matching condition was tuned and the resulting difference frequency signal was varied from 1.035 μm to 1.19 μm.

Sohler and Suche managed to reduce their waveguide losses sufficiently so that they were able to measure signal gain. In the initial experiments [20], the signal source was a cw HeNe laser (λ = 1.15 μm) and the pump was derived from a tunable pulsed dye-laser operating in the visible ($\lambda \simeq 0.65$ μm). The best single-pass parametric gain observed was 1.75 obtained with a peak pump power of about 200 W and a TE_{10}(pump) $\rightarrow$ TM_{00}(signal, idler) coupling. Much improved performance was obtained in a later experiment [20] with a 20-μm-wide channel, 48-mm long. With 150 W of peak power, a gain of 16 dB, which corresponds to a signal amplification of 43, was achieved.

The quality of the 10-μm-wide channel waveguides made by Sohler and Suche was so good that parametric oscillation was obtained when the ends of the channels were coated to a reflectivity of 0.96 at the signal wavelength [20]. At high enough pump powers, the signal and idler gain surpasses the resonator losses, and simultaneous oscillation at the signal and idler frequencies will start from noise. This will, of course, happen only under phase-matching conditions. The results for the signal and idler power generated are shown in Fig. 7 as a function of pump power (at λ = 0.598 μm). For a pump power of 14 W, 1 W is converted into the sum of signal and idler power. Varying the input pump wavelength results in tuning of the signal idler wavelengths as phase-matching is maintained. This is illustrated in Fig. 8. By also varying temperature, the signal and idler wavelengths can be tuned from 0.587 μm to 0.616 μm.

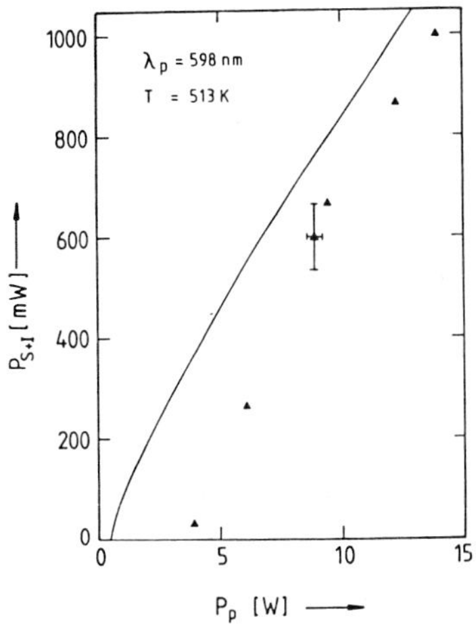

Fig. 7. Sum of the signal and idler powers versus pump wavelength for optical parametric oscillation in Ti:in-diffused $LiNbO_3$ channel waveguides. The solid line is the theoretical curve [20].

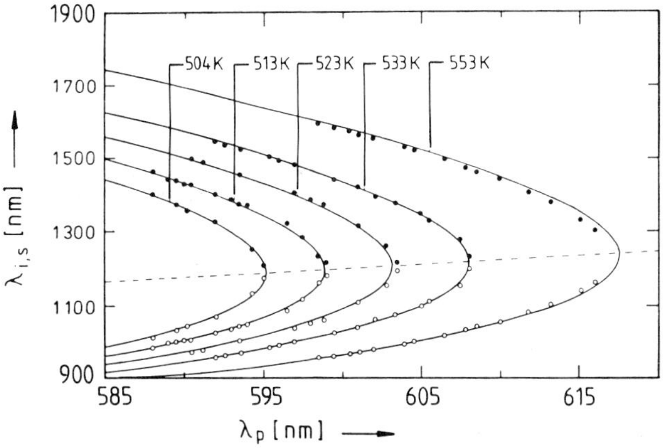

Fig. 8. The signal and idler wavelengths versus pump wavelength for optical parametric oscillation in a Ti:in-diffused $LiNbO_3$ channel waveguide at a series of temperatures [20].

7. Degenerate Four-Wave Mixing

Degenerate four-wave mixing involves three input beams and one output beam, all at the same frequency. It has been studied extensively with plane waves, as well as in fibers [3]. It is unique because it is automatically phase-matched in this interaction geometry, which is shown for two guided-wave cases in Fig. 9. Wavevector conservation is given by $\beta^{(m,1)}\mathbf{k}_1 + \beta^{(m',2)}\mathbf{k}_2 + \beta^{(m'',3)}\mathbf{k}_3 + \beta^{(n,4)}\mathbf{k}_4 = 0$, with $k_1 = k_2 = k_3 = k_4$. If the two input beams 1 and 2 are exactly contra-directional and have the same mode number $m = m'$, $\beta^{(m,1)}\mathbf{k}_1 = -\beta^{(m,2)}\mathbf{k}_2$. Therefore, if $m'' = n$, then $\beta^{(n,4)}\mathbf{k}_4 = -\beta^{(n,3)}\mathbf{k}_3$ and the signal beam 4 is generated backward along the direction of beam 3, independent of the incidence direction of beam 3. Note that this process requires that beams 1 and 2 have the same mode order m, and that beam 4 is radiated into the same mode index n as beam 3, and m and n are not necessarily equal [25].

The nonlinear polarization field corresponding to degenerate four-wave mixing can be obtained from Eqs. (5). For an isotropic material with equal mode indices for all of the input beams,

$$P^{NL}_i(z,\omega) = 4\varepsilon_0\chi^{(3)}_{1122}(-\omega,\omega,-\omega,\omega)\, E_i^{(n,3)}(z)E^*_j{}^{(n,2)}(z)E_j^{(n,1)}(z) + 2\varepsilon_0\chi^{(3)}_{1221}(-\omega,\omega,-\omega,\omega)\, E^*_i{}^{(n,1)}(z)E_j^{(n,2)}(z)E_j^{(n,3)}(z)\ . \tag{16}$$

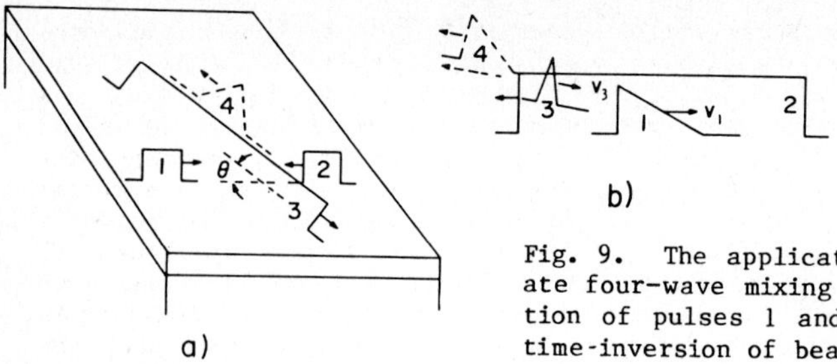

Fig. 9. The application of degenerate four-wave mixing to (a) convolution of pulses 1 and 2, and (b) the time-inversion of beam 1.

This expression simplifies far from any resonant behavior in the $\chi^{(3)}$ terms and $\chi^{(3)}{}_{1122} = \chi^{(3)}{}_{1221} = \chi^{(3)}{}_{eff}$. For guided wave fields,

$$P^{NL}{}_{\gamma i}(z,\omega) = 2c\varepsilon_0{}^2 n_\gamma{}^2 n_{2\gamma}[\tfrac{2}{3}E_{\gamma i}{}^{(n,3)}(z)E_{\gamma j}{}^{(n,1)}(z)E^*{}_{\gamma j}{}^{(n,2)}(z) + \tfrac{1}{3}E_{\gamma j}{}^{(n,3)}(z)E_{\gamma j}{}^{(n,2)}(z)E^*{}_{\gamma i}{}^{(n,1)}(z)] \,, \tag{17}$$

in each medium (labeled γ) where $n_{2\gamma} \neq 0$. Here we have defined $3\chi^{(3)}{}_{eff} = n_{2\gamma}n_\gamma{}^2\varepsilon_0 c^2$, the justification for which will be discussed later. Assuming that TE beams 1 and 2 propagate along x', which is oriented at an angle θ to the x-axis along which TE beams 3 and 4 are traveling,

$$\mathbf{E}^{(n,1 \text{ or } 2)}(z) = E^{(n,1)}(z)\,[\sin\theta,\cos\theta,0] \;;\; E^{(n,3 \text{ or } 4)}(z) = E^{(n,3)}(z)\,[0,1,0]. \tag{18}$$

In the limit of no attenuation or depletion for the input beams, and assuming that the beams 1, 2, and 3 are introduced into the interaction region at x' = 0, x' = L', and x = L, respectively, substituting into Eq. (7) and integrating the signal beam from 0 → L gives

$$a^{(n,4)}(L) = (kL)\; A\; a^{(n,1)}(0)a^{(n,2)}(L')a^{(n,3)}(0) \,, \tag{19}$$

$$A = \frac{ic^2\varepsilon_0{}^2}{2}\,\frac{2+\cos^2\theta}{3}\int_{-\infty}^{\infty} dz\, n_\gamma(z)^2 n_{2\gamma}(z) \times E_\gamma{}^{(n,1)}(z)E_\gamma{}^{(n,2)}(z)^* E_\gamma{}^{(n,3)}(z)E^*{}_\gamma{}^{(n,4)}(z). \tag{20}$$

In the general case, it is necessary to take incident beam depletion into account, which leads to a series of coupled mode equations between the amplitudes of the various beams.

One experiment has been reported to date [26]. The nonlinear medium was CS_2 used as a cladding medium, inside a cell optically contacted to the surface of a thin-film glass waveguide, as shown in Fig. 10. Three coupling prisms were required for the four beams needed in the interaction, leading to difficult alignment problems. Since the nonlinear mixing occurred via the evanescent tail of the guided waves, and because the nonlinearity of CS_2 is small ($n_{2c} = 3\times10^{-18}$ m^2/W), the reflectivity (fractional conversion of beam 3 into beam 4) was only 10^{-9}. Nevertheless, the cubic dependence of the four-wave mixing signal was verified experimentally for 15-ns-long pulses from a Q-switched, doubled Nd:YAG laser.

This interaction has some possible applications to real-time all-optical signal processing [27], two of which are illustrated in Fig. 9. The two input signals to be convolved are beams 1 and 2, which have pulse envelopes $U_1(t - x/v)$ and $U_2(t + x/v)$. Whether the convolution process occurs or not can be controlled

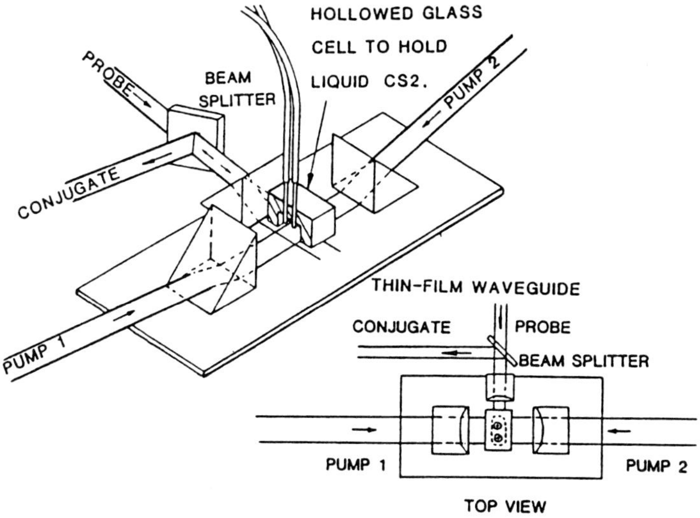

Fig. 10. Sample and coupling degenerate four-wave mixing geometry for CS_2 on a thin-film glass waveguide. The probe beam intersects the pump beams at 90° under the liquid cell [26].

by beam 3, which is assumed to have a constant amplitude during the overlap of pulses 1 and 2 inside the waveguide. Assuming that $\theta = \pi/2$ for this case and that L is small relative to the characteristic pulse size (to preserve detail in the convolved signal), the 4'th beam signal radiated is directly proportional to the instantaneous overlap of beams 1 and 2. Mathematically, the time evolution of the signal beam can be written as $U_4(t) \propto \int U_1(t-\tau)U_2(\tau)U_3\, d\tau$, where it has been explicitly assumed that the control beam is "turned on." For this application, the input signal length cannot be larger than the waveguide dimension, which limits the process to pulses of 100 ps or less.

Another potential application of four-wave mixing to real-time processing is time-reversal of an optical waveform, as illustrated in Fig. 9b. Here, $\theta \simeq 0$ and beams 1 and 2 are the waveform to be reversed and a control beam of constant amplitude, respectively. Beam 3 is a δ-function in time (very short pulse) and 4 is the time-reversed signal. A crucial factor here is the fact that beams 1 and 3 can be propagated in different modes. If the mode number of beam 3 is larger than that of beam 1, beam 3 can overtake beam 1 from "behind." Thus beam 4 is "read out" from trailing edge to leading edge, which corresponds to a time reversal of the pulse envelope.

8. Coherent Anti-Stokes Raman Scattering (CARS)

This interaction has been used with plane waves to obtain Raman spectra, and has recently been applied to the Raman spectroscopy of thin films [28] and monolayers deposited on thin films. It has the potential of becoming a very powerful, albeit specialized, experimental technique for studying surface species on a picosecond time-scale.

The basic concept of CARS [2] is summarized in Fig. 11a. Two laser beams of frequency and wavevector (ω_a, k_a) and (ω_b, k_b) with $\omega_a \geq \omega_b$ illuminate a sample containing molecules characterized by a vibrational frequency Ω_r. When the difference between the two laser frequencies is tuned to approximately the resonance frequency, that is $\omega_a - \omega_b \simeq \Omega_r$, the molecules are excited to vibrate at this difference frequency. This results in a coherent (in space and time)

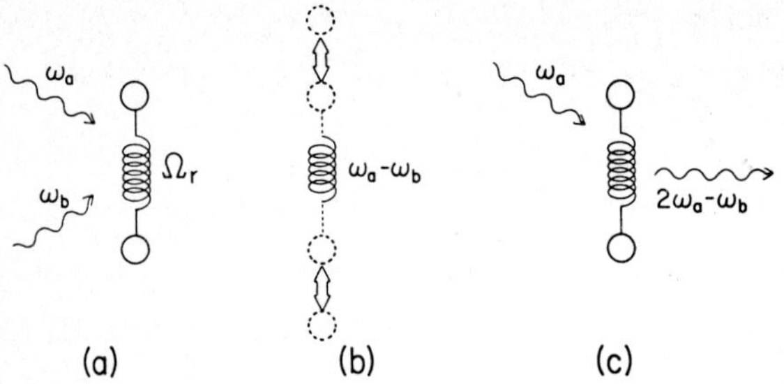

Fig. 11. Schematic of the CARS interaction. (a) Two lasers of frequency ω_a and ω_b mix at a molecule to excite it at the difference frequency $\omega_a-\omega_b$. (b) Laser "a" is Raman scattered by the excited molecule to produce a signal at $2\omega_a-\omega_b$.

excitation of the vibrational normal mode, which can be expressed as a contribution to the linear susceptibility of the material of the form

$$\chi_{ij}(\mathbf{r},t) = \frac{1}{2}\frac{\eta\varepsilon_0}{4m[\omega_a - \omega_b - \Omega_r - i\Gamma_r]} \alpha_{ik}\alpha^*_{j\ell}\, E_k^{(m,a)}(z)E^*_\ell{}^{(m',b)}(z)$$

$$\times\, e^{i[(\omega_a - \omega_b)t - (\beta^{(m,a)}\mathbf{k}_a - \beta^{(m',b)}\mathbf{k}_b).\mathbf{r}]} + \text{c.c.}\,, \qquad (21)$$

where η is the density of molecules, m is the electron mass, and α_{ik} is the Raman susceptibility tensor. Information about the orientation of the molecules relative to the coordinate system is contained in the α_{ik} terms. Light incident onto the molecules is now scattered via the usual Raman process, as shown in Fig. 11c. Since the induced nonlinear polarization is just the product of the susceptibility with the incident fields, nonlinear polarization fields oscillating at the frequencies $\omega_s = 2\omega_a - \omega_b$ (CARS) and $\omega_s = 2\omega_b - \omega_a$ (CSRS - Coherent Stokes Raman Scattering) are created. For CARS the spatial periodicity is given by $\beta_p\mathbf{k}_s = 2\beta^{(m,a)}\mathbf{k}_a - \beta^{(m',b)}\mathbf{k}_b$. The corresponding CARS nonlinear polarization field is

$$P_i^{NL}(z,2\omega_a-\omega_b) = \varepsilon_0\left[\chi_b{}^{(3)}{}_{ijk\ell} + \sum_r \chi_r{}^{(3)}{}_{ijk\ell}\frac{\Gamma_r/\pi}{[\omega_a - \omega_b - \Omega_r - i\Gamma_r]}\right]$$

$$\times\, E^{2(m,a)}(z)E^{*(m',b)}(z)\; a^{2(m,a)}(x)a^{(m',b)}(x)\,. \qquad (22)$$

The $\chi_b{}^{(3)}$ term corresponds to the background third-order susceptibility term due to the electronic degrees of freedom and can be considered to be independent of frequency. The details of the resonant $\chi_r{}^{(3)}$ terms can be obtained from Eq. (21).

The amplitude of the guided waves generated at the frequencies $2\omega_a - \omega_b$ and $2\omega_b - \omega_a$ are calculated from coupled mode theory via Eq. (7). Assuming an interaction distance L, a CARS guided-wave signal field (mode number n), which starts with zero amplitude at x = 0, and assuming no depletion of the incident beams

$$|a^{(n,s)}(L)|^2 = |\, k_sL\, |^2\, |K_{2a-b}|^2\, \frac{\sin^2\phi}{\phi^2}\, |a^{2(m,a)}(0)|^2\, |a^{(m',b)}(0)|^2\,, \qquad (23)$$

$$K_{2a-b} = \frac{c\varepsilon_0}{4}\int_{-\infty}^{\infty} dz\left[\chi_b{}^{(3)}{}_{ijk\ell} + \sum_r \chi_r{}^{(3)}{}_{ijk\ell}\frac{\Gamma_r/\pi}{[\omega_a-\omega_b-\Omega_r-i\Gamma_r]}\right] E^{2(m,a)}(z)E^{*(m',b)}(z), \qquad (24)$$

with $\phi = (1/2)(\beta^{(n,s)} - \beta_p)\mathbf{k}_s \cdot \mathbf{r}$. Phase-matching occurs for $\phi = 0$ and clearly can happen for only the CARS or SCRS case, but not for both simultaneously. Typically the angles between the two incident beams are only a few degrees.

How this phenomenon can be used for Raman spectroscopy is clear from Eqs. (23) and (24). The CRS signal, CARS and CSRS, are both proportional to $|K|^2$. If the background signal is negligible, or can be eliminated, then the CRS power is

$$P_s \propto \sum_r |\chi_r^{((3))}|^2 \frac{(\Gamma_r/\pi)^2}{[(\omega_a-\omega_b-\Omega_r)^2 + \Gamma_r^2]} . \tag{25}$$

The signal power is a maximum whenever $\omega_a - \omega_b = \Omega_r$. Therefore, by tuning the frequency difference $\omega_a - \omega_b$ it is possible to map out all of the vibrational transitions. Even when the background term cannot be eliminated, the signal power undergoes dispersion of some kind with frequency difference, and the characteristic Ω_r can be identified.

Numerical calculations [28] for thin films based on realistic materials indicate that the signal power on a strong resonance can be very large. For example, for a 2.0-μm-thick polystyrene film deposited on a glass substrate, for the 1000-cm^{-1} ring vibration, and for laser pulse energies of less than 0.1 mJ in pulses of ≈15-ns duration, efficiencies of ≈0.2% are predicted for the conversion of the incident beams into the CARS signal. Such laser energies are easily available, making this approach very promising for nonlinear spectroscopy. The initial experiments were carried out on 2-μm-thick polystyrene waveguides using two tuneable dye lasers with 100 ps long pulses [29]. The power densities in the two beams were 30 and 60 MW/cm^2 which corresponded to pulse energies of hundreds of nanojoules. The conversion efficiency was 0.2%, in excellent agreement with the theoretical calculations.

The remarkable efficiency of this process suggests applications to the investigation of monolayers deposited on film surfaces [28]. The problem is that the background term for the film can lead to a large signal that can mask the desired monolayer signal. In other words, under normal conditions the guiding film can be thousands of Angstroms thick, whereas a monolayer has a thickness of say 5 Å. The problem, therefore, is to eliminate the background term in K_{2a-b} as completely as possible. The solution lies in the coherent nature of this process, and the advantages of using guided waves. As noted before for second-harmonic generation, the overlap integral term is reduced if guided waves with unequal mode numbers are used for the interacting fields. If at least one of the input fields corresponds to a higher order mode, that is TE_m with $m > 0$, interference effects occur, and hence the film contribution can be minimized leaving the monolayer term as the dominant contribution. This approach has been used under vacuum conditions to observe the bonding of Ti atoms to oxygen bonds on the surface of a Nb_2O_5 film. The result is shown in Fig. 12 in which two TiO_2 vibrational modes were observed.

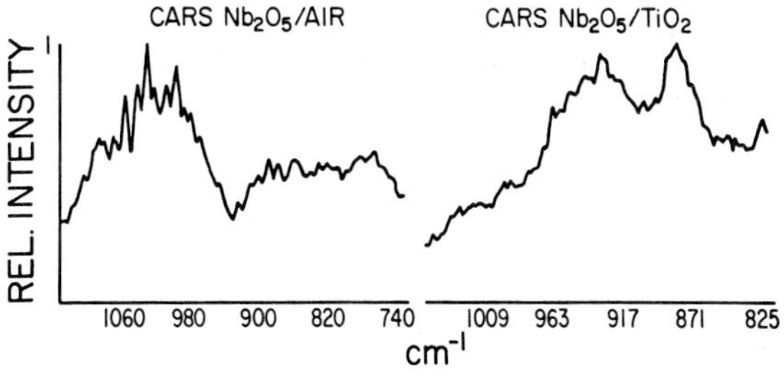

Fig. 12. Waveguide CARS signal from species on the surface of a Nb_2O_5 film. (a) Spectrum of contaminants on surface in laboratory environment. (b) Spectrum of Ti-O vibration in vacuum from monolayer of Ti atoms.

This new technique could prove valuable for surface science. Since a variety of different polarizations can be used for the incident and signal guided-wave modes, a large number of the nonlinear susceptibility terms can be measured. Their relative magnitudes will yield direct information about the orientation of surface species. Furthermore, this remarkable projected sensitivity may also find application in identifying both the nature and location of impurities in thin films. For example, different mode combinations can be used for the input and output guided waves, and the interference effects obtained in K_{2a-b} can then be used to locate the impurity sources as a function of depth into the film.

9. Intensity-Dependent Refractive Index and Dielectric Tensor

The optical properties, specifically the refractive index and relative dielectric constant of a medium, can be changed by the presence of a high-power optical beam. For plane waves, this leads to self-focusing or self-defocusing of the beam, depending on the sign of the nonlinearity. Starting from Eq. (5b) for a single guided wave, the pertinent nonlinear polarization for an isotropic medium is [30,31]

$$P^{NL}_i(z,\omega) = 2\varepsilon_0\chi^{(3)}_{1122}(-\omega,\omega,-\omega,\omega)\, E_i^{(n)}(z)E^*_j{}^{(n)}(z)E_j^{(n)}(z) + \varepsilon_0\chi^{(3)}_{1221}(-\omega,-\omega,\omega,-\omega)\, E^*_i{}^{(n)}(z)E_j^{(n)}(z)E_j^{(n)}(z)\,. \tag{26}$$

Far from any resonant behavior, Kleinman symmetry is valid and $\chi^{(3)}_{1122} = \chi^{(3)}_{1221} = \chi^{(3)}_{eff}$. For plane waves, Eq. (26) simplifies, and, including the linear susceptibility, the total polarization is

$$P_i(\omega) = \varepsilon_0[\chi^{(1)}_{ii} + 3\chi^{(3)}_{eff}|E_j(\omega)|^2]\, E_i(\omega)\,. \tag{27}$$

Noting that $\chi^{(1)}_{ii} = n_\gamma^2 - 1$ and expressing $|E_j(\omega)|^2$ in terms of the intensity $S = (1/2)c\varepsilon_0 n_\gamma|E_j(\omega)|^2$, the refractive index can be written as

$$n = n_\gamma + n_{2\gamma}S\,; \quad n_{2\gamma} = \frac{3\chi^{(3)}_{eff}}{n_\gamma\varepsilon_0 c}\,. \tag{28}$$

For guided-wave fields, the expression for the nonlinear polarization is more complex than that for the plane wave case. It is given by [31]

$$P^{NL}_{\gamma i}(z,\omega) = c\varepsilon_0{}^2 n_\gamma{}^2 n_{2\gamma}\left[\frac{2}{3}\, E_{\gamma i}^{(n)}(z)|E_{\gamma j}^{(n)}(z)|^2 + \frac{1}{3}\, E^2{}_{\gamma j}^{(n)}(z)E^*{}_{\gamma i}^{(n)}(z)\right]. \tag{29}$$

For TE this simplifies to

$$P^{NL}_{\gamma y}(z,\omega) = c\varepsilon_0{}^2 n_\gamma{}^2 n_{2\gamma}|E_{\gamma y}^{(n)}(z,\omega)|^2 E_{\gamma y}^{(n)}(z,\omega)\,. \tag{30}$$

The situation with TM waves is more complex, because there are two field components E_x and E_z. The difficulty arises from the $E^2{}_{\gamma j}^{(n)}(z)E^*{}_{\gamma i}^{(n)}(z)$ term whose value depends on the relative phase [31] between the fields for products with $i\neq j$. For example, writing

$$E_{\gamma x}^{(n)}(z) = |E_{\gamma x}^{(n)}(z)|\,; \quad E_{\gamma z}^{(n)}(z) = e^{i\Psi}\,|E_{\gamma z}^{(n)}(z)|\,, \tag{31}$$

then

$$P^{NL}_{\gamma x}(z,\omega) = c\varepsilon_0{}^2 n_\gamma{}^2 n_{2\gamma}\left[|E_{\gamma x}^{(n)}(z)|^2 + \left(\frac{2}{3} + \frac{e^{2i\psi}}{3}\right)|E_{\gamma z}^{(n)}(z)|^2\right] E_{\gamma x}^{(n)}(z)\,, \tag{32a}$$

$$P^{NL}_{\gamma z}(z,\omega) = c\varepsilon_0{}^2 n_\gamma{}^2 n_{2\gamma}\left[|E_{\gamma z}^{(n)}(z)|^2 + \left(\frac{2}{3} + \frac{e^{-2i\psi}}{3}\right)|E_{\gamma x}^{(n)}(z)|^2\right] E_{\gamma z}^{(n)}(z). \tag{32b}$$

Fortunately, in media for which $\beta^{(n)} > n_\gamma$, $\Psi = \pi/2$ and Eqs. (32) simplify to

$$P^{NL}_{\gamma x}(z,\omega) = \varepsilon_0\alpha_{xx}|E_{\gamma x}^{(n)}(z)|^2 + \varepsilon_0\alpha_{xz}|E_{\gamma z}^{(n)}(z)|^2 , \quad (33a)$$

$$P^{NL}_{\gamma z}(z,\omega) = \varepsilon_0\alpha_{xz}|E_{\gamma x}^{(n)}(z)|^2 + \varepsilon_0\alpha_{zz}|E_{\gamma z}^{(n)}(z)|^2 , \quad (33b)$$

$$\alpha_{ii} = c\varepsilon_0 n_\gamma^2 n_{2\gamma} \qquad \alpha_{ij}(i\neq j) = \frac{1}{3}c\varepsilon_0 n_\gamma^2 n_{2\gamma} . \quad (33c)$$

Before we proceed to guided-wave problems involving nonlinear media, it is useful to consider the effects of nonlinearities on the propagation of plane-wave beams with a finite aperture. The case $n_{2\gamma} > 0$ is called a self-focusing nonlinearity for reasons that are evident in Fig. 13a. Because the phase velocity $v = c/n$ decreases with increasing power, regions of high intensity along a wavefront propagate a shorter distance than regions of low intensity. This leads to curvature in the wavefront equivalent to passing through a focusing lens, and the beam self-focuses under the action of its own high intensity. Conversely, $n_{2\gamma} < 0$ leads to self-defocusing of the wavefront, and hence this case is termed a self-defocusing nonlinearity.

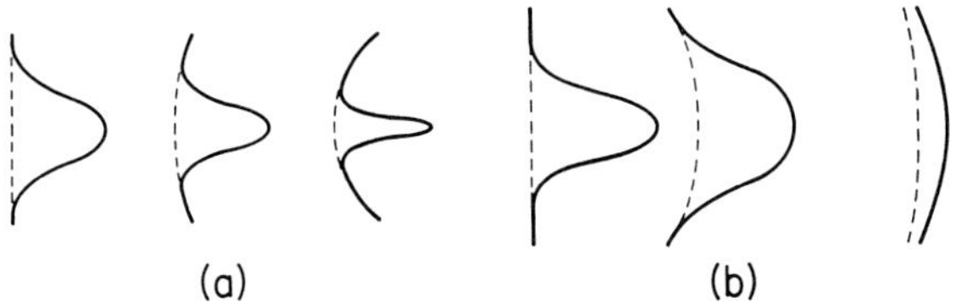

Fig. 13. (a) Evolution of plane-wave field into a self-focused beam in a self-focusing medium. (b) Defocusing of a plane wave in a self-defocusing medium.

10. Nonlinear Waves - Single Interface

It is well known that waves cannot be guided by the interface between two dielectric media whose refractive indices do not depend on intensity. (TM polarized surface plasmon polaritons can be guided by the interface between a dielectric and a metal by virtue of the negative dielectric constant of the metal: Here we do not consider surface plasmon polaritons because of their high attenuation coefficients.) However, when at least one of the two dielectric media exhibits a self-focusing nonlinearity, a guided wave can exist above a well-defined power threshold [34-36].

(a) TE Waves

The single-interface guided-wave geometry is shown in Fig. 1 with h = 0. Including the nonlinearities discussed in the preceding section, the nonlinear wave equation for TE polarized (the only case to be solved exactly to date) waves is [32]

$$\nabla^2 E_{\gamma y}(\mathbf{r},t) + \frac{\omega^2}{c^2}[\varepsilon_\gamma + \alpha_\gamma|E_{\gamma y}(z)|^2]E_{\gamma y}(\mathbf{r},t) = 0 , \quad (34)$$

where $\alpha_\gamma = \alpha_{yy}$ and $\varepsilon_\gamma = n_\gamma^2$. The most general form for the field is

$$E_{\gamma y}(\mathbf{r},t) = \frac{1}{2}E_\gamma(z)\exp i[\beta kx - \omega t + \phi(z)] + c.c. , \quad (35)$$

where the $E_\gamma(z)$ is real. Substituting into Eq. (34) yields, after equating real and imaginary parts separately to zero,

$$\frac{d^2}{dz^2}E_\gamma(z) - E_\gamma(z)\left[\frac{d\phi}{dz}\right]^2 - q_\gamma{}^2k^2E_\gamma(z) + k^2\alpha_\gamma E^3{}_\gamma(z) = 0 \ ; \qquad \frac{d}{dz}\left[E^2{}_\gamma(z)\frac{d\phi}{dz}\right] = 0 \ . \tag{36}$$

The last equation expresses the conservation of energy flux in the nonlinear medium and integrates to a constant K [33]. Therefore the complete equation for $E_\gamma(z)$ is

$$\frac{d^2}{dz^2}E_\gamma(z) - q_\gamma{}^2k^2E_\gamma(z) - \frac{K^2}{E^3{}_\gamma(z)} + k^2\alpha_\gamma E^3{}_\gamma(z) = 0 \ . \tag{37}$$

For the two semi-infinite bounding media, the fields must decay to zero at $\pm\infty$ so that $K = 0$ for this case and the fields are completely real. (For thin film media, or cases in which there is a radiation component that extends to infinity, K need not be zero and this case must be considered carefully.) For a nonlinear upper (cladding, $\gamma \equiv c$, $z \leq 0$) medium, the field solutions are [32]

$$E_c(z) = \sqrt{\frac{2}{\alpha_c}}\ \frac{q_c}{\cosh[q_ck(z_c - z)]} \qquad n_{2c} > 0 \ , \tag{38a}$$

$$E_c(z) = \sqrt{\frac{2}{|\alpha_c|}}\ \frac{q_c}{\sinh[q_ck(z_c - z)]} \qquad n_{2c} < 0 \ , \tag{38b}$$

where $q_c{}^2 = \beta^2 - n_c{}^2$. For $n_{2\gamma} > 0$, $z_c < 0$ corresponds to a self-focused field maximum in the cladding region. For $n_{2\gamma} < 0$, $z_s < 0$ leads to a divergent field in the cladding, a solution that is rejected as unphysical. Therefore the parameters z_c, which depend directly on the guided-wave power, play pivotal roles in defining the field distributions. Similar field solutions are obtained for the substrate medium ($z \geq 0$, $\gamma \equiv s$) with q_c replaced by q_s, and $z_c - z$ by $z + z_s$. For a linear substrate

$$E_s(z) = A_s \exp(-q_skz) \qquad n_{2s} = 0 \ , \tag{39}$$

with $q_s{}^2 = \beta^2 - n_s{}^2$. The parameter z_s, which is related to z_c via the continuity of the tangential electric field, identifies where field maxima occur, and whether the field solutions are physically meaningful.

Matching the tangential electric and magnetic fields across the boundary leads to the dispersion relation [36]

$$q_c \tanh(kq_cz_c) = -q_s \tanh(kq_sz_s) \ ; \quad \alpha_c,\alpha_s > 0 \ . \tag{40}$$

Using continuity of the E_y field across $z = 0$ and the dispersion relation, one can show that

$$E_c{}^2(0) = 2\frac{n_s{}^2 - n_c{}^2}{\alpha_c - \alpha_s} \ , \tag{41}$$

which predicts the interesting result that the field at the surface is a constant, independent of guided-wave power. For $n_s > n_c$ (which also implies $q_c > q_s$), the condition $\alpha_c > \alpha_s$ must be satisfied. Furthermore, if $\alpha_\gamma < 0$, then $\tanh(kq_\gamma z_\gamma) \rightarrow \text{cotanh}(kq_\gamma z_\gamma)$ from which all four possibilities are obtained. The limit corresponding to $\alpha_\gamma \rightarrow 0$ (a refractive index that is independent of power) is obtained formally by taking the limit $z_\gamma \rightarrow 0$. Taking into account Eqs. (40) and (41), solutions exist for the following cases (assuming $n_s > n_c$)

$\alpha_c > 0$ and $\alpha_s > 0$ with $\alpha_c > \alpha_s$; $z_c < 0$, $z_s > 0$ <u>or</u> $z_c > 0$, $z_s < 0$

$\alpha_c > 0$ and $\alpha_s < 0$; $z_c < 0$ and $z_s > 0$ $\qquad$ $\alpha_c > 0$ and $\alpha_s = 0$; $z_c < 0$, $z_s \rightarrow \infty$.

In each case, at least one of the media must have a self-focusing nonlinearity.

The pertinent experimental quantity is the guided-wave power per unit length along the wavefront (y-axis) in Watts per meter or milliwatts per meter. This is obtained from

$$P = \int_{-\infty}^{\infty} \mathbf{ExH}\, dz = P_c + P_s \tag{42}$$

$$P_\gamma = \frac{\beta q_\gamma}{k n_\gamma^2 n_{2\gamma}} \left[1 - [\tanh(k q_\gamma z_\gamma)]^{\pm 1}\right], \tag{43a}$$

where the + and - signs refer to $\alpha_\gamma > 0$ and $\alpha_\gamma < 0$, respectively. If $\alpha_s = 0$ and $\alpha_c > 0$,

$$P_s = \frac{\beta q_c^2}{2 k q_s n_c^2 |n_{2c}| \cosh^2(k q_c z_c)}, \tag{43b}$$

and the $\alpha_c < 0$ case is obtained by replacing cosh with sinh. Note that these expressions contain the parameter z_γ explicitly. Therefore the position of the self-focused peak depends on the guided-wave power.

To date, no exact, analytical, closed-form solution of the nonlinear wave equation which includes loss has been reported. It is, however, very useful as a guideline for experiments to estimate in a simple way the effect of the linear attenuation coefficient on nonlinear guided waves. Assuming that the field distribution obtained in the lossless case will still be valid if the attenuation per wavelength is small, it is possible to calculate the nonlinear guided-wave attenuation coefficient approximately from the imaginary component of the dielectric constant [36]. Expanding $\nabla\cdot(\mathbf{ExH})$ and substituting from Maxwell's equations for $\nabla\mathbf{xE}$ and $\nabla\mathbf{xH}$ leads to

$$\nabla\cdot(\mathbf{ExH}) = \frac{1}{2}\omega\varepsilon_0\varepsilon_{\gamma I}\mathbf{E}(z)\cdot\mathbf{E}^*(z), \tag{44}$$

where $\varepsilon_{\gamma I}$ is the imaginary component of the dielectric constant in the γ'th medium. Integrating over a small volume element, applying the divergence theorem, and assuming small losses gives

$$\beta_I = \Sigma\varepsilon_{\gamma I} P_\gamma / 2P\beta_R, \tag{45}$$

where β_I is the imaginary component of the effective index $\beta = \beta_R - i\beta_I$.

Sample calculations are shown in Figs. 14 and 15 for β versus guided wave power [36]. For a single nonlinear (self-focusing) medium, there is a minimum

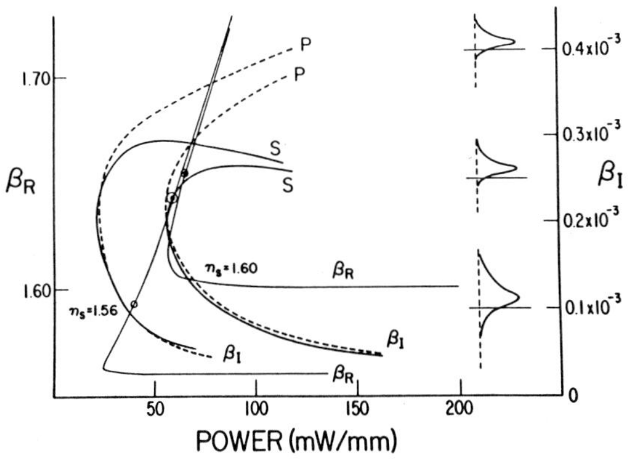

Fig. 14. The effective index $\beta = \beta_R - i\beta_I$ versus power for a wave guided by the interface between a linear and a nonlinear dielectric [36]. Here $\varepsilon_c = 2.4025 - 0.001i$, $n_{2c} = 10^{-9}$ m^2/W, and $n_s = 1.56$ or 1.6 for s- and p-polarized $(\varepsilon_z(E_z^2))$ waves. The evolution in field distribution with β_R is also shown for $n_s = 1.56$. The single and double circles correspond to $\Delta n = 0.1$ and 0.2 respectively [36].

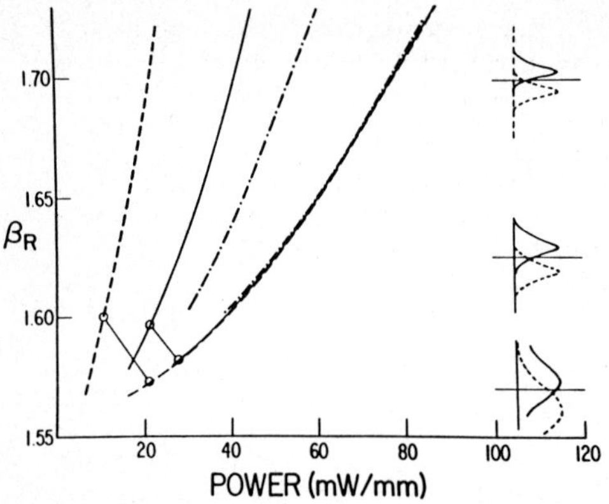

Fig. 15. The effective index versus guided-wave power for TE polarized waves guided by a single dielectric interface separating two self-focusing nonlinear media. Here $n_c = 1.55$, $n_s = 1.56$ and $n_{2s} = 2\times10^{-9}$ m^2/W. The dashed, solid, and dash-dotted curves correspond to $n_{2c} = 4\times10^{-9}$, 2×10^{-9} and 1.5×10^{-9} m^2/W respectively. The separated curves are for $z_c > 0$, and the clustered curves for $z_s > 0$. The field distributions are for the $n_{2c} = 4\times10^{-9}$ m^2/W case [36].

threshold power required for a guided wave to exist. The larger the difference in indices $(n_s - n_c)$ between the two media, the higher the threshold power. This is to be expected, since the index of the cladding near the interface must be increased to the point that it is larger than that of the substrate, which requires higher powers for larger initial values of $n_s - n_c$. As $\beta \to n_s$ (cutoff condition), the field in the substrate degenerates into a plane wave and the substrate power diverges, see Eq. (43b). For two self-focusing media, there are two separate branches, each corresponding to a self-focused field maximum in a different medium. In both figures, the higher the guided-wave power (past the minimum power point), the narrower the self-focused peak.

The key quantity that will limit the experimental utilization of these nonlinear guided waves is Δn_{sat}, the maximum change in refractive index that can be induced optically in a real medium. This quantity is typically less than 0.01, but can be as large as 0.1 for some exceptional cases such as orientational nonlinearities in liquid crystals or semiconductors near their band gaps. As shown in Figs. 14 and 15, this material constraint limits the initial index difference $n_s - n_c$.

(b) TM Waves

The analysis of this case is complicated greatly by the existence of two field components, both of which can contribute to field-dependent dielectric constants and refractive indices. The best starting point for deriving the pertinent nonlinear wave equations are the Maxwell's equations $\nabla\times\mathbf{H} = i\omega\varepsilon_0\varepsilon\mathbf{E}$ and $\nabla\times\mathbf{E} = -i\omega\mu_0\mathbf{H}$. For TM waves, this leads to

$$E_{\gamma x} = \frac{i}{\omega\varepsilon_0\varepsilon_{xx}}\frac{d}{dz}H_{\gamma y} \qquad E_{\gamma z} = -\frac{\beta}{c\varepsilon_0\varepsilon_{zz}}H_{\gamma y} \qquad -i\omega\mu_0 H_{\gamma y} = \frac{d}{dz}E_{\gamma x} + i\beta k E_{\gamma z}\ , \tag{46}$$

from which nonlinear wave equations can be constructed in terms of $E_{\gamma x}$, $E_{\gamma z}$, or $H_{\gamma y}$. However, no analytic solutions have been found yet to these equations without some form of simplifying approximations. The most common one has been to assume $\alpha_{\gamma xz} = \alpha_{\gamma zx} = \alpha_{\gamma zz} = 0$ [37] which we call $\varepsilon_{xx}(|E_x|^2)$. The second has been $\alpha_{\gamma xz} = \alpha_{\gamma zx} = \alpha_{\gamma xx} = 0$ [38], identified as $\varepsilon_{zz}(|E_z|^2)$. Yet a third approach with a different set of approximations has been suggested. Here we will concentrate on the first two, with the purpose of justifying the second one in terms of typical material parameters.

We discuss first the nonlinear wave equation for $E_{\gamma x}$ in the $\varepsilon_{xx}(|E_x|^2)$ approximation [37]. Eliminating $H_{\gamma y}$ and $E_{\gamma z}$ and assuming $K = 0$,

$$\frac{d^2}{dz^2}E_{\gamma x}(z) - q_\gamma{}^2k^2E_{\gamma x}(z) - \frac{k^2q_\gamma{}^2}{\varepsilon}\,\alpha_\gamma E^3{}_{\gamma x}(z) = 0\ , \tag{47}$$

which has solutions of the form discussed previously for the TE case. There is, however, one very important difference, namely the sign of the nonlinear term is opposite to that of Eq. (37) for TE waves. For a self-focusing nonlinearity ($\alpha_\gamma > 0$) in the cladding, the field solution is proportional to $\sinh[q_c k(z_c - z)]^{-1}$, which cannot lead to self-focusing of a field in the nonlinear medium. If $z_c < 0$, the solution fields diverge and are unphysical. Therefore, the behavior at high powers is quite dissimilar from the TE case, whose interpretation was consistent with previous experience in self-focusing nonlinear optics.

The elimination of E_x and E_z does lead to an equation for H_y that cannot be rigorously solved [38]. In the approximation $\varepsilon_{zz}(|E_z|^2)$,

$$\frac{d^2}{dz^2}H_{\gamma y}(z) - q_\gamma{}^2k^2H_{\gamma y}(z) + \frac{\beta^4}{c^2\varepsilon_0{}^2\varepsilon_{zz}{}^3}\alpha_\gamma|H_{\gamma y}(z)|^2H_{\gamma y}(z) = 0. \tag{48}$$

Although the sign of the nonlinear term is consistent with that found previously for TE waves, no analytical solutions are possible because the $1/\varepsilon_{zz}$ prefactor depends on the power, and hence on $H_{\gamma y}(z)$. On the other hand, since Δn_{sat} is small for any realistic material, the change in $\varepsilon_{zz}{}^3$ with power will be small, and hence this prefactor constitutes a small correction to a nonlinear term, which is small anyway. Setting $\varepsilon_{zz} = \varepsilon$ and $\beta \simeq \varepsilon$, which is accurate to the same level of approximation, leads for the self-focusing cladding case to

$$H_{\gamma y}(z) = \sqrt{\frac{2c^2\varepsilon_0{}^2n_c{}^2}{\alpha_c}}\ \frac{q_c}{\cosh[q_c k(z_c - z)]}, \tag{49}$$

and to self-focusing behavior similar to that obtained for the TE case.

Based upon the previous experience with the TE case, it is possible to justify why the $\varepsilon_{zz}(|E_z|^2)$ approximation is the preferable one. From Eqs. (46) and (49)

$$\frac{|E_{\gamma x}|^2}{|E_{\gamma z}|^2} = \frac{1}{\beta^2k^2}\frac{\left|\frac{dH_{\gamma y}}{dz}\right|^2}{|H_{\gamma y}|^2} = \frac{q_c{}^2}{\beta^2}\tanh^2[q_c k(z_c - z)] \leq \frac{\beta^2 - n_c{}^2}{\beta^2} < \frac{2(\beta - n_c)}{\beta}. \tag{50}$$

It was shown in Figs. 14 and 15 that the value of Δn_{sat} limits the maximum index difference $\beta - n_c$ to less than Δn_{sat}. As mentioned previously, Δn_{sat} is typically 0.01 or less, and can be as large as 0.1 in very special cases only. Therefore $|E_{\gamma z}|^2 \gg |E_{\gamma x}|^2$ and the dominant TM nonlinearity will involve $\varepsilon_{zz} = \varepsilon + \alpha_{zz}|E_z|^2$, which suggests that the $\varepsilon_{zz}(|E_z|^2)$ approximation is the most realistic of the two.

The dispersion relations are obtained by matching boundary conditions, and the <u>same</u> relations [36] are obtained for both forms of the uniaxial approximation, namely

$$\frac{q_c}{n_c{}^2}\tanh(kq_cz_c) = -\frac{q_s}{n_s{}^2}\tanh(kq_sz_s) \tag{51}$$

for $\alpha_c > 0$ and $\alpha_s > 0$. If $\alpha_\gamma < 0$, then $\tanh(kq_\gamma z_\gamma) \to \operatorname{cotanh}(kq_\gamma z_\gamma)$, from which all four possibilities are obtained. The power flow relations for this case can be shown to be exactly the same as for TE waves and are given by Eqs. (45). Some typical power-β curves are shown in Fig. 14. For small Δn_{sat}, $n_s \simeq n_c$ and the dispersion relations Eqs. (40) and (51) are effectively identical.

11. Nonlinear Waves - Thin Film Guided

Thin film waveguides (Fig. 1), in which some combination of the film, the cladding, or the substrate are nonlinear, exhibit many interesting power-dependent characteristics. They are more useful than the single interface nonlinear waves of section 10 because the solutions have low power limits. In addition, self-focused fields are simultaneously possible in more than one medium, which leads to multiple guided-wave branches. This area was first pioneered by Akhemediev [39], and many authors have subsequently contributed [34,40-43], especially for TE cases dealing with self-focusing nonlinearities.

The mathematical analysis for this case follows exactly the procedures outlined for the single interface problem, with the added complication of the thin film. The pertinent nonlinear wave equation for any of the media is Eq. (37). K must be equal to zero for the bounding media, which means that many of the formulas can be used from the preceding section with only small modifications. Since a nonlinear film is a bounded medium, solutions with $K \neq 0$ are also possible: To date only the case $K = 0$ has been analyzed for the nonlinear film case.

(a) Nonlinear Bounding Media

The most general case of two nonlinear bounding media will be treated here, since the limits of a vanishingly small nonlinearity can easily be obtained from the most general results. The solutions for the field equations are by now well known [32-43]. The general form was defined previously by Eqs. (1) and (2). It turns out to be convenient to set $a^{(m)}(x) = 1$. Furthermore, since the fields now become power dependent, the solutions are no longer normal modes and the usual orthogonality conditions for normal modes no longer hold. An additional consequence is that the solutions discussed here are valid for only one guided wave present in the waveguide at a time. (This is self-evident, because the presence of any one wave alters the refractive index profile of the waveguide by virtue of the intensity-dependent refractive index.)

Many of the analytical formulas for the field distributions can be taken over directly from the previous section. For example, the fields in the cladding media are given directly by Eqs. (38) with the small change in notation that $E_c(z)$ is replaced by $E_c^{(m)}(z)$ to bring in the dependence on the mode number. Equations (39) describe the field distributions in the substrate with z replaced by $z - h$ to take into account the film thickness, and the superscript (m) added to denote the mode number. The fields inside the film are written in the usual way as a superposition of sine and cosine functions ($n_f > \beta$) or sinh and cosh functions (for $\beta > n_f$, which is now allowed for some cases) with argument $k\kappa z$ where $\kappa^2 = |\beta^{(m)2} - n_f^2|$. For $n_{2c} > 0$ and $\beta^2 < n_f^2$ they are

$$E_f^{(m)}(z) = E_c^{(m)}(0)[\cos(k\kappa z) + \frac{q_c}{\kappa}\tanh(kq_cz_c)\sin(k\kappa z)] \;. \tag{52}$$

For $\beta^2 > n_f^2$, cos → cosh and sin → sinh respectively. When $n_{2c} < 0$, $\tanh(kq_cz_c)$ is replaced by $\text{cotanh}(kq_cz_c)$ with the obvious changes in $E_c^{(m)}(0)$. Continuity of the tangential electric field at $z = h$ relates z_s directly to z_c. The dispersion relations are obtained by matching the tangential magnetic fields at $z = h$. For $n_f^2 > \beta^2$,

$$\tan(k\kappa h) = \frac{\kappa[q_c\tanh(kq_cz_c) + q_s\tanh(kq_sz_s)]}{\kappa^2 - q_cq_s\tanh(kq_cz_c)\tanh(kq_sz_s)} \;. \tag{53}$$

For $\beta^2 > n_f^2$, $\tan(\kappa kh) \to \tanh(\kappa kh)$ and $\kappa^2 \to -\kappa^2$. For $n_{2\gamma} < 0$, it is straightforward algebra to show that $\text{cotanh}(kq_\gamma z_\gamma)$ replaces $\tanh(kq_\gamma z_\gamma)$ in Eq. (53).

What remains is to determine the constant z_c (and z_s) from the guided wave power, or vice versa. The guided wave power per unit length along the y-axis is given by Eq. (42) with an additional term P_f to include the power guided by the film. Equation (43a) holds for the power guided by the cladding (P_c) and substrate (P_s). For the film with $n_{2c} > 0$

$$P_f = \frac{\beta^{(m)} q_c^2}{2 n_c^2 |n_{2c}|} \frac{1}{\cosh^2(k q_c z_c)} \Big[h\Big(1 \pm \frac{q_c^2}{\kappa^2} \tanh^2(k q_c z_c)\Big)$$

$$- \frac{\sin(2k\kappa h)}{2k\kappa}\Big(-1 \pm \frac{q_c^2}{\kappa^2} \tanh^2(k q_c z_c)\Big) \pm \frac{q_c}{\kappa^2 k} \tanh(k q_c z_c)(1 - \cos(2k\kappa h))\Big] . \quad (54)$$

For $n_f^2 > \beta^2$, the upper sign is used. For $\beta^2 > n_f^2$, the lower sign is appropriate, and, in addition $\sin(2k\kappa h) \to \sinh(2k\kappa h)$ and $\cos(2k\kappa h) \to \cosh(2k\kappa h)$. For $n_{2c} < 0$, one needs to replace $\cosh(kq_cz_c)$ by $\sinh(kq_cz_c)$, and $\tanh(kq_cz_c)$ by $\mathrm{cotanh}(kq_cz_c)$.

The other case of interest here is a substrate whose refractive index is independent of power. This corresponds to taking the limit $z_s \to \infty$ in all of the formulas just given for the two nonlinear media case. For example, in the dispersion relations given by Eqs. (53), $\tanh(kq_sz_s) \to 1$ and $\mathrm{cotanh}(kq_sz_s) \to 1$. The substrate field is given by Eq. (39) with $A_s \to E_f^{(m)}(h)$ and z replaced by z - h. The power flow P_s is given by Eq. (45b) multiplied by the field ratio $|E_f^{(m)}(0)|^2/|E_c^{(m)}(0)|^2$.

Single Nonlinear Bounding Medium

Typical numerical calculations [43] for the effective index β versus the guided wave power for a self-focusing cladding are shown in Fig. 16. The variation in field distribution with increasing β in Fig. 17 indicates the progressively stronger self-focusing that occurs. For large values of β the TE_0 wave

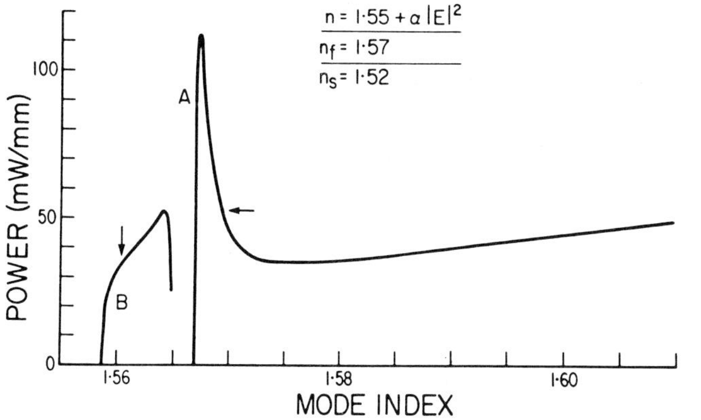

Fig. 16. (a) The guided wave power versus the effective mode index for TE_0 (A) and TE_1 (B) waves guided by a linear film bounded by a self-focusing cladding and a linear substrate. The waveguide is a 2.0-μm film of Corning 7059 glass on a soda lime glass substrate with a liquid crystal MBBA cladding [43].

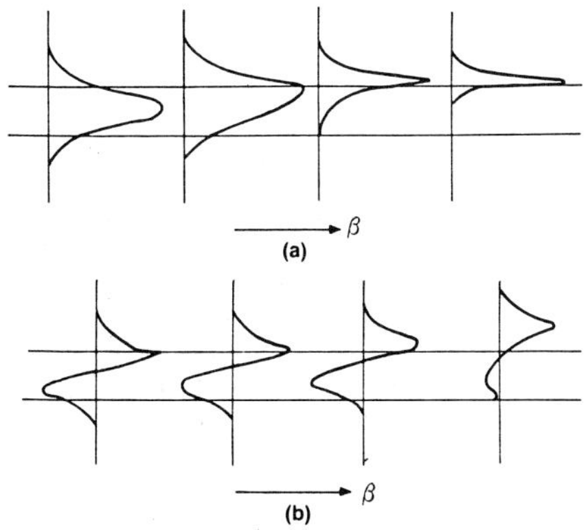

Fig. 17. Field distributions associated with (a) TE_0 and (b) TE_1 nonlinear guided waves for a linear film, linear substrate, and nonlinear cladding. The field evolution with increasing β is shown. Here $n_c = n_s = 1.55$, $n_f = 1.57$, $n_{2c} = 10^{-9}$ m^2/W and h = 2.0 μm [49].

degenerates into a single-interface nonlinear guided wave of the type discussed in the preceding section. A maximum is obtained in the guided wave power because the power required to sustain a single-interface wave is less than that carried by the film when self-focusing starts to dominate. (For thinner films, no maximum is obtained and the TE_0 wave evolves monotonically into the single-interface solution.) For the TE_1 solution, one of the extrema must remain inside the film, which implies that the solutions in the film are oscillatory, and hence that $n_f > \beta$. Therefore the TE_1 cannot degenerate into a nonlinear single-interface wave, and the branch must terminate for some value β, as seen in Fig. 16. Since there are two possible values for β at some guided-wave power levels, the possibility exists of switching, and perhaps bistability under the appropriate conditions.

The power dependence of the solutions has been verified experimentally [44,45] for the sample geometry shown in Fig. 18. A guided wave was propagated through a transverse boundary from a linear into a nonlinear waveguide region, and then through a second transverse boundary into another linear waveguide. Prisms were used to couple radiation into and out of the linear waveguide sections. The nonlinear medium was liquid crystal MBBA which has a thermal nonlinearity with $n_{2c} \simeq 10^{-9}$ m^2/W [45]. With reference to Fig. 16, the guided wave power available was intermediate between the TE_0 and TE_1 maxima.

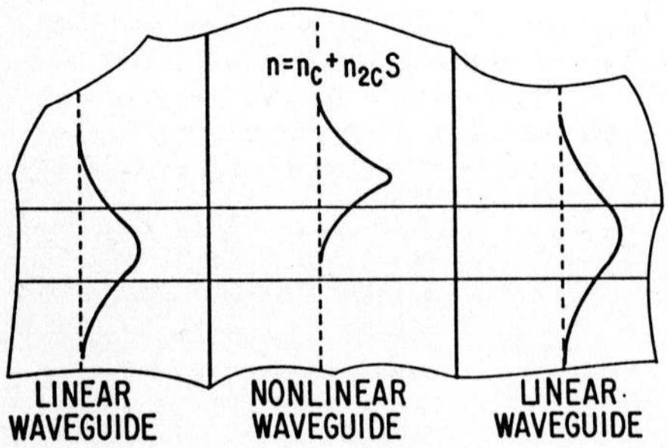

Fig. 18. Excitation of nonlinear guided waves by a linear guided wave incident onto a transverse boundary between a linear and nonlinear waveguide.

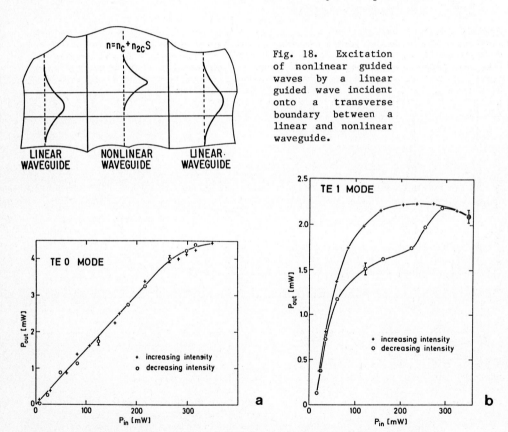

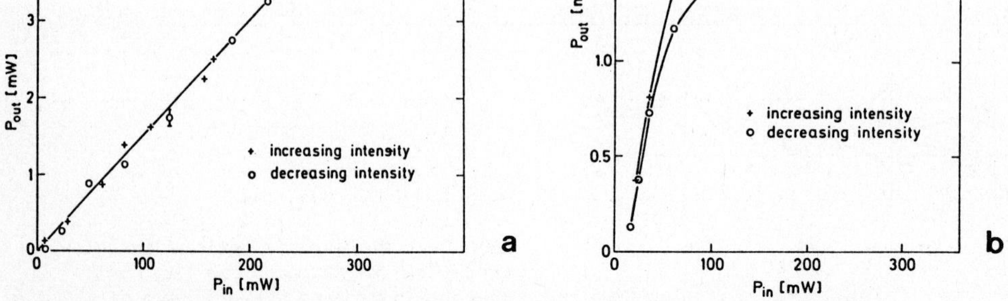

Fig. 19. The guided wave power for TE_0 (a) and TE_1 (b) waves transmitted through a waveguide with a nonlinear cladding. The waveguide consisted of the liquid crystal MBBA as the cladding ($n_c = 1.55$, $n_{2c} \simeq 10^{-9}$ m^2/W), a 1.0-μm film of borosilicate glass ($n_f = 1.61$) and a soda lime glass substrate ($n_s = 1.52$) [44].

The transmitted versus incident power for TE_0 and TE_1 shown in Fig. 19 can be interpreted as follows. The field maximum moves out of the film into the cladding with increasing TE_1 guided wave power, and hence the transmission coefficient for both transverse boundaries decreases due to field mismatch. Furthermore, the net transmission coefficient decreases as a progressively larger fraction of the guided wave power is propagated in the lossy MBBA cladding. Thus the transmission for the TE_1 wave decreases with increasing power until the transmitted power becomes a constant when the maximum for TE_1 in Fig. 16 is reached. For subsequent decreases in power, both branches of the TE_1 curve are excited. Since the high β branch corresponds to higher losses, hysteresis is expected and observed [44]. For the TE_0 wave, there is insufficient power to significantly move the field extremum towards the boundary, and the transmission remains linear with no hysteresis, as observed [44]. A similar experiment has been carried out with CS_2 as the nonlinear medium [45].

The increase in β with guided wave power suggests that cutoff may be power dependent for asymmetric ($n_c \neq n_s$) waveguides. That is, for a waveguide of thickness less than that required for low-power TE_0 wave propagation, an optical field can be used to increase the effective index past cutoff. A typical variation [46] in β with guided wave power is shown in the inset of Fig. 20. The variation in the threshold power with film thickness is shown in Fig. 20. This corresponds to a lower threshold device.

For a self-defocusing cladding medium, β decreases monotonically with guided wave power [46], see the inset of Fig. 21. If $n_s > n_c$, cutoff occurs when the field in the substrate degenerates into a plane wave and the guided wave power diverges. However, for $n_c > n_s$, cutoff occurs at a finite power, as shown in Fig. 21. From Eq. (43b), the substrate power remains finite since $q_s \rightarrow (n_c^2 - n_s^2)^{1/2} \neq 0$. As $q_c \rightarrow 0$, $P_c \rightarrow [k^2 z_c n_c |n_{2c}|]^{-1}$ with $z_c > 0$, and the cladding power also remains finite. This phenomenon can be used for upper threshold devices in which the cutoff power can be tuned, for example by tuning the cladding index, see Fig. 21.

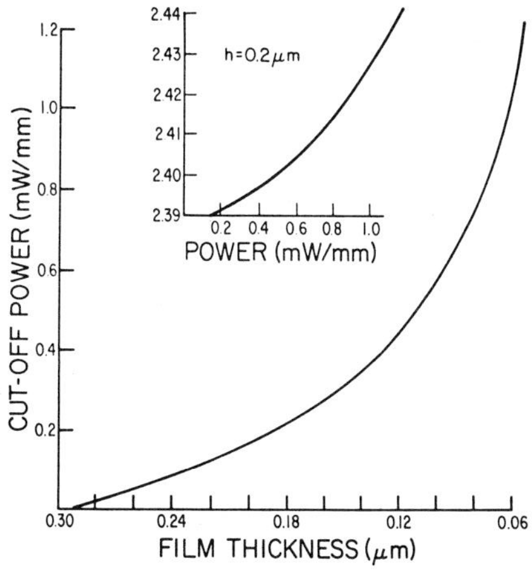

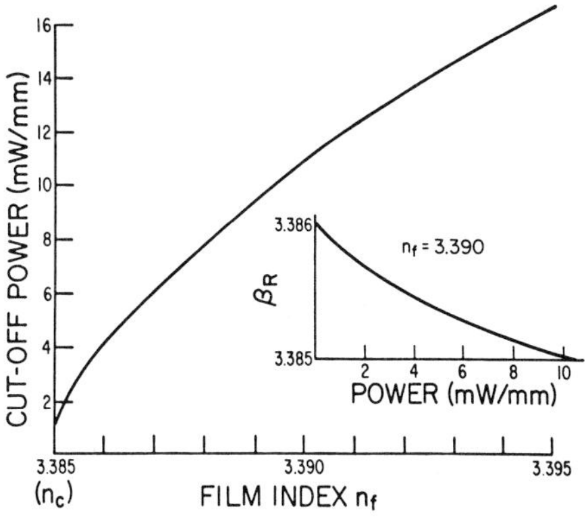

Fig. 20. The cut-off power versus film thickness for the TE_0 wave for a nonlinear self-focusing cladding and a linear film and substrate. Here $n_{2c} = 3\times10^{-11}$ m^2/W, $n_c = 2.39$, by $n_f = 2.40$ and $n_s = 2.38$. The inset shows the variation in effective index β with guided wave power for a film thickness of 0.2 μm. At low powers, the TE_0 mode for this structure is cut off at a film thickness of ≈0.30 μm [46].

Fig. 21. The maximum TE_0 guided-wave power that can be propagated versus index difference between a linear film and a self-defocusing cladding ($GaAs-GaAl_xAs_{1-x}$. Inset is the effective index versus guided wave power for $n_f = 3.39$ and film thickness of 1.07 μm [46].

Two Nonlinear Bounding Media

The most interesting cases here occur for both media with self-focusing nonlinearities [39-43,46-49]. The possibilities are that self-focused fields can occur in either or both of the bounding media. This implies three separate asymptotic curves that degenerate into single-interface waves at high powers.

Numerical results [43,49] are shown in Fig. 22 for a totally symmetric waveguide, that is $n_c = n_s$ and $n_{2c} = n_{2s}$. There are three TE_0 branches, two of which are degenerate with respect to power, and have power thresholds. As shown in Fig. 23, the fields associated with branch A remain symmetric with respect to the film center. For large β, this branch degenerates into single-interface surface waves, one guided by each film boundary. On the other hand, for branch B, a single-field maximum remains and moves out into one of the bounding media. Thus the symmetry of the waveguide is broken by the high-power optical field, a rather unique result [39]. Both branches B degenerate at high powers into single-interface surface guided waves.

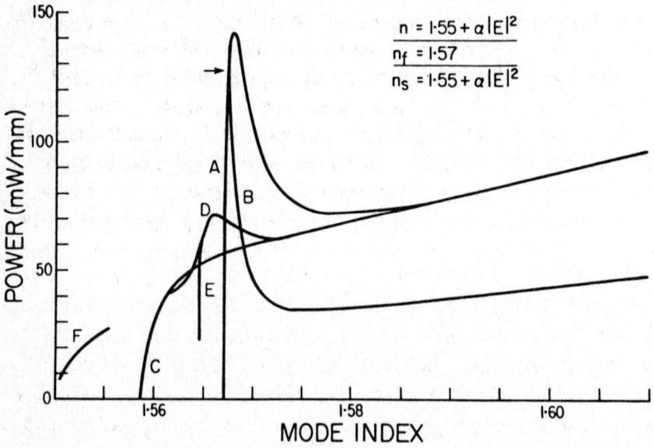

Fig. 22. The mode index β versus guided wave power for a linear film bounded by identical self-focusing cladding media. There are TE_0-like waves (A,B), TE_1-like waves (C,D,E) and TE_2-like waves (F) for a glass film with $n_f = 1.57$ with $n_c = n_s = 1.55$ and $n_{2c} = n_{2s} = 10^{-9} m^2/W$ [44].

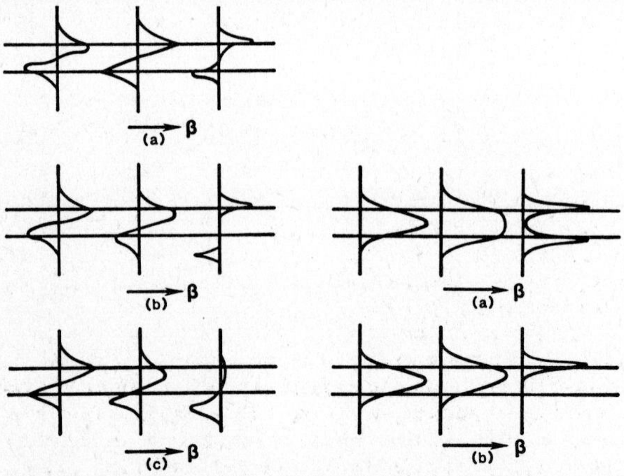

Fig. 23. TE_0 guided-wave field distributions for (a) branch A and (b) branch B. Here $n_c = n_s = 1.55$, $n_f = 1.57$, $h = 2.0$ μm and $n_{2c} = n_{2s} = 10^{-9}$ m^2/W. Field distributions associated with TE_1 guided waves with both bounding media nonlinear; $n_c = n_s = 1.55$, $n_{2c} = n_{2s} = 10^{-9}$ m^2/W and $h = 2.0$ μm. Asymmetric branches C, D and E correspond to (a), (b) and (c) respectively [49].

There are also multiple branches associated with the TE_1 solutions. Branch C retains symmetry about the film center, and both field extrema move symmetrically into the bounding media and degenerate into two out-of-phase single-interface nonlinear waves for high β. Branch D has a power threshold and consists of two degenerate curves. The two field extrema both move into the bounding media, but asymmetrically as indicated by the field distributions in Fig. 23. Again, for large β the fields degenerate into two single-interface surface waves. Branch E consists of two degenerate curves, each with a field extremum in the film and in one of the self-focusing media. It corresponds to $k\kappa h = \pi$ so that in Eq. (53) $\tan(k\kappa h) = 0$ and $z_c = -z_s$. For the parameters chosen, the TE_2 solution is cut off at low powers. However, at high powers, the branch F can be excited over a finite range of powers. The corresponding field distributions are shown in Fig. 23.

The profusion of solutions indicates that this combination of nonlinear media will be useful only if the film thickness is small enough to ensure only TE_0 solutions. When either the linear or nonlinear symmetry of the sample geometry is broken ($n_c \neq n_s$, $n_{2c} \neq n_{2s}$), two disconnected branches are obtained for TE_0, as shown in Fig. 24 [46,48,49]. We discuss first the case $n_c = n_s$ and $n_{2c} > n_{2s}$. In the limit of large β ($\beta > n_f$), the asymptotic curves in order of increasing power correspond to self-focusing in the most nonlinear medium (cladding), in the least nonlinear medium (substrate), and self-focusing in both bounding media. If in addition $n_c \neq n_s$, it becomes possible to obtain two branches to the dispersion curves that do not overlap in power. That is, there exists a range of powers over which guided waves cannot propagate. Note that the origin of the maxima in the branches is the same as discussed for the single nonlinear cladding in the previous section.

The curves shown in Fig. 24 appear promising for switching or bistability. For the $n_c = n_s$ case, increasing guided-wave power past the maximum of the lower curve can be accomplished only by switching to the upper curve. This may occur smoothly, or with a discontinuous jump in transmitted power. A detailed calculation of the field transmission through a transverse boundary of the type shown in Fig. 18 will be necessary to settle this issue. As the power is decreased along the upper branch, switching back to the lower branch is required at a much lower power than for the reverse process. This is a prime candidate for bistability. With $n_c \neq n_s$, the gap between the two curves implies a discontinuous jump in power when the guided wave switches from the lower to the upper branch, and vice versa. Bistability could occur, and these characteristics could be useful for an all-optical switch.

There is one other case of potential interest. If both media have self-defocusing nonlinearities, increasing power always leads to cutoff at finite powers [50].

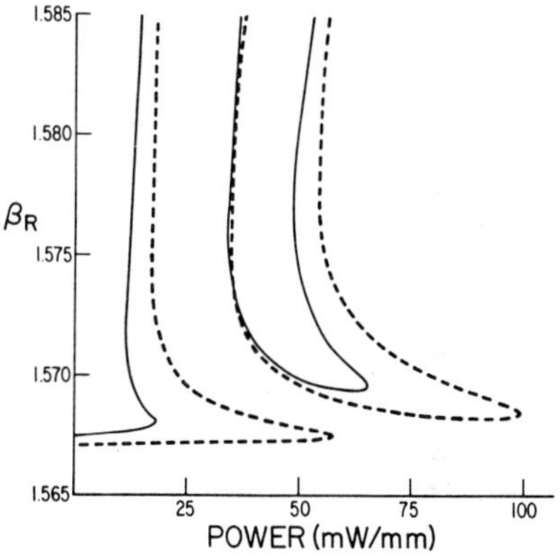

Fig. 24. The effective index versus TE_0 guided wave power for a linear film bounded by two dissimilar self-focusing media. Solid lines are for $h = 2.0$ μm, $n_f = 1.57$, $n_c = 1.56$, $n_{2c} = 2\times10^{-9}$ m^2/W, $n_s = 1.55$ and $n_{2s} = 10^{-9}$ m^2/W. Dashed lines are for $h = 2.0$ μm, $n_f = 1.57$, $n_c = 1.55$, $n_{2c} = 2\times10^{-9}$ m^2/W, $n_s = 1.55$ and $n_{2s} = 10^{-9}$ m^2/W [49].

(b) Nonlinear Film

This case is considerably more complicated than for the nonlinear bounding media and has not been explored fully yet because of two complicating factors [50-53]. The first is that K is not necessarily zero, since the fields exist only over a bounded region of space. The second is that the fields in the linear film are standing waves, and the generalization of this to the nonlinear case involves Jacobi elliptic integrals. For these reasons we give only a brief summmary, that elucidates the key features for a self-focusing film bounded by linear media.

For a self-focusing nonlinearity ($\alpha_f > 0$), the appropriate field solution to Eq. (37) in the approximation K = 0 is [53]

$$E_f^{(m)}(z) = \sqrt{\frac{2(\delta^2 + \kappa^2)}{\alpha_f}}\ \mathrm{cn}[\delta k(z + z_f)|p] \ , \tag{55a}$$

$$\delta^2 = \sqrt{\kappa^4 + 2\alpha_f E^2_o\left[n_f^2 - n_c^2 + \frac{1}{2}\alpha_f E^2_o\right]} \ , \tag{55b}$$

and $p = (\delta^2 + \kappa^2)/2\delta^2$ where $\mathrm{cn}(r|p)$ is the Jacobian elliptic function of order p and argument r, and $E_f^{(m)}(0)$ has been abbreviated by E_o. Assuming that the fields in the cladding and substrate are of the form

$$E_c^{(m)}(z) = E_o \exp[kq_c z] \ ; \quad E_s^{(m)}(z) = E_h \exp[-kq_s(z-h)], \tag{56}$$

with $E_h = E_f^{(m)}(h)$, matching the boundary conditions across both film interfaces leads to a dispersion relation of the form

$$\mathrm{cn}(\delta kh|p) = \frac{2E_oE_h(\delta^2 - q_cq_s)}{q_c^2E_o^2 + q_s^2E_h^2 + \delta^2(E_o^2 + E_h^2) + \frac{1}{2}\alpha_f(E_o^2 - E_h^2)} \ . \tag{57}$$

The power flow is given by

$$P_c = \frac{\beta^{(m)}\varepsilon_0 cE^2_o}{4kq_c} \ ; \quad P_s = \frac{\beta^{(m)}\varepsilon_0 cE^2_h}{4kq_s} \ , \tag{58a}$$

$$P_f = \frac{\beta^{(m)}\varepsilon_0 cE^2_o}{2k}\int_0^h dz\ \frac{\delta}{k}\left[\frac{\frac{\delta}{k}\mathrm{cn}\left[\frac{\delta}{k}|p\right] + \frac{q_c}{k}\mathrm{sn}\left[\frac{\delta}{k}|p\right]\mathrm{dn}\left[\frac{\delta}{k}|p\right]}{\frac{\delta^2}{k^2}\mathrm{dn}^2\left[\frac{\delta}{k}|p\right] + \frac{\alpha_f E^2_o}{2}\mathrm{sn}^2\left[\frac{\delta}{k}|p\right]}\right] \ . \tag{58b}$$

Numerical calculations for TE_0 waves have been performed for thin films with self-focusing nonlinearities. For waveguides with thickness chosen above cutoff for TE_0, the effective index β increases monotonically with power as self-focusing occurs inside the film, as expected [52]. For films whose thickness is below low-power waveguide cutoff, there is a minimum power above which the TE_0 wave can propagate, as shown in Fig. 25 [53]. As β increases past its cutoff value of n_s, the substrate guided-wave field changes from a plane wave to a field progressively more localized near the film-substrate boundary. Therefore the power associated with the substrate field decreases. As the field becomes progressively more self-focused in the film, it presumably degenerates into a single-interface surface wave, and β increases with increasing power. These two effects combine to produce a minimum in the guided-wave power. This phenomenon could find application as a lower threshold device.

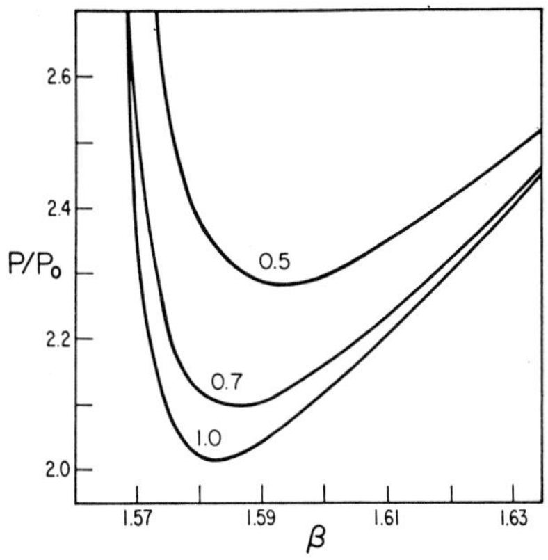

Fig. 25. Normalized power flow for the TE_0 wave guided by a self-focusing thin film versus effective index β. The curves are labelled by D = kh [53].

12. Intensity-Dependent Wavevector Devices

In the previous section we discussed nonlinear waveguide media excited at high enough power levels to produce power-dependent field distributions. At lower power levels, it can be seen from Figs. 16, 22, and 24 that the variation in effective index with power is linear. (The same result can be obtained from first-order perturbation theory or coupled mode theory [31].) That is,

$$\beta^{(m)} = \beta_0^{(m)} + \Delta\beta_0^{(m)}|a^{(m)}(x)|^2 . \tag{59}$$

Since we are usually dealing with only one guided-wave order m, we drop all the superscripts for simplicity. Such a power-dependent wavevector can affect a synchronous coupling condition for a dual-channel coherent coupler or prism coupler. For a grating imbedded in a nonlinear medium, the Bragg condition becomes power dependent.

(a) Nonlinear Prism Coupling

When a high-index prism is placed at a distance of less than a wavelength above a thin film, waveguide modes can be excited if the projection of the optical wavevector onto the base of the prism matches $\beta^{(m)}k$ for a guided wave. If the guided-wave power increases as the guided wave grows under the base of the prism, β changes, synchronism with the external field is lost, and the coupling efficiency is reduced. This can be expressed mathematically by the coupling equation [54]

$$\frac{d}{dx}a(x) = \hat{t}a_{inc}(x)e^{i(n_p\sin\theta - \beta_0 - \Delta\beta_0|a(x)|^2)kx} - (\ell^{-1} + \alpha)a(x) , \tag{60}$$

where $\hat{t}$ is the coupling coefficient, n_p is the prism index, θ is the angle of incidence measured from normal to the surface, ℓ is the characteristic distance for reradiation of the guided-wave field back into the prism and α is the waveguide absorption coefficient. The incident field amplitude varies as $a_{inc}(x)$ when projected onto the base of the prism. If the incidence angle is adjusted for maximum coupling at low powers ($\beta_0 = n_p\sin\theta$), then the guided-wave field falls out of phase with the generating field for high powers, and the coupling efficiency is reduced.

This phenomenon has been verified experimentally [55]. The gap between a coupling prism and a thin-film waveguide was filled with the liquid crystal MBBA, which now acts as a nonlinear cladding. Results for the in-coupled versus incident power are shown in Fig. 26. Note that the in-coupled power starts out linear with incident power, and then saturates at high powers. These results are in good agreement with theoretical calculations.

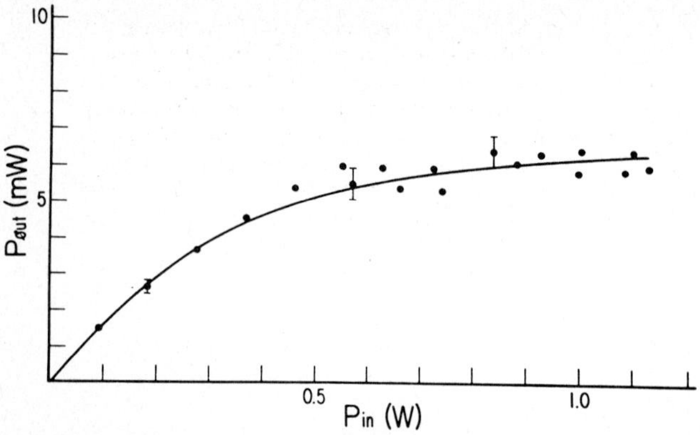

Fig. 26. The power prism-coupled into a nonlinear waveguide versus incident power for the TE_1 wave. The liquid crystal MBBA (n_c = 1.55, n_{2c} = 10^{-9} m^2/W) was the nonlinear cladding, the film was 1.7-μm Corning 7059 glass (n_f = 1.57) on a Pyrex substrate [55].

(b) Nonlinear Gratings

Guided waves can be deflected efficiently by a grating, provided that the Bragg condition is satisfied. For gratings defined either by a surface corrugation or a bulk index modulation with periodicity ℓ, the associated wavevector whose direction is orthogonal to the "grooves" is $\kappa_r = 2\pi r/\ell$ with r = 1 for sinuosoidal gratings. The Bragg condition $\beta^{(n)}\mathbf{k}_s = \beta^{(m)}\mathbf{k}_i + \kappa_r$ is an expression of wavevector conservation where the subscripts "s" and "i" identify the incident and scattered guided-wave beams. Here it is assumed that m = n and r = 1 since the best coupling coefficients Γ are usually obtained for this case. For simplicity, we drop these subscripts in the subsequent discussion.

Since the effective index of a guided-wave mode depends on power [Eq. (59)], the Bragg condition can be tuned optically by using waveguide media with self-focusing or self-defocusing nonlinearities. Three ways in which this tuning can be effected are shown in Fig. 27. The "control" beams 1, 2, and 3 can change the Bragg condition for the wave incident at an angle to the grating. In addition, beam 1 can tune its own Bragg condition. Most of these cases can be summarized by the coupling equations [56]

$$i\frac{d}{dx}a_s(x) = \Gamma e^{-i\Delta\beta kx}a_i(x) + \Delta\beta_0 k[a^2{}_s(x) + 2a^2{}_i(x)]a_s(x) + 2\Delta\beta_0 ka^2{}_c a_s(x) \ , \qquad (61a)$$

$$-i\frac{d}{dx}a_i(x) = \Gamma e^{i\Delta\beta x}a_s(x) + \Delta\beta_0 k[a^2{}_i(x) + 2a^2{}_s(x)]a_i(x) + 2\Delta\beta_0 ka^2{}_c a_i(x) \ , \qquad (61b)$$

where the initial offset from the Bragg condition is given by $\Delta\beta = 2\beta_0\cos\theta - \kappa$ and θ is the angle between the deflected beam and the grating wavevector. For the surface corrugation $u = u_0\sin(\kappa x)$ centered on the z = 0 interface, $\Gamma = \omega\varepsilon_0 u_0/8\ [n_c{}^2 - n_f{}^2]E_{ic}(z \to 0)\cdot E_{sf}(z \to 0)$. Here $a^2{}_c$ is the control beam power for cases 2 and 3. (The situation when beam 1 controls the deflection of the second weak beam is more complicated and requires four coupled equations.)

Bistable switching can occur when beam 1 is the only incident field present, that is $a_i(x) \neq 0$, $a_s(x) \neq 0$ and $a_c = 0$. The equations have solutions in terms of Jacobi elliptic functions [56]. Defining $|a_{sw}(0)|^2 = 2\beta_0/3\Delta\beta_0 kL$, $I = |a_i(0)|^2/|a_{sw}(0)|^2$ and $J = |a_i(L)|^2/|a_{sw}(0)|^2$ where I and J are normalized incident and transmitted powers,

$$T = J/I = 2[1 + nd(2\sqrt{(\Gamma^2L^2 + J^2)}|[1 + J^2/\Gamma^2L^2]^{-1})]^{-1} \ , \qquad (62)$$

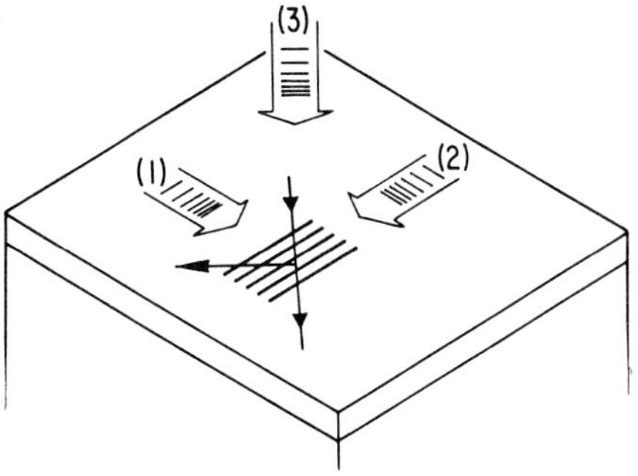

Fig. 27. General nonlinear grating geometry for switching guided waves. The grating can be tuned (in a bistable fashion) by a high-power guided wave incident (for example) along the grating axis, by a control guided wave incident parallel to the lines of the grating, or by illumination from above or below.

where L is the grating length (along κ). The variation in transmitted versus incident power shown in Fig. 28a is the classic one for optical bistability. Switching powers for InSb and polydiacetylene channel waveguides can be as small as 10's of nanowatts and 100's of milliwatts respectively [57].

Smooth tuning of the grating reflectivity can be achieved with control beams 2 or 3. For a weak incident signal beam, the terms in the [ ] brackets in Eqs. (61) can be neglected and the reflectivity is

$$R = \frac{4\Gamma^2 \sinh^2(\mu L)}{4\mu^2 \cosh^2(\mu L) + \Delta k^2(a^2_c)\sinh^2(\mu L)} , \tag{63a}$$

$$\mu = \sqrt{\Gamma^2 - \Delta k^2(a^2_c)} \; ; \qquad \Delta k(a^2_c) = k[\Delta\beta_0 + 4\Delta\beta_0 |a_c|^2] \; . \tag{63b}$$

Therefore the grating reflectivity can be tuned as shown in Fig. 28b. This property is useful for implementing a variety of all-optical logic functions. For example, if two input beams of power P_s are incident as control beams with the grating initially offset from maximum reflectivity (for a third weak signal beam) by $-P_s$, an XOR operation is obtained for a weak signal beam. Similarly, operations such as AND may also be implemented.

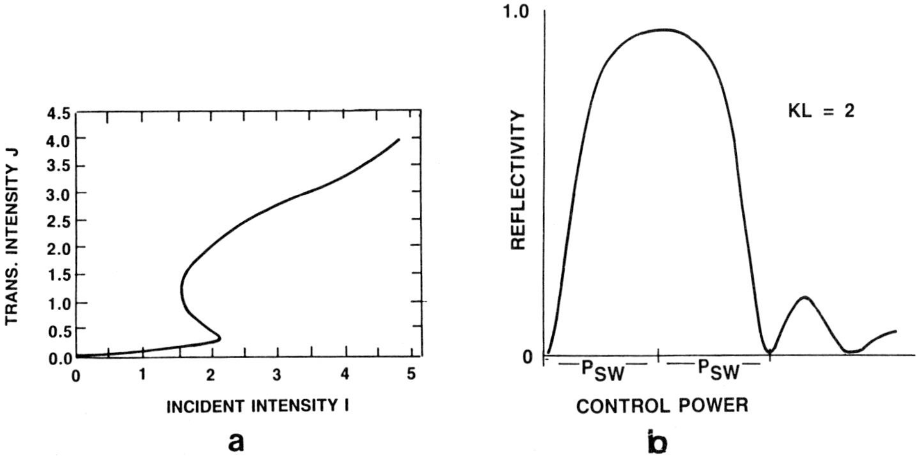

Fig. 28. (a) The transmitted versus incident power for distributed-feedback guided-wave bistability with $\Gamma L = 2$ [56]. (b) Grating reflectivity of a nonlinear grating versus beam power for a control beam incident parallel to the grooves [46].

(c) Nonlinear Coherent Coupler

An intensity-dependent refractive index can be used to alter the phase-matching condition between two coupled channel waveguides. For example, consider the channel waveguide directional coupler shown in Fig. 29. When light is injected into one channel, the overlap of that guided-wave field with the adjacent channel waveguide results in power transfer into the second waveguide. If one of the media in which the guided-wave fields exist is characterized by an intensity-dependent refractive index, the cross-coupling conditions become power-dependent and new signal-processing devices become possible [58,59]. That is, the response of the device depends on the intensity of the input waves. Assuming identical waveguides with guided-field amplitudes described by $a_1(x)$ and $a_2(x)$ respectively and neglecting attenuation, the interaction can be described by the coupled-wave equations [58]

$$-i\frac{d}{dx}a_1(x) = \Gamma a_2(x) + [\Delta\beta_0 a_1(x)^2 + 2\Delta\beta_0' a_2(x)^2]a_1(x) \qquad (64a)$$

$$-i\frac{d}{dx}a_2(x) = \Gamma a_1(x) + [\Delta\beta_0 a_2(x)^2 + 2\Delta\beta_0' a_1(x)^2]a_2(x). \qquad (64b)$$

The parameter $\Delta\beta_0'$ quantifies the nonlinear effect of a strong field in one channel on the propagation characteristics of the neighboring channel.

The solutions to these equations as obtained by Jensen [58] are shown in Fig. 30 for a single input channel excited at different power levels. They suggest a variety of possible applications. For example, if the device is set to produce complete transfer into the neighboring channel at low powers, at appropriately high input power levels the transfer can be minimized. This corresponds to an intensity-dependent optical switch. A series of such switches can be used to demultiplex signals that are intensity coded. Alternatively, if the signals are

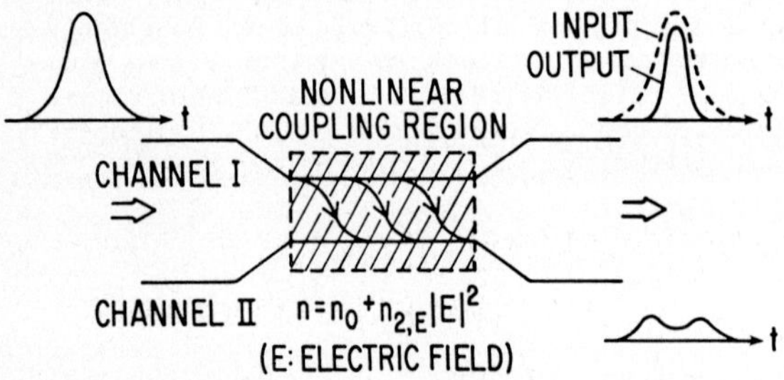

Fig. 29. Schematic of two coupled channel waveguides with a nonlinear material in the coupling region.

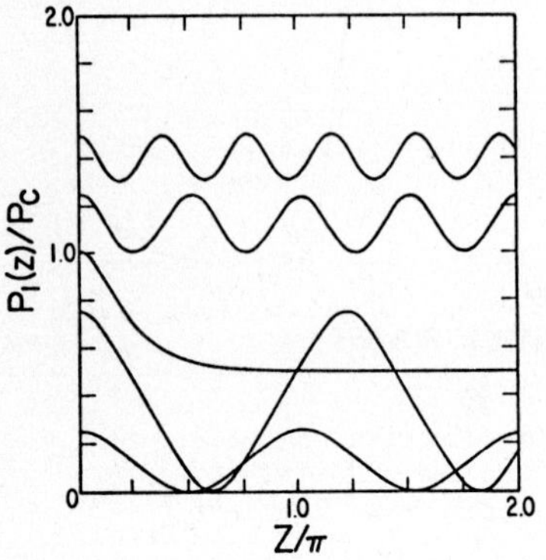

Fig. 30. The amount of power remaining in waveguide 1 as it propagates along a nonlinear coherent coupler. Here z/π is the normalized propagation coordinate, $P_1(z) \rightarrow a_1^2(x)$ and P_c is the typical switching power [58].

of approximately equal intensity but are wavelength multiplexed, the signals can be demultiplexed if the material nonlinearity is strongly wavelength-dependent.

Intensity-dependent power transfer has recently been demonstrated using in-diffused channel waveguides in $LiNbO_3$ [59], and in $GaAs-Ga_{0.7}Al_{0.3}As$ multiple-quantum-well (MQW) channel waveguides [60]. Although $n_{2\gamma}$ is quite small for $LiNbO_3$ and in fact was measured in this experiment, a measurable transfer with picosecond time-resolution has been observed by Haus and colleagues [59]. The recent report [60] of nonlinear switching in the MQW structures demonstrated partial switching of laser diode radiation at waveguide power levels of only 1 mW. In that case the two channel waveguides were strain-induced by depositing two gold stripes on top of a planar MQW waveguide. Although this device operated at a temperature of 180 K, it represents a very new and exciting development in this field.

Acknowledgments

This research was supported by the National Science Foundation (DMR-8300599 and ECS-8304749), the Army Research Office (DAAG-29-85-K-0026), the Air Force Office of Scientific Research (AFOSR-84-0277), the Joint Services Optics Program (ARO, AFOSR and BMSDC) and the NSF-Industry Optical Circuitry Cooperative.

References

1. Y.R. Shen, The Principles of Nonlinear Optics (Wiley, New York, 1984).
2. M.D. Levenson, Introduction To Nonlinear Laser Spectroscopy, (Academic Press, New York, 1982).
3. Reviewed in the book Optical Phase Conjugation, R.A. Fisher, ed. (Academic Press, New York, 1982).
4. P.K. Tien, Rev. Mod. Phys. 49, 361 (1977).
5. In Surface Polaritons, Electromagnetic Waves at Surfaces and Interfaces, V.M. Agranovich and D.L. Mills, eds. (North-Holland, Amsterdam, 1982).
6. W. Sohler, in New Directions in Guided Wave and Coherent Optics, D.B. Ostrowsky, ed. (Plenum Press, New York, 1983).
7. C.T. Seaton, G.I. Stegeman, W.M. Hetherington III and H.G. Winful, in Integrated Optics, H.P. Nolting and R. Ulrich, eds. (Springer-Verlag, New York, 1985) p. 178.
8. H. Kogelnik, Integrated Optics, Vol 7 of Topics in Applied Physics, T. Tamir, ed. (Springer-Verlag, Berlin, 1975), p. 66.
9. D. Marcuse, Theory of Dielectric Optical Waveguides, (Academic Press, New York, 1974).
10. J.E. Sipe and G.I. Stegeman, in Surface Polaritons, D.L. Mills and V.N. Agranovich, eds., (North-Holland, New York, 1982), p. 661.
11. N. Uesugi, Appl. Phys. Lett. 36, 178 (1980).
12. J.P. van der Ziel, M. Ilegems, P.W. Foy, and R.M. Mikulyak, Appl. Phys. Lett. 29, 775 (1976).
13. H. Ito and H. Inaba, Opt. Lett. 2, 139 (1978).
14. Y. Suematsu, Y. Sasaki, K. Furuya, K. Shibata, and S. Ibukuro, IEEE J. Quant. Electron. 10, 222 (1974).
15. G.H. Hewig and K. Jain, Optics Commun. 47, 347 (1983).
16. K. Sasaki, T. Kinoshita, and N. Karasawa, Appl. Phys. Lett. 45, 333 (1984).
17. R.V. Schmidt and I.P. Kaminow, Appl. Phys. Lett. 25, 458 (1974).
18. N. Uesugi and T. Kimura, Appl. Phys. Lett. 29, 572 (1976).
19. W. Sohler and H. Suche, Appl. Phys. Lett. 33, 518 (1978).
20. W. Sohler and H. Suche, in Integrated Optics III, L.D. Hutcheson and D.G. Hall, eds., Proc. SPIE 408, 163 (1983).
21. R. Normandin and G.I. Stegeman, Opt. Lett. 4, 58 (1979).
22. R. Normandin and G.I. Stegeman, Appl. Phys. Lett. 40, 759 (1982).
23. N. Uesugi, K. Daikoku, and M. Fukuma, J. Appl. Phys. 49, 4945 (1978).
24. A.T. Reutov and P. P. Tarashchenko, Opt. Spectrosc. 37, 447 (1974).
25. C. Karaguleff and G.I. Stegeman, IEEE J. Quant. Electron. QE-20, 716 (1984).

26. C. Karaguleff, G.I. Stegeman, R. Zanoni, and C.T. Seaton, Appl. Phys. Lett. 7, 621 (1985).
27. G.I. Stegeman, J. Opt. Commun. 4, 20 (1983).
28. G.I. Stegeman, R. Fortenberry, C. Karaguleff, R. Moshrefzadeh, W.M. Hetherington III, N.E. Van Wyck, and J.E. Sipe, Opt. Lett. 8, 295 (1983).
29. W.M. Hetherington III, N.E. Van Wyck, E.W. Koening, G.I. Stegeman, and R.M. Fortenberry, Opt. Lett. 9, 88 (1984).
30. P.D. Maker and R. Terhune, Phys. Rev. 137, A801 (1964).
31. G.I. Stegeman, IEEE J. Quant. Electron. QE-18, 1610 (1982).
32. A.E. Kaplan, Sov. Phys. JETP 45, 896 (1977).
33. A.E. Kaplan, IEEE J. Quant. Electron. QE-17, 336 (1981).
34. A.A. Maradudin, in Optical and Acoustic Waves in Solids-Modern Topics, M. Borissov, ed. (World Scientific Publ., Singapore, 1983), p. 72.
35. W.J. Tomlinson, Opt. Lett. 5, 323 (1980).
36. J. Ariyasu, C.T. Seaton, G.I. Stegeman, A.A. Maradudin, and R.F. Wallis, J. Appl. Phys., in press.
37. V.M. Agranovich, V.S. Babichenko, and V.Ya. Chernyak, Sov. Phys. JETP Lett. 32, 512 (1981).
38. C.T. Seaton, J.D. Valera, B. Svenson, and G.I. Stegeman, Opt. Lett. 10, 149 (1985).
39. N.N. Akhemediev, Sov. Phys. JETP 56, 299 (1982).
40. A.D. Boardman and P. Egan, J. Physique Colloq. C5, 291 (1984).
41. F. Fedyanin and D. Mihalache, Z. Physik B 47, 167 (1982).
42. F. Lederer, U. Langbein, and H.-E. Ponath, Appl. Phys. B 31, 69 (1983).
43. G.I. Stegeman, C.T. Seaton, J, Chilwell, and S.D. Smith, Appl. Phys. Lett. 44, 830 (1984).
44. H. Vach, C.T. Seaton, G.I. Stegeman, and I.C. Khoo, Opt. Lett. 9, 238 (1984).
45. I. Bennion, M.J. Goodwin, and W.J. Stewart, Electron. Lett. 21, 41 (1985).
46. C.T. Seaton, Xu Mai, G.I. Stegeman, and H.G. Winful, Opt. Eng. 24, 593 (1985).
47. A.D. Boardman and P. Egan, IEEE J. Quant. Electron., in press.
48. A. Boardman and P. Egan, Phil. Trans. Roy. Soc. London A313, 363 (1984).
49. C.T. Seaton, J.D. Valera, R.L. Shoemaker, G.I. Stegeman, J. Chilwell, and S.D. Smith, IEEE J. Quantum Electron. QE-21, 774 (1985).
50. N.N. Akhemediev, K.O. Boltar and V.M. Eleonskii, Opt. Spektrosk. 53, 906 and 1097 (1982).
51. V.K. Fedyanin and D. Mihalache, Z. Phys. B 47, 167, (1984).
52. U. Langbein, F. Lederer, and H.E. Ponath, Opt. Commun. 46, 167 (1983).
53. A.D. Boardman and P. Egan, IEEE J. Quant. Electron., in press.
54. C. Liao, G.I. Stegeman, C.T. Seaton, R.L. Shoemaker, J.D. Valera, and H.G. Winful, J. Opt. Soc. Am. A 2, 590 (1985).
55. J.D. Valera, C.T. Seaton, G.I. Stegeman, R.L. Shoemaker, Xu Mai, and C. Liao, Appl. Phys. Lett. 45, 1013 (1984).
56. H.G. Winful, J.H. Marburger, and E. Garmire, Appl. Phys. Lett. 35, 379 (1979).
57. G.I. Stegeman, C. Liao, and H.G. Winful, in Optical Bistability II, C.M. Bowden, H.M. Gibbs, and S.L.McCall, eds. (Plenum Press, New York, 1984) p. 389.
58. S.M. Jensen, IEEE J. Quant. Electron. QE-18, 1580 (1982).
59. A. Lattes, H.A. Haus, F.J. Leonberger, and E.P. Ippen, IEEE J. Quant. Electron. QE-19, 1718 (1983).
60. P. Li Kam Wa, J.E. Sitch, N.J. Mason, J.S. Roberts, and P.N. Robson, Electron. Lett. 21, 26 (1985).

Nonlinear Interactions and Excitonic Effects in Semiconductor Quantum Wells

D.S. Chemla

AT & T Bell Laboratories, Holmdel, NJ 07733, USA

1. INTRODUCTION

The optimization of the nonlinear optical response of solids is a key element in the development of future optical switching and signal processing systems. Demonstrations of device feasibility have been obtained in laboratories. However, the characteristics of the materials utilized so far do not meet all the stringent requirements imposed by implementation in real systems. It is most likely that new concepts have to be introduced in nonlinear optics and that new physical mechanisms have to be explored in order to manufacture the materials that will satisfy these requirements.

Recent advances in crystal-growth techniques [1] enable us now to manufacture artificial semiconductor microstructures with atomically smooth interfaces and precisely controlled composition. These novel materials, because of quantum size effects, exhibit properties not shown by the parent compounds in the bulk [2]. The new degrees of freedom in the control of sample microscopic size and geometry provided by modern material science open new opportunities to engineer locally the band structure and thus to tailor the optical properties of semiconductors.

Quantum size effects appear whenever the spatial dimensions of an object become comparable or eventually smaller than a characteristic length that governs some quantum mechanical process within that object. For example, optical transitions in semiconductors near the band gap are governed by excitonic effects [3]. The characteristic length for electron-hole (e-h) correlation is the exciton Bohr radius, that in the case of III-V material ranges from 10 Å to 500 Å. Consequently, samples with dimensions of the order of 100 Å exhibit optical properties that are qualitatively different from those of bulk compounds. Recent investigations have shown that quantum well structures (QWS) indeed exhibit optical nonlinearities and present new electro-absorption, not encountered in three-dimensional semiconductors, which have promising potential applications to optical signal processing.

In this set of lectures, we will review the nonlinear optical processes and electro-absorption effects that are seen in QWS. In Section 2 we discuss linear optical effects in QWS; in Section 3 the modification of refractive index and absorption by cw, picosecond and femtosecond excitations are presented and analyzed. Finally in Section 4 we describe the electro-absorption of QWS for static field perpendicular and parallel to the layers.

2. EXCITONIC EFFECTS IN SEMICONDUCTOR QUANTUM WELLS

QWS consist of stacks of ultrathin layers with specific chemical and physical compatibility but different composition, grown by epitaxy one on another [2]. Since the band gap depends on the composition, it is modulated in the direction perpendicular to the layer. If the discontinuities of the valence and conduction bands are large enough and if the barrier (large gap compound) layers

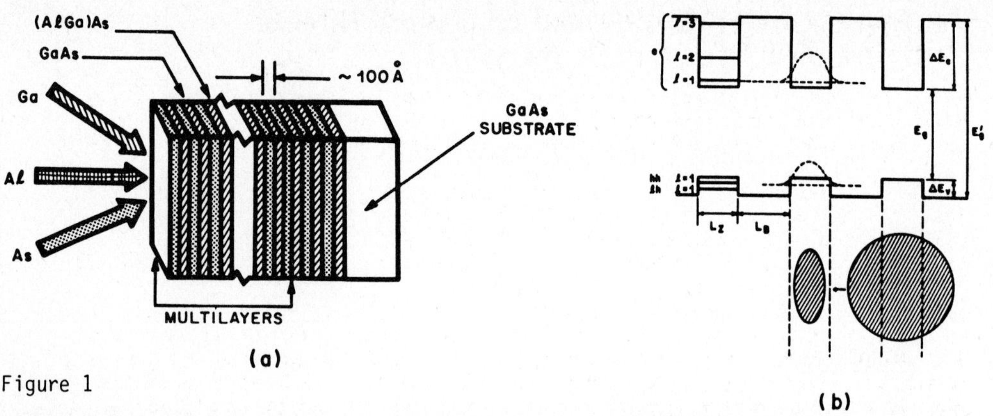

Figure 1

Quantum-well structure and corresponding real-space energy band structure. The schematic diagram in a shows compositional profiling in thin layers. The circle in b represents an exciton in the bulk compound, and the ellipse represents an exciton confined in a layer with a low band gap.

are wide enough, the carriers are confined in the quantum well (low gap compound) [4]. The quantization of motion along the modulation axis produces a set of discrete states, Fig. 1, whereas the carriers can move freely along the plane of the layers.

The band structures of QWS consist of a set of subbands with two-dimensional density of states [5]. The electron subbands that originate from the nondegenerate conduction band of the parent compound are relatively simple. Conversely for the hole, the confinement splits the degenerate valence bands of III-V bulk compounds and produces a double set of complex QW-subbands [6,7]. The valence subbands are labeled according to the hole masses perpendicular to the layer i.e., heavy and light hold bands as shown in Fig. 2. Simple symmetry arguments indicate dichroic behavior with relative oscillator strength of 3/4 and 1/4 or 0 and 1 for light polarized along or perpendicular to the layers respectively.

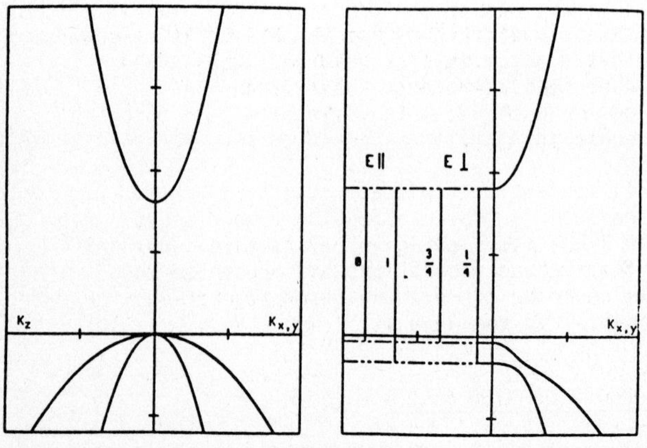

Figure 2

Schematic of the band structure of a 3-D semiconductor (left) and of an ultrathin layer of the same material (right). The selection rules shown in the figure, for optical transitions with fields parallel and perpendicular to the normal to the layers, assume no band mixing.

As in the case of three-dimensional (3-D) semiconductors, Coulomb interaction produces bound e-h pairs states or excitons. Excitons in two-dimensional systems (2-D) are well described by a natural extension of Elliot's theory [8,9]. Extensive studies of excitons spectroscopy in GaAs/AlGaAs QWS at low temperature have been performed [10], as well as investigations of exciton localization [11] due to fluctuations in the layer thickness [12].

The principal effect of the confinement is to artificially decrease the e-h distance and thus to increase the exciton binding energy. In a pure 2-D semiconductor, the binding energy would be four times larger than in 3-D [9], but in layers of finite thickness the enhancement ranges only between 2 and 3 [13,14]. Associated with the increased binding is a large augmentation of the oscillator strength and of the contrast with the transitions to the continuum, as illustrated in Fig. 3 [15]. Interaction of phonons with QWS excitons has been investigated by resonant Raman scattering [16]. It is found that, because of the strong confinement, low-lying excitons interact only with the QW phonons whereas the higher-lying states can interact both with the QW and barrier layer vibrations. The reduced symmetry of the QWS also causes significant modifications of the scattering selection rules [16].

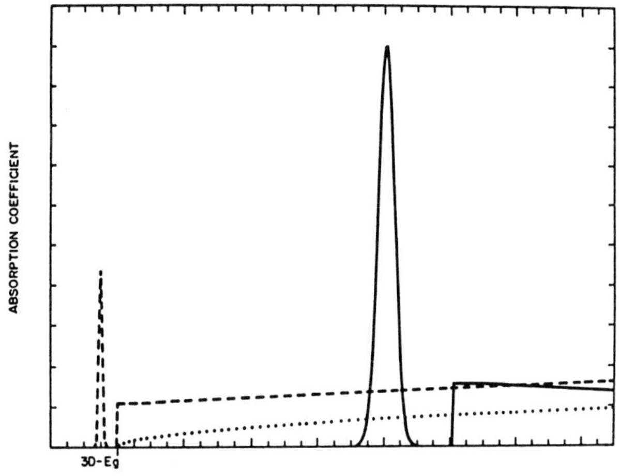

Figure 3

Comparison of the band-edge absorption spectra in two dimensions (solid lines) and in three dimensions (dashed lines). The dotted line shows the interband absorption in the absence of excitonic enhancement in three dimensions. The theoretical infinitely marrow exciton peaks have been replaced by more realistic Gaussian profiles. For clarity, only the 1S peaks have been represented.

An important consequence of the increased binding of e-h bound states in QWS is that excitonic resonances are observed at room temperature in a number of III-V QWS [17,18,19]. Resonances can be resolved if the binding energy significantly exceeds the line-width. In the polar III-V semiconductors, the major cause of exciton line-broadening is due to the interaction with the LO-phonons. In QWS the well-confined low-lying excitons interact only with the phonons of the QW layers. And if the confinement produces any effect on the temperature-broadening it is only a small reduction due to the selection rules diminution of the interaction channels with vibrations. On the other hand, the binding energy increases in proportion to the ratio of the Bohr radius to the QW layer thickness. Since the Bohr radius in inversely proportional to the 3-D binding energy that scales with the band gap, for the same well thickness the confinement produces a larger compression on small-gap-compound excitons.

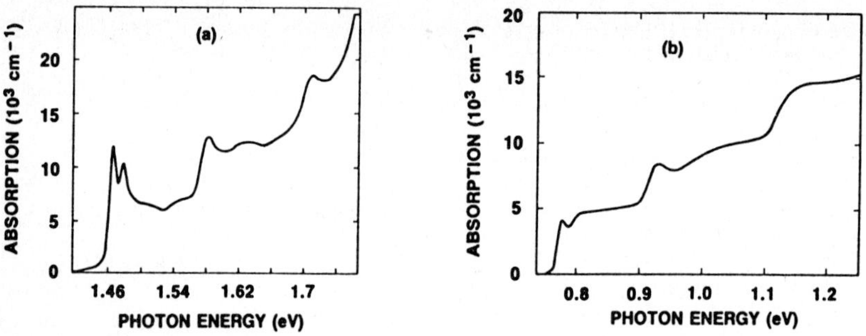

Figure 4

Room-temperature absorption spectra of (a) GaAs-AlGaAs and (b) GaInAs-AlInAs MQWS's showing the strong exciton resonances and the characteristic plateaux of the 2-D continua. In both cases the quantum-well thickness is about 100 Å.

Altogether, for a number of III-V materials in QW about 100 Å thick, the enhancement of the binding is large enough to compensate the temperature line-broadening, and well-resolved resonances can be observed even above room temperature. This is shown in Fig. 4 where the room-temperature absorption spectra of GaAs and GaInAs QWS are shown.

However, even if clear resonances are seen in absorption, the excitons are not stable against LO-phonon collision. In all III-V QWS the LO-phonon energy is several times larger than the quasi 2-D exciton binding energy, so that the only possible outcome of collision between an exciton and an LO-phonon is a free e-h pair with a substantial excess energy. For example, in the case of GaAs QWS about 100 Å thick the heavy-hole exciton binding energy is 9 meV whereas the LO-phonon energy is 37 meV. Study of the temperature dependence of the exciton resonance in GaAs and GaInAs QWS show that the line-width is well described by a constant term that represents the low-temperature inhomogeneous width originating from the layer fluctuations [11,12] plus a term proportional to the density of LO-phonons [17-18]. If the temperature-broadening is interpreted as a reduction of lifetime by thermal LO-phonons ionization, it is found that the mean time for ionization is of the order of half a picosecond in GaAs QWS and a quarter of a picosecond in GaInAs QWS. Thermodynamic arguments show that the probability for an exciton to reform from the free e-h pair is extremely small at room temperature [20]. Thus, when a photon at resonance with an exciton is absorbed in a room-temperature QWS, it generates a bound e-h pair that lives only a fraction of a picosecond before being ionized by the thermal vibrations.

3. ROOM-TEMPERATURE EXCITONIC NONLINEAR EFFECTS IN QWS

The nonlinear optical effects associated with excitons in semiconductors are qualitatively different from those encountered in atomic or molecular systems [21]. First the photo-generation of e-h pairs, bound or unbound, produce a phase-space filling, which because of the exclusion principle inhibits further optical transitions toward the occupied states and thus results in a reduction of the absorption. This effect is well known in the case of free e-h pairs, for which it is called state-filling or band-filling [21]. However, bound states also produce phase-space filling because of the composite nature of excitons; this aspect of generation of excitons is less well documented [21]. In addition, free e-h and excitons produce a screening of the Coulomb interaction that comprises exchange and direct screening. It modifies the energy of the single particle states (band gap renormalization) and of bound states. The effect of renormalization on the energy of the bound states is strongly reduced owing to their electric neutrality. Thus, when e-h pairs are generated in a semiconductor, the band gap reduces whereas the energy of the exciton remains almost constant. However, the binding energy of the bound states with respect to the renormalized gap diminishes. The exciton wave function blows up in real space and the oscillator strength decreases. In 3-D semiconductors the charged plasma screening is much more effective that that of the neutral exciton gas [21].

The nonlinear optical effects due to room-temperature excitons in QWS are further more complicated for two reasons [15]. First, the irreversible transformation of excitons into free e-h pairs produces transient exciton and plasma populations which have different effects on the optical transitions. Because of the very short ionization time, these effects need ultrafast excitation and probe to be resolved. Then, because electrostatic screening depends critically on the density of states, the relative importance of the long-range direct screening and of the short-range exchange interaction is significantly modified by the reduced dimensionality of QWS.

Saturation of QWS absorption under cw, ps and fs excitation has been extensively studied [15,17,20,22-26]. For excitation with cw sources or with pulses long compared to the exciton ionization time, the effects are only governed by the number of free carriers generated in the sample. For resonant excitation these originate from the exciton ionization by thermal phonons, for nonresonant excitation they are directly created above the band gap. Typical saturation intensities are of the order of 500W/cm^2 for GaInAs QWS and 1 kW/cm^2 for GaInAs QWS. The index of refraction changes induced by free carriers correlates very precisely with the absorption saturation, as shown by simultaneous four-wave mixing/saturation measurements, Fig. 5. The maximum values of the change of absorption coefficient and refractive index induced by one e-h pair are $\sigma_{eh} \sim 7 \times 10^{-14}$ cm^2 and $n_{eh} \sim 3.7 \times 10^{-19}$ cm^{-2}, respectively. This corresponds to huge third-order nonlinearities $\chi^3 \sim 6 \times 10^{-2}$ esu that has been used to demonstrate 0.1% efficiency for wave mixing in a 1.25 μm thick sample using as sole light source the 17 W/cm^2 output of a cw commercial laser diode [23].

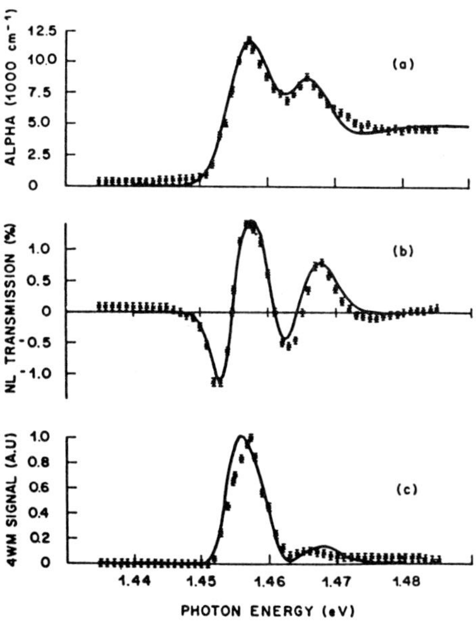

Figure 5

Comparison of the experimental spectra (crosses) for (a) linear absorption, (b) nonlinear absorption, and (c) degenerate four-wave mixing with the theorectical spectra (solid lines).

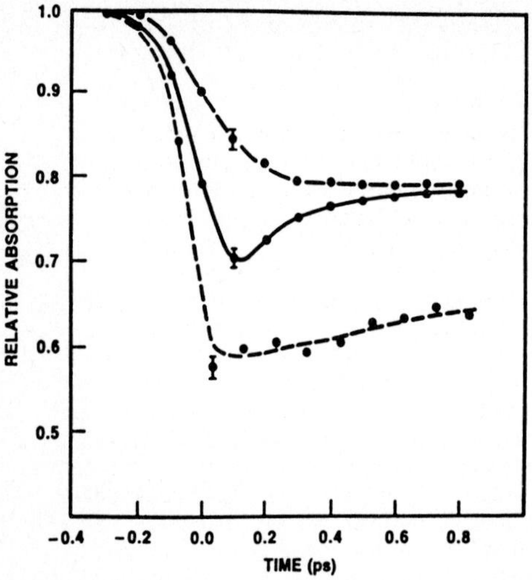

Figure 6

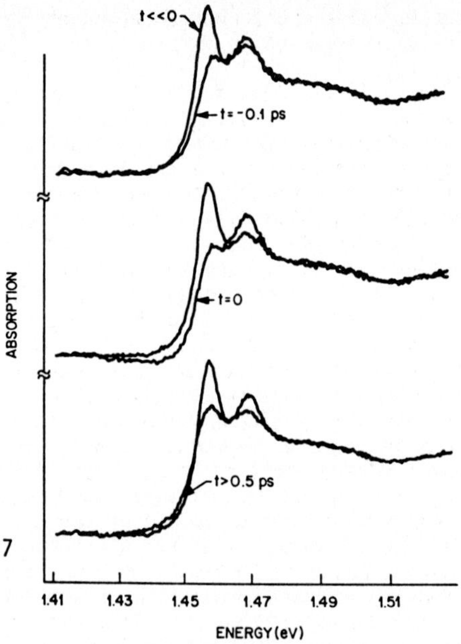

Figure 7

Figure 6
Dynamics of the heavy-hole exciton-peak absorption under femtosecond excitation at room temperature for resonant pumping (solid line) and nonresonant pumping (long-dashed line) showing the exciton ionization and the stronger saturation induced at short times by excitons. The effect of density of thermal phonon on the ionization is shown by the short-dashed line, which gives the dynamics of the exciton peak under resonant pumping at a temperature of T = 15 K.

Figure 7
Comparison of the absorption spectrum of an unexcited MQWS to the spectra after excitation by an ultra short pump pulse at resonance with the heavy-hole exciton.

Under ultrashort (100 fs) excitation the behavior is quite different for nonresonant and resonant excitation. When free e-h pairs are directly generated by a nonresonant fs-pump pulse, the effects are similar to those observed with longer pulses. The fast time resolution only enables one to see the switching on of the exciton bleaching. It follows roughly the integral of the pump pulses and stabilizes after one ps to the level and profile observed with cw excitation. Conversely, when bound e-h pairs are resonantly generated, the low-energy side of the exciton peak saturates preferentially, and a strong ultrafast bleaching is observed for the first 300 fs. During this phase the absorption at the exciton peak is significantly larger than when free e-h pairs are created. Then the absorption recovers partially to the same level as for nonresonant excitation, as shown on Fig. 6, and the absorption profile also evolves toward that observed when free e-h pairs are directly excited, Fig. 7. This result has been interpreted as the effects of selectively generated excitons that transform in 300 fs in free e-h pairs. It is considered as the direct time-resolved observation of the excitons ionization by thermal phonons. The surprising observation that the neutral bound states can produce a diminution of the excitonic absorption about twice larger than the charged e-h plasma has been recently explained [27]. It is due to the reduced strength of the long-range direct screening in 2-D compared to the short-range effects of the exclusion principle i.e., phase-space filling and exchange, and to the large temperature of the free e-h pairs directly generated above the gap or released by the exciton ionization.

The excitons selectively generated by a resonant excitation occupy the lowest excited states of the QWS. The exciton gas temperature (if it can be defined) is determined by the distribution of the absorbed photons. Thus, for the very short time between the generation of the excitons and the first scattering with the thermal phonons, the exciton gas is at a very low temperature. The modifications of the optical transitions induced by this cold gas are due to the filling of the single particle states from which the bound states are constructed [21], and to exchange interaction which also concern like-states. The excitonic long-range screening again is strongly reduced by dimensionality effects and by the neutrality of the excitons. The blocking of excitonic transitions due to the exclusion principle is extremely effective, precisely because excitons utilize preferentially the single particle states that are less than one binding energy above the gap. Conversely, the e-h plasma that is released by excitons-ionization has a rather high energy because the LO-phonon energy is about four times larger the exciton binding energy. Then the plasma equilibrates to the lattice temperature in a few carrier-carrier and carrier-phonon collisions. For the same reasons the free e-h pairs that are directly generated above the gap by high-energy photons also have a large energy and quickly equilibrate. The distribution of thermalized high-temperature e-h plasma follow a Boltzman law with a long tail in the high-energy part of the density of states. The occupation close to the gap, at the location of the single particle states from which excitons are built is small, and thus the phase-space filling and exchange effects they produce are smaller than those due to a cold gas of exciton with the same density. The mechanism that governs exciton beaching by charged plasma in 3-D semiconductors is the long-range direct screening, however, in QWS it is strongly reduced owing to the lower dimensionality. Altogether at room temperature excitons are more efficient in reducing the absorption close to the gap than thermalized free e-h pairs [27]. If the carriers are cold too, then it can be shown that they become more efficient than excitons [27].

This is an unusual example of reverse relaxation, where an optically active species is observed before it can interact with the thermal reservoir and relax toward another species less optically active and lying higher in energy. It confers to QWS very large optical nonlinearities with double components. One component has a very fast recovery time ($\tau < 0.5$ ps) and is about twice as large as the component with a longer recovery time (100 ps $< \tau <$ 20 ns). A recent theory [28] has speculated that this feature plays an important role in the pulse-shaping mechanisms of passively modelocked diode lasers that use QWS as saturable absorbers [29].

4. EFFECTS OF STATIC ELECTRIC FIELDS ON THE ABSORPTION OF QWS

When static electric fields are applied to 3-D semiconductors, the band edge is shifted and broadened by the Franz-Keldysh effect [30]. At low temperature where exciton resonances are observed, the dominant effect of a static electric field is to broaden the exciton peak owing to the life-time shortening by field ionization [31,32]. There is also a Stark shift of the ground-state but it never exceeds a small fraction of the binding energy and is hardly observed before the resonance is washed out by the field ionization.

Simple symmetry considerations show that the application of a static electric field to a QWS will produce different effects according to the orientation of the field with respect to the plane of the layers. Both geometries have been extremely studied and theoretical models in good agreement with experiment have been developed [33,34]. An interesting aspect of QWS electro absorption is that the energies of the levels involved in the optical transitions, i.e., single-particle state confinement energies and bound state binding energies, are small (between ten and a hundred meV) and the corresponding wave functions are extended over large volumes (for the single particle envelope wave functions, the characteristic length is the well thickness and for the bound state wave functions it is the exciton Bohr radius, both of which are of the order of 100 Å). Thus, the application of small fields of the order of a few V/μm can cause significant changes of energies and of the absorption spectra, the description of these modifications requires theories with applicability well beyond that of the perturbation approximation. In addition, the electro absorption mechanisms are intrinsically fast, since they are limited only by the response time of the envelope wave functions (in practice, circuit considerations will limit the devices speed before).

When the field is applied parallel to the plane of the layer, the effects that are observed are similar to those seen in 3-D semiconductors. As shown in Fig. 8, the resonances broaden and eventually they become no more resolvable. No significant shift is observed even on the very sensitive differential spectra. The change of absorption at the exciton peak induced by field of the order of some V/μm can be as large as a few thousand cm^{-1}. It is rather difficult to compare the experimental spectra under applied field to pure 2-D or 3-D theories because of the finite thickness of the wells, the proximity of the light-hole and heavy-hole exciton resonances and the inhomogeneous broadening due to interface roughness. However, a reasonable qualitative agreement is found with the pure 2-D theory [34], with excitons more difficult to field-ionize than pure 3-D excitons [35].

In this configuration the capacitance of the samples is small because the thickness of the electrodes is limited to the sample thickness i.e., about one μm. Very fast light modulations with switching on and off times of about 30 ps and a few percent modulation depth have been demonstrated using this geometry [36].

When the electric field is applied perpendicular to the layers qualitatively, different effects are seen [33,34]. The exciton resonances shift to lower

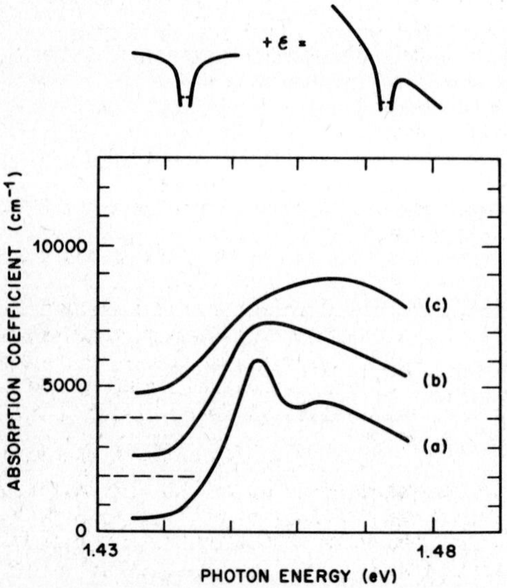

Figure 8

Absorption spectra at various electric fields for the parallel-field sample. (a) 0 V/cm; (b) 1.6×10^4 V/cm; (c) 4.8×10^4 V/cm. The insert shows schematically the distortion of the Coulomb potential of electron and hole with applied field. The zeros are displaced as shown by the dashed lines for clarity.

Figure 9

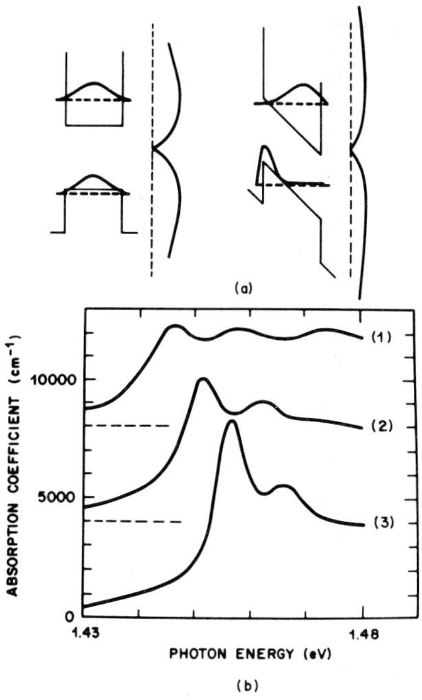

Excitonic wave functions without and with an applied electric field (a), and the quantum-confined Stark shift in an absorption spectrum (b). The wave functions illustrate how the walls of a quantum well hold an electron and hole in a bound state, even at applied fields much stronger than the classical ionization field. The absorption spectra are those of a quantum-well structure under three different static electric fields applied normal to the layers. The fields are 10^4 V/cm to the layers. The fields are 10^4 V/cm (bottom curve), 5 X 10^4 V/cm (middle curve) and 7.5 X 10^4 V/cm (top curve).

energies with little broadening and the height of the peaks decreases as the applied field is increased. A typical set of absorption spectra measured at various applied fields is shown in Fig. 9a. Most remarkable is the fact that this behavior is observed up to very large fields. In fact, with this geometry it is possible to resolve excitonic resonances at applied field several orders of magnitude larger than the classical ionization field [34,37]. The explanation for such process is that the perpendicular field pushes the electron and the hole in opposite directions, but the walls of the quantum will prevent them flying apart until the magnitude of the field is large enough to induce tunneling out of the well. The well thickness is smaller than the 3-D Bohr radius and the well depth, that is controlled by the aluminum content of the barrier layers, can be made much larger than the exciton binding energy. Hence, even under very large fields the electron and the hole are constrained to remain close enough one to another that they can still form bound states. This behavior is illustrated in Fig. 9a. Since the Stark shift observed in this configuration is only seen when the carriers are confined in very thin layers, it has been named the Quantum Confined Stark Effect (QCSE).

The experimental results are in good agreement with a theory that accounts for the aforementioned mechanisms [33,34] as shown by Fig. 10 where the observed shifts of the excitonic peaks are compared to the theory.

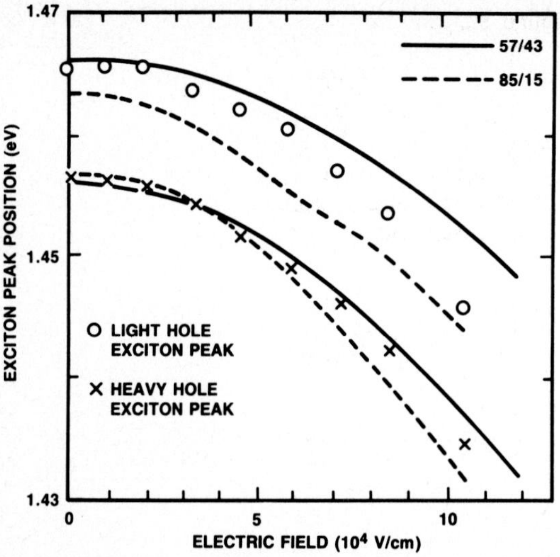

Figure 10

Positions of the exciton peaks with applied field perpendicular to the layers. The points are experimental. The lines are theoretical calculations using the known room-temperature GaAs band gap and strain shift of the excitons, and the shifts of the exciton peaks as evaluated for the 57:43 split (solid lines) and 85:15 (dashed lines) of the band discontinuities. There are no fitted parameters in the theory.

In order to easily apply the field perpendicular to the layer and avoid large currents and thermal effects, the sample design requires some care, QWS is usually contained within the depletion region of a p-i-n diode and the field across the QWS is applied by reverse biasing the diode. The QCSE has been applied to high-speed optical modulation using both the conventional geometry with light-propagating perpendicular to the layers [38,39] or in waveguides with light propagating along the layers [37,40]. Fast (100 ps) and efficient (2 dB for the usual configuration and 10 dB in waveguides) QWS modulators have been demonstrated. These devices represent a new approach to modulation; they are small (a few cubic microns) hence, they require only moderate driving voltages and avoid the problems of electric and optical velocity mismatch. They are also compatible with laser diodes' wavelength and semiconductor electronic power.

Because the QWS is in a p-i-n diode, the sample is also a photodetector. In fact, excellent detection with unity quantum efficiency have been observed. The responsivity of the photodiode closely follows the absorption coefficient, and thus provides information upon the optical properties of the sample in an electric form that can be used in an external circuit to produce a feedback. With a positive feedback, low switching energy optical bistable devices have been demonstrated [41], with negative feedback it has been shown that the QWS p-i-n diodes can act as optical level shifter and as self-linearized modulator [42].

The use of the QCSE for optical switching deserves some comments. The p-i-n diode is connected to a voltage supply through a resistor, a potential is applied to the QWS in order to shift the absorption profile to lower energy and light at the wavelength of maximum absorption for zero-electric field is sent through the sample. First, due to the field on the QWS, the absorption is small and the sample is relatively transparent. As the intensity of the light source increases, the photocurrent induces a drop of potential across the resistor, so that the field on the QWS decreases. This causes the absorption to increase, which in turn increases the photocurrent ... and so on until the QWS switches to

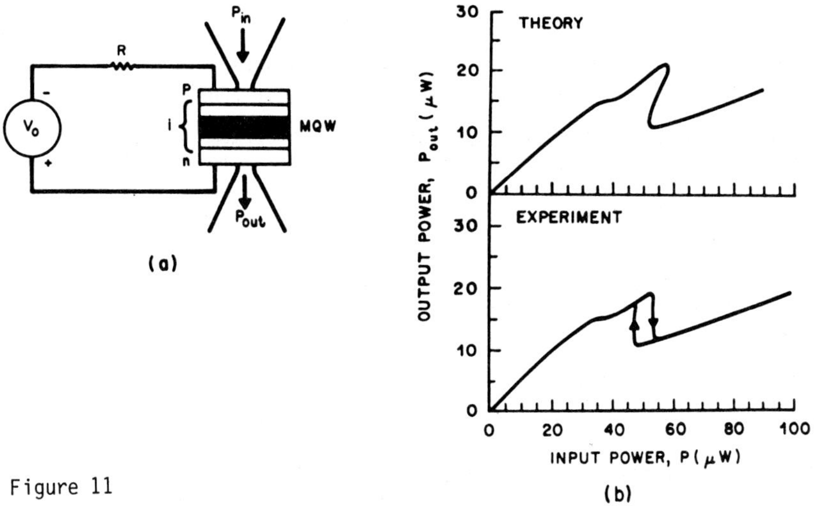

Figure 11

Self-electro-optic device and plots showing its optical bistability. The device in the schematic diagram shows optical bistability when it is connected to a simple resistive load.

the state of maximum absorption. Thus, the positive feedback leads to the switching from a transparent to an absorptive state. Such operation is shown in Fig. 11. The optical bistable devices that use this effect do not require cavities, they need very low energy to switch (optical power 4 fJ/μm^2, total power 20 fJ/μm^2) and they can present fast switching times (30 ns for 100 μm diameter samples have been observed and faster operation is expected with smaller samples). In addition, they do not require coherent light sources, they are compatible with semiconductor technology and can be manufactured in large arrays. They are the first example of optical switches that have reasonable chances of being used in real systems.

5. CONCLUSION

In this lecture we have presented a quick survey of the novel nonlinear optical properties and electro-absorption effects which can be seen in QWS. The possibility of manufacturing artificial semiconductor microstructures gives us new degrees of freedon to tailor the optical and electric properties of condensed matter. Our main studies were performed on GaAs/AlGaAs QWS but recent progress in crystal growth opens new opportunities to extend these investigations to other materials. The III-V semiconductors represent a natural step for infrared applications and II-VI compounds, which offer, in addition, the possibility of including magnetic inpurities in the QWS, open new perspectives in magneto-opto-electronics. In the future, 3-D patterned growth will make it possible to fabricate other structures, such as 1-D quantum wires and 0-D quantum dots. There is no doubt that they will possess fascinating physical properties, and will find novel applications in opto-electronics.

ACKNOWLEDGEMENTS

The work summarized in these notes has been performed at AT&T Bell Laboratories in close collaboration with many colleagues. I am pleased to acknowledge the invaluable contributions of D.A.B. Miller, A.C. Gossard, P.W. Smith, T.C. Damen, T.H. Wood, W.H. Knox, J.S. Weiner, C.V. Shank, R.L. Fork, A.Y. Cho, C.A. Burrus and W.T. Tsang.

REFERENCES

1. L. L. Chang and K. Ploog, eds., "Molecular Beam Epitaxy and Heterostructures," (Nijhoff, Dordrecht, The Netherlands, 1985).

2. R. Dingle, ed. "Device and Circuit Applications of III-V Semiconductor Superlattice and Modulation Doping," (Academic, New York, 1985).

3. E. I. Rashba and M. D. Sturge, eds., "Excitons," Vol. 2 of Modern Problems in Condensed Matter Science, (North-Holland, Amsterdam 1982).

4. R. Dingle, W. Wiegmann, and C. H. Henry, "Quantum states of confined carriers in very thin AlGaAs-GaAs heterostructures," Phys. Rev. Lett. 33 827 (1974).

5. G. Bastard, Phys. Rev. B25, 7584 (1982).

6. U. Ekenberg and M. Altarelli, Phys, Rev. B30, 3569 (1984).

7. A. Fasolino and M. Altarelli, in "Heterostructures and Two-Dimensional Electron Systems in Semiconductors," (Mauendorf, Austria, 1984).

8. R. J. Elliott, Phys. Rev. 106, 1384 (1957).

9. M. Shinada and S. Sugano, J. Phys. Soc. Jpn. 21, 1936 (1966).

10a. C. Weisbuch, R. C. Miller, R. Dingle, and A. C. Gossard, Solid State Commun. 37, 219 (1981).

10b. R. C. Miller, D. A. Kleinman, W. A. Norland, and A. C. Gossard, Phys. Rev. B22, 863 (1980).

10c. R. C. Miller, D. A. Kleinman, J. Lumin 30, S20 (1985).

11. J. Hegarty and M. D. Sturge, J. Opt. Soc. Am. B2, 1143 (1985).

12. C. Weisbuch, R. Dingle, A. C. Gossard, and W. Wiegmann, J. Vac. Sci. Technol 17, 1128 (1980).

13. G. Bastard, E. E. Mendez, L. L. Chang, and L. Ezaki, Phys. Rev. B26, 1974 (1982).

14a. R. L. Greene and K. K. Bajaj, Solid State Commun. 45, 831 (1983).

14b. R. L. Greene, K. K. Bajaj, and D. E. Phelps, Phys. Rev. B29, 1807 (1984).

15. D. S. Chemla, D. A. B. Miller, J. Opt. Soc. Am. B2, 1155 (1985).

16a. J. E. Zucker, A. Pinczuk, D. S. Chemla, A. C. Gossard, and W. Wiegmann, Phys. Rev. Lett. 51, 1293 (1983).

16b. J. E. Zucker, A. Pinczuk, D. S. Chemla, A. C. Gossard, and W. Wiegmann, Phys. Rev. Lett. 53, 1280 (1984).

16c. J. E. Zucker, A. Pinczuk, D. S. Chemla, A. C. Gossard, and W. Wiegmann, Phys. Rev. B29, 7065 (1984).

17. D. A. B. Miller, D. S. Chemla, P. W. Smith, A. C. Gossard, and W. T. Tsang, Appl. PHys. B28, 96 (1982).

18. J. S. Weiner, D. S. Chemla, D. A. B. Miller, T. H. Wood, D. Sivco and A. Y. Cho, Appl. Phys. Lett. 46, 619-621 (1985).

19. H. Temkin, M. B. Panish, P. M. Petroff, R. A. Hamm, J. M. Vandenberg, S. Sumski, Appl. Phys. Lett. 47, 394 (1985).

20. D. S. Chemla, D. A. B. Miller, P. W. Smith, A. C. Gossard, and W. Wiegmann, IEEE J. Quantum Electron. QE-20, 265 (1984).

21. H. Haug and S. Schmitt-Rink, Prog. Quantum Electron. 9, 3 (1984).

22. D. A. B. Miller, D. S. Chemla. P. W. Smith, A. C. Gossard, and W. T. Tsang, Appl. Phys. Lett. 41, 679 (1982).

23. D. A. B. Miller, D. S. Chemla, D. J. Eilenberger, P. W. Smith, A. C. Gossard, and W. Wiegmann, Appl. Phys. Lett. 42, 925 (1983).

24. W. H. Knox, R. F. Fork, M. C. Downer, D. A. B. Miller, D. S. Chemla, and C. V. Shank, in "Ultrafast Phenomena IV, D. H. Auston and K. B. Eisenthal, eds. (Springer-Verlag, Berlin, 1984), p. 162.

25. W. H. Knox, R. F. Fork, M. C. Downer, D. A. B. Miller, D. S. Chemla, and C. V. Shank, Phys. Rev. Lett. 54, 1306 (1985).

26. J. S. Weiner, D. S. Chemla, D. A. B. Miller, H. Haus, A. C. Gossard, W. Wiegmann, and C. A. Burrus, Appl. Phys. Lett. 47, 664 (1985).

27. S. Schmitt-Rink, D. S. Chemla, D. A. B. Miller, Phys. Rev. 32 (1985).

28. H. A. Haus, Y. Silberberg, J. Opt. Soc. Am. B2, 1237 (1985).

29. P. W. Smith, Y. Silberberg, D. A. B. Miller, J. Opt. Soc. Am. B2, 1228 (1985).

30a. W. Franz, Z. Naturforsch 13a, 484 (1958).

30b. L. V. Keldysh, Zh. Eksp. Teor. Fiz. 34, 1138 (1958) [Sov. Phys. - JETP 7, 788 (1958)].

31. J. D. Dow and D. Redfield, Phys. Rev. B1, 3358 (1970).

32. Q. H. F. Vrehen, J. Phys. Chem. Solids 29, 129 (1968).

33. D. A. B. Miller, D. S. Chemla, T. C. Damen, A. C. Gossard, W. Wiegmann, T. H. Wood, and C. A. Burrus, Phys. Rev. Lett. 53, 2173 (1984).

34. D. A. B. Miller, D. S. Chemla, T. C. Damen, A. C. Gossard, W. Wiegmann, T. H. Wood, and C A. Burrus, Phys. Rev. B32, 1043 (1985).

35. F. L. Lederman and J. D. Dow, Phys, Rev. 13, 1633 (1976).

36. D. A. B. Miller, W. H. Knox, D. S. Chemla, T. C. Damen, P. M. Downey, J. E. Henry, Paper TUK2, Conference on Lasers and Electro-optics, Baltimore, MA (1985).

37. J. S. Weiner, D. A. B. Miller, D. S. Chemla, T. C. Damen, C. A. Burrus, T. H. Wood, A. C. Gossard, W. Wiegmann, To be published in Appl. Phys. Lett.

38. T. H. Wood, C. A. Burrus, D. A. B. Miller, D. S. Chemla, T. C. Damen, A. C. Gossard, and W. Wiegmann, Appl. Phys. Lett. 44, 16 (1984).

39. T. H. Wood, C. A. Burrus, D. A. B. Miller, D. S. Chemla, T. C. Damen, A. C. Gossard, and W. Wiegmann, IEEE J. Quantum Electron. QE-21, 117 (1985).

40. T. H. Wood, C. A. Burrus, R. S. Tucker, J. S. Weiner, D. A. B. Miller, D. S. Chemla, T. C. Damen, A. C. Gossard, W. Wiegmann, Electronics Letters 21, G93 (1985).

41. D. A. B. Miller, D. S. Chemla, T. C. Damen, A. C. Gossard, W. Wiegmann, T. H. Wood, and C. A. Burrus, Appl. Phys. Lett. 45, 13 (1984).

42. D. A. B. Miller, D. S. Chemla, T. C. Damen. T. H. Wood, C. A. Burrus, A. C. Gossard, and W. Wiegmann, Opt. Lett. 9, 567 (1984).

Part II

Ultrafast Charge Carrier Dynamics in Semiconductors

Femtosecond Lasers and Ultrafast Processes in Semiconductors

C.L. Tang

School of Electrical Engineering, Cornell University, Ithaca, NY 14853, USA

I. Introduction

Rapid progress has been made in the last few years[1] on pico-second and femto-second lasers and related experimental techniques for studying ultrafast processes in various materials, including the dynamics of hot carriers in semiconductors. Of particular interest are GaAs and related compounds and structures, which are important for modern high-speed electronic devices, for example.

The intraband relaxation of nonequilibrium carriers in the conduction and valence bands of semiconductors is known to be in the subpicosecond time domain. The corresponding rates determine the intrinsic lifetimes of the relevant Bloch states. There have been numerous theoretical studies of such relaxation processes, but experimental confirmation has been difficult and often indirect. The development of lasers with pulse lengths less than 100 femtoseconds opens up the possibility of studying these processes directly by optical means. The experimental problems involved are still formidable, however, because the femtosecond lasers and related measurement techniques developed so far are still somewhat limited in the choice of wavelengths, pulse repetition rates, and pulse lengths.

For GaAs and related compounds and structures, the time-constants of interest can be in the 50 to 100 fs range,which is comparable to the shortest pulse lengths that can be achieved conveniently in the laboratory for measurement purposes. This makes the interpretation of the data very complicated. Rapid progress is being made in reducing the pulse length, however. Pulses as short as 27 fs at 10^8 Hz rate[2] and 8 to 12 fs at 1 KHz rate[3,4] have recently been reported. These lasers will undoubtedly be used for applications related to semiconductors eventually. At the moment, however, most of the experimental studies of the ultrafast dynamics of hot carriers in semiconductors reported in the literature are based upon lasers in the 80 - 100 fs range. For these cases, complicated theories are, unfortunately, needed to extract lifetime information from the experimental results.

The available laser wavelengths are also seriously limited. The energy of the femtosecond photons generated directly from the laser are generally limited to about 2.02 eV at 10^8 Hz rate. It is possible to Raman-shift such photons in, for example, water vapor to a continuum at lower wavelengths at about 10 Hz. The loss of repetition rate could present a problem in some experiments. For

GaAs and related compounds, the conduction band states that can be excited with 2 eV or less energetic photons are, therefore, limited to about 0.6 eV above the band edge or below. One would like to be able to study conduction band states further above the band edge, however, because the intraband relaxation time is a function of excitation energy.

Several types of relaxation rates are of interest: the energy relaxation rate of the electrons in the initially excited states, which is on the order of 50 - 100 fs; the time it takes, on the order of a picosecond, the electrons to reach quasi-equilibrium at the bottom of the band at a temperature which may be higher than that of the lattice; finally, the rate at which this quasi-equilibrium distribution cools down to the lattice temperature, which may take many picoseconds. Because these rates and the excitation energies of the carriers involved are very different, the measurement techniques needed are also different.

In this article, a brief review is given of our recent studies[5-7] of the relaxation dynamics of hot carriers in GaAs, $Al_{0.32}Ga_{0.68}As$, and related multiple quantum well (MQW) structures using femtosecond and picosecond lasers.

II. Femtosecond Relaxation of Hot Carriers in Semiconductors

Figure 1 shows the energy schematic near the Brillouin zone center of a direct bandgap semiconductor. The lifetime of the excited states can be studied with appropriately chosen photons as shown in Figure 1. The basic scheme is to first photo-excite the carriers into the states of interest with a short laser pulse. The rate of decay of the population in the initially excited states can then be estimated by measuring the absorption of successively delayed probe pulses at the same wavelength.

In applying this basic pump-and-probe technique to GaAs and related compounds, a slightly improved procedure, the so-called equal-pulse correlation technique[8], is used. The schematic of the experimental setup is shown in Figure 2. Instead of distinguishing the pump and probe pulses, two trains of orthogonally polarized and colinearly propagating pulses of equal intensity but with a variable delay,

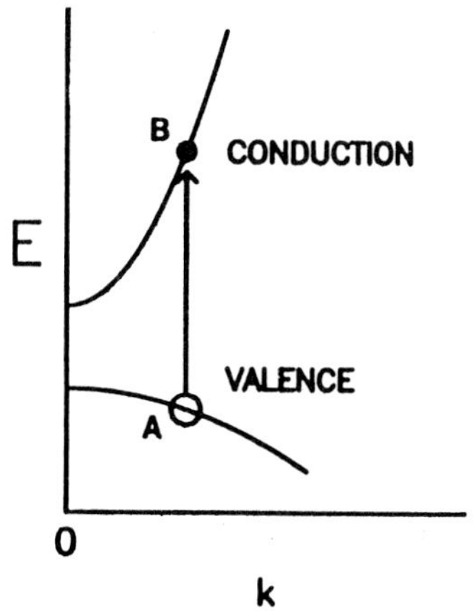

Figure 1-Schematic band structure of a direct-bandgap semiconductor and the absorption of a photon to create a hole and an electron (A and B, respectively).

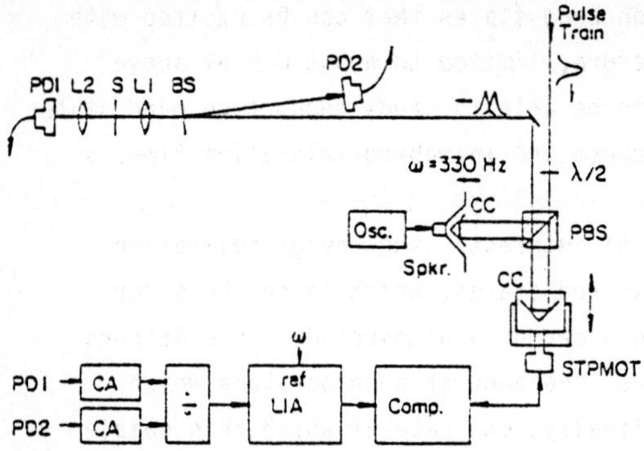

Fig.2. Optical setup for the equal-pulse correlation technique. PBS, BS, 50% polarizing and 4% ordinary beam splitters, respectively; N2, half-wave plate; CC, retroreflecting corner cubes; SPKR, speaker; STPMOT, stepping-motor-driven linear actuator; PD1, PD2, P-I-N photodiodes. L1, L2, 40 x and 20 x microscope objectives, respectively; S, sample; CA, current amplifier; :, dividing circuit; LIA, lock-in amplifier; Comp., computer

τ, between them are used. The time-averaged total transmission of the pulses is measured as a function of the delay between the two sets of pulses. In this case, reversing the sign of τ reverses the role of the two pulses as pump or probe. Symmetrizing the two pulses simplifies the interpretation of the experimental results when the relaxation process being measured is superimposed on a slower relaxation process that also affects the transmission of the probe pulses[8].

When the delay between the two pulses is longer than the lifetime of the carriers in the initially excited states, the absorption or transmission of the two pulses are independent of each. As the delay is varied in this range, the time-averaged total transmission is independent of the delay. As the delay becomes comparable or shorter than the carrier lifetime, T_0, the absorption of one pulse is reduced, because the optically coupled conduction and valence band states are partially occupied by carriers excited by the other pulse. This leads to a mutual saturation effect,and the total transmission increases with decreasing delay between them, reaching a maximum at zero-delay. The shape of this transmission correlation peak [TCP(τ)] of the two pulses contains information on the lifetime of the excited

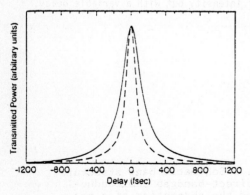

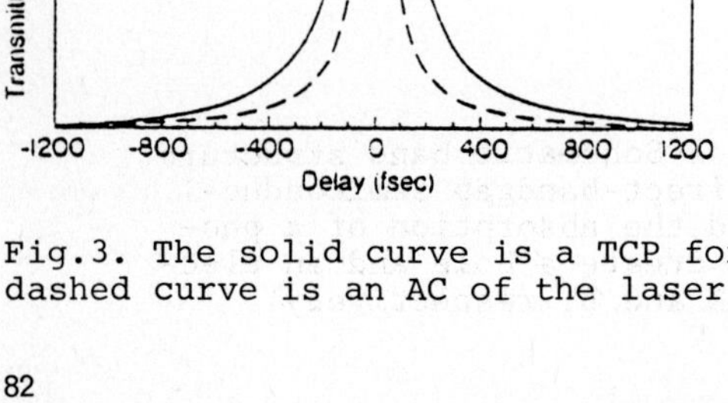

Fig.3. The solid curve is a TCP for a sample of AlGaAs, and the dashed curve is an AC of the laser pulse

states. If this is the only effect present, the TCP is the convolution of the autocorrelation peak of the excitation pulses and the population decay function of the excited states[5,8]. A typical measured TCP(τ) for $Al_{0.32}Ga_{0.68}As$ along with the corresponding autocorrelation trace of the pulses used are shown in Figure 3. The increased width of the TCP from that of the autocorrelation trace is a measure of the energy relaxation time T_0 of the excited carriers.

Because the incoherent saturation effect is a nonlinear optical effect proportional to at least the square of the light intensity I^2 or the fourth power of the E-field, the possible existence of coherent interference[9] when the two pulses overlap directly in time should also be considered. If the excited carrier distribution is isotropic, coherent interference between the two pulses can be avoided when the two pulses are othogonally polarized.

In the case of GaAs excited by 2-eV photons, the excitation is not isotropic[10]. To understand this, one has to consider the band structure of GaAs in more detail. Figure 4 shows a schematic of the energy diagram near the Brillouin zone center of GaAs and $Al_{0.32}Ga_{0.68}As$. For 2-eV photons, the transitions from the heavy-hole, light-hole, and split-off bands are all allowed. Based on energy-momentum conservation considerations and the known effective masses of the conduction and valence bands, the directly excited electronic states (marked h, l, and s in Figure 4) are at 1.93, 1.87, and 1.57 eV and the corresponding hole states are at -0.086, -0.152, and -0.45 eV, respectively. The bandwidth of the electronic states excited by a 100 fs pulse of 2-eV photons is on the order of 45 meV, taking into account the estimated energy-relaxation time of the optically coupled states. Thus, the excited electronic states from the heavy-hole, light-hole, and split-off bands are all well separated. The corresponding electronic distribution in the momentum space will depend upon the selection rules. For electric dipole transitions from the heavy-hole state and in the spherical approximation using Kane's model for III-V compounds[11], the selection rule near the zone center is such that the magnetic quantum number changes by 1 ($\Delta m = \pm 1$). The k-vector of the excited electrons is, therefore, perpendicular to the E-field direciton, or the electron distribution is anisotropic and oriented primarily normal to the E-field. For the transitions

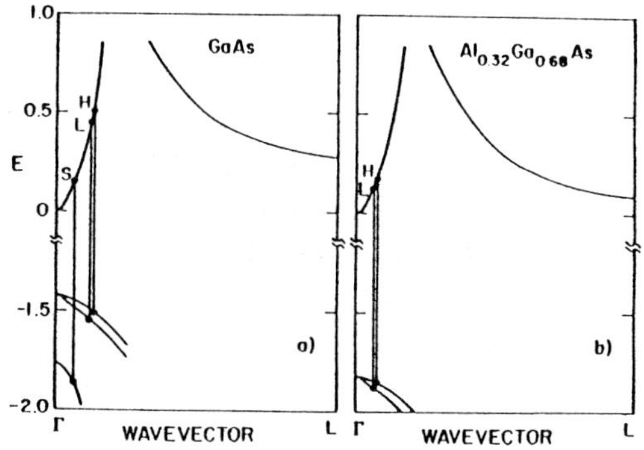

Fig.4-Band structure of GaAs and AlGaAs and allowed transitions for 2.02 eV photons from the heavy-hole (H), light-hole (L), and split-off bands (S).

from the light-hole, it is primarily in the direction of the E-field. For the split-off transition, it is isotropic. From density of states considerations, transitions from the split-off band contribute only about 15% of the absorption of the 2-eV photons, while the heavy-hole, and light-hole transitions are comparable and contribute the remaining 85%. Thus, for GaAs and 2-eV photons, there can be a coherent artifact contribution in the measured transmission correlation peak while the pulses overlap.

The extent of the coherent artifact contribution [CA(τ)] to the measured TCP(τ) depends upon the orientation relaxation time T_θ relative to the pulse width τ and the energy relaxation T_0. Clearly, if T_θ is much less than the pulse width, the anisotropy of the electron distribution will be wiped out over most of the pulse, and the coherent artifact contribution will be negligible. If the orientation relaxation time is not negligible compared to the pulse width, but the pulse width is much shorter than the energy-relaxation time, the coherent artifact, which appears only where the two pulses overlap, will appear as a sharp peak on a much broader TCP whose width is determined by the energy-relaxation time. The two can then be clearly separated in this case. The complication comes when the orientation relaxation time is comparable to both the pulse width and the energy-relaxation time. An additional piece of experimental information is then needed to determine the fractional contribution:

$$FCA = CA(0)/TCP(0) \tag{1}$$

of the coherent artifact at $\tau = 0$. It has been shown[5], based upon detailed and very tedious numerical calculations based upon Kane's model for the relevant wavefunctions and the theory of Ref.12, that this fractional contribution can be determined from the experimentally measured ratio, R, of the TCP(0)'s with parallel and perpendicular polarizations for the two pulses:

$$R = [TCP(0)-TCP(\infty)]_{parallel}/[TCP(0)-TCP(\infty)]_{perpendicular} \tag{2}$$

It leads to the following numerical relationship:

$$FCA(0) \simeq \begin{cases} 0.732(2-R) & \text{for GaAs} \\ 0.581(2-R) & \text{for } Al_{0.32}Ga_{0.68}As \end{cases} \tag{3}$$

for 2 eV-photons. Theoretically, the value of R for GaAs ($Al_{0.32}Ga_{0.68}As$) can vary from 1.38(1.45) to 2 depending upon the orientational relaxation time relative to the energy-relaxation time of the carriers, giving a maximum fractional contribution of the coherent artifact to the TCP of 45%(38%).

The same theory shows that the measured R also gives an estimate of the orientation-relaxation time T_θ from the energy-relaxation time T_0 according to the following numerical relationship:

$$T_\theta/T_o \simeq \begin{cases} (2-R)/(1.58R-2.2) & \text{for GaAs} \\ (2-R)/(1.6R-2.33) & \text{for } Al_{0.32}Ga_{0.68}As \end{cases} \qquad (4)$$

for 2-eV photons. Experimentally, the measured R value is approximately 1.75 for both materials. The implied orientational relaxation time for both materials is then approximately half of the energy-relaxation times. The corresponding fractional contribution, FCA, of the coherent artifact to the transmission correlation peak is according to Eq. (3) no more than 18% for both cases, which is not significant.

Making use of these results, the intraband relaxation times of electrons excited by 2.02 eV photons in GaAs, $Al_{0.32}Ga_{0.68}As$, and a GaAs/AlGaAs multiple quantum well structure have been measured by the equal-pulse correlation technique. The results are shown in Figure 5. Several points are of particular interest.

First, in the low carrier density limit, the energy-relaxation times for GaAs and AlGaAs are respectively, 45 and 80 femtoseconds. These are among the fastest relaxation times ever measured. Table I shows a summary of the calculated relaxation rates[5] using existing theories for various scattering processes out of the initially excited states. These calculated results neglect the carrier-density effects, such as carrier-carrier scattering, carrier-plasmon scattering, and Debye screening effects on the electron- phonon scattering rates, which are less important in the low density limit. As can be seen, the measured results

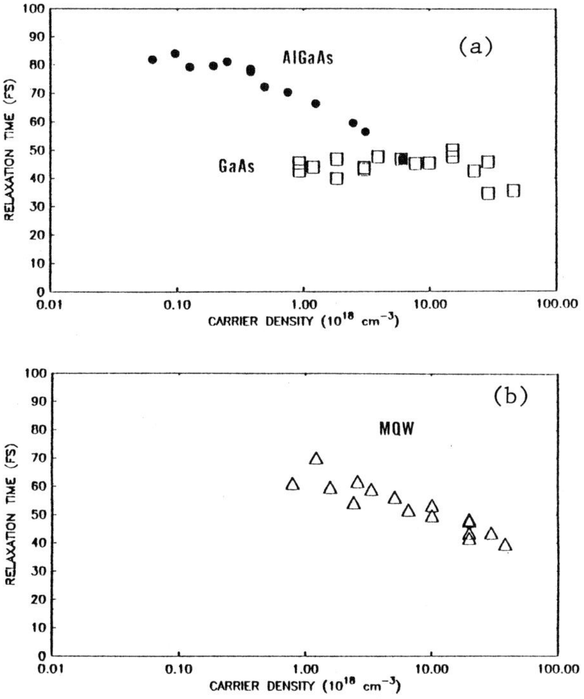

Fig. 5 - Relaxation time T_o vs. photo-generated carrier density for a) GaAs and AlGaAs and b) multiple quantum well.

Table I - Calculated scattering rates[a] for the heavy-hole and light-hole optically coupled states

Type of Scattering	Electron OCR Heavy	Electron OCR Light	Hole OCR Heavy	Hole OCR Light
GaAs				
R_{POP}^+	1.7	1.7	5.5	5.5
R_{POP}^-	6.4	6.4	18.5	18.5
R_{ACS}	0.6	0.6	4.7	6.2
R_{ALLOY}	o	o	o	o
R_{IV}	25(12)	19(0)	o	o
AlGaAs				
R_{POP}^+	2.2	2.2	6.8	6.4
R_{POP}^-	7.9	7.9	0	19.8
R_{ACS}	0.5	0.4	3.5	6.3
R_{ALLOY}	1.0	1.0	0.4	0.4
R_{IV}	15(9)	9(0)[f]	0	0

[a] Rates are in units of $10^{12}sec^{-1}$. $R_{POP}^{+,-}$ is polar optical scattering for absorption (+) and emission (-). R_{ACS} is acoustic and includes emission and absorption of a phonon, R_{ALLOY} is random potential allow scattering. Note that the latter two rates do not affect the carriers energy but only its momentum; thus they affect only T_0 rather than T_θ. R_{IV} is $\Gamma \rightarrow L$ IV deformation-potential scattering by emission and absorption of a phonon using a deformation potential of 1×10^9 eV/cm. The values in parentheses are the $\Gamma \rightarrow X$ rates.

give a reasonable confirmation of the calculated results. The calculated rates for the individual processes show that the dominant relaxation mechanism for the 0.6 eV electrons in the initially excited states is electron-phonon scattering into the satellite valleys.

A second point of interest is that, in the case of AlGaAs, there is a clear carrier-density dependence in the energy-relaxation rate above about 10^{18} cm^{-3}. Straight forward application to this problem of conventional theories for carrier-carrier scattering and screening effects did not yield any consistent results. However, these theories assume a thermal equilibrium carrier distribution. In the femtosecond domain, the carrier distribution is far from equilibrium but is closer to a δ-function excitation in energy space. To our knowledge, no screening theory has yet been developed for such a highly nonequilibrium distribution. This should be an interesting theoretical problem.

The third point of interest is that in the femtosecond time domain, the initial lifetimes of the excited carriers in the quantum-well structure and the bulk materials are comparable at room temperature. This shows that the initial relaxation of the electron distribution is independent of the dimensionality of the electron gas. This is very different from the liquid nitrogen case discussed in Section IV where there is a very significant difference between the cooling rates of the quasi-equilibrium distributions down to the lattice temperature for the bulk materials and quantum structures.

III. Dynamic Burstein-Moss Shift in GaAs and Related Compounds

After the photo-excited electrons leave the initial states, primarily through phonon-scattering into the satellite valleys, the subsequent steps of relaxation of the carriers in the satellite valleys are the following: relax to the bottom of the satellite valleys through phonon emission, scatter from there back to the central valley, and finally from there through polar optical- phonon scattering down the central valley. As the carriers accumulate at the bottom of the central valley, the corresponding quasi-Fermi level shifts upward, leading to a shift of the absorption band edge. This is known as the Burstein-Moss shift. For high enough carrier densities, this shift can lead to increasing saturation of the transition from the split-off band (marked s in Figure 4) in the absorption of 2-eV photons by GaAs. This transition can, therefore, be used as a convenient probe for the time-dependence of the carrier distribution near the bottom of the central valley. It gives the time it takes for the carriers to equilibrate near the bottom of the band after leaving the initially excited states (marked h and l in Figure 4).

The dynamic Burstein-Moss shift following femtosecond pulse excitation in GaAs has been observed[6]. In $Al_{0.32}Ga_{0.68}As$, energy- momentum conservation does not allow absorption of 2-eV photons from the split-off band; indeed, no Burstein-Moss shift was observed. Figure 6 shows the transmission correlation trace for GaAs at higher intensity and, hence, carrier density levels than the case shown in Figure 3. The results show that, in addition to the central transmission correlation peak, there is a rising wing on each side of the peak.

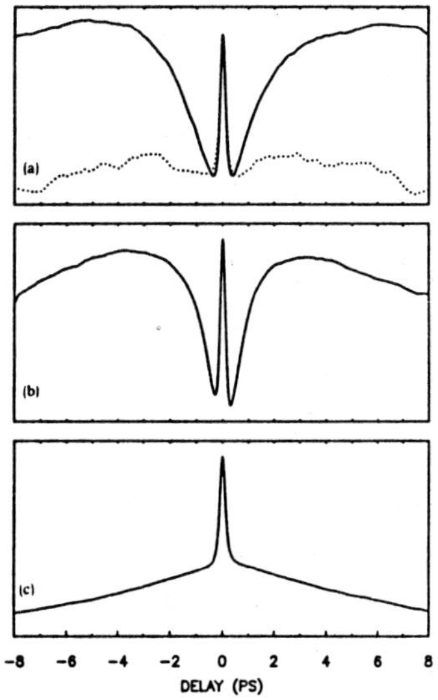

Fig.6-(a)Experimental scans of transmission vs τ for GaAs at P=o.6 mW(solid curve)and P=0.2 mW (dotted). The wing rise time for the solid curve is 1.6±0.2 psec. (b)Experimental scan for MQW at P=0.6 mW and with a wing rise time of 1±0.2 psec. (c) Experimental scan for $Al_{0.32}Ga_{0.68}As$ at P=0.6 mW.

According to Ref.6, this is due to the saturation of the split-off transition or the dynamic Burstein-Moss shift. The approximately 1 picosecond risetime of the wings corresponds to the filling up of the bottom of the band and the concomitant rise of the quasi-Fermi level.

This time-constant is considerably longer than the corresponding calculated LO phonon emission rate ($6.4 \times 10^{12} s^{-1}$) within the central valley given in Table I. The explanation is that the density of states in the satellite valleys is much higher than that of the central valley in the same energy range. There is, therefore, a bottle neck for draining of the carriers in the satellite valleys through scattering back to the central valley. Numerical considerations[6] based upon the relevant densities of states show that it is not unreasonable to expect an almost 10-fold increase from the phonon emission rate in the time it takes to drain all the carriers in the satellite valleys to the bottom of the central valley. A more detailed numerical modeling[6] of this problem confirmed this explanation of the observed rising wings. The numerical results also show that the carrier temperature near the bottom of the band is already close to the lattice temperature in the case when the lattice temperature is at room temperature. This is not the case if the lattice temperature is at liquid nitrogen temperature or lower; in this case, the electron temperature following short pulse excitation can be considerably higher than the lattice temperature, as will be discussed in the following section.

IV. Picosecond Relaxation of Hot Carriers in Highly Photoexcited Bulk GaAs and GaAs/AlGaAs Multiple Quantum Wells

The final question is the rate of cooling of a hot carrier distribution in the central valley near the bottom of the conduction band following femtosecond or picosecond photoexcitation.

Following femtosecond photoexcitation with the lattice at room temperature, by the time the carriers relax to the bottom of teh band and reach quasiequilibrium the carrier temperature is already close to the lattice temperature[6]. The question of cooling to the lattice temperature does not arise in this case. With the lattice at liquid nitrogen temperature, the carrier distribution remains at a higher temperature many picoseconds after short pulse excitation. To study the cooling rate, one can use, therefore, picosecond photoexcitation. To determine the cooling rate of the distribution, one would need to look at the carriers at different energies, however. For this, different optical techniques from that discussed above must be used.

Instead of the equal-pulse correlation technique which probes the initially excited states, one must now measure the time-dependence of various states below the initially excited states. A convenient technique for doing so is the nonlinear hot luminescence correlation technique[13,7].

In this technique, the time-averaged total luminescence of the sample is measured as a function of delay between two excitation pulses with the pulse length shorter than the lifetime of the carriers. If the carriers generated by the two pulses see each other, there will be additional luminescence due to recombination of electrons from one pulse with holes from the other and vice versa. Thus, as the delay between the two pulses are varied, there should be a luminescence correlation peak at each wavelength of the luminescence. The shape of the correlation peak at each wavelength contains information about the carriers in the corresponding optically coupled states. This technique has been used in particular to measure the rate of cooling of the carrier distribution following picosecond photoexcitation in bulk GaAs and GaAs/AlGaAs multiple quantum wells with the lattice at liquid nitrogen temperature. The results are summarized in Figure 7.

There are three rather interesting points about these results[7]. First, there is a strong carrier density-dependence in the relaxation rates for both the bulk material and the quantum well in the carrier density-range shown (above $10^{17}cm^{-3}$). Second, the relaxation rates for both vary with the carrier energy, which is not unexpected. Third, the most striking and surprising point is that there is a drastic difference in the relaxation rates between the bulk material and the quantum well in this density range. These results were first reported for GaAs and GaAs/AlGaAs MQW and were confirmed by a detailed study of the hot luminescence tails for the two cases using picosecond pulse and cw excitations by Xu and Tang[7]. They have also been independently confirmed shortly afterward by Ryan, et al[14].

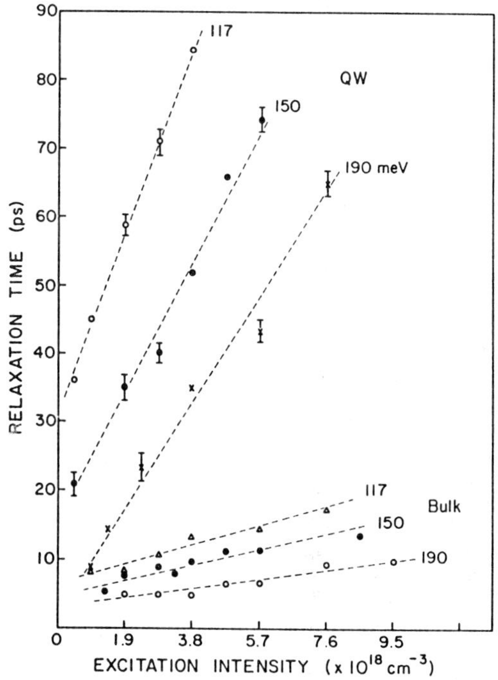

Fig. 7- Hot-carrier relaxation time vs. excitation intensity for both QW and bulk GaAs at different energies: $E-E_{1e}$=117, 150, 190 meV, respectively. The error bar reflects the uncertainties in the deduced relaxation time due to noise and asymmetry in the recorded traces. Dashed lines serve as a guide to the eye only.

Thus, the relaxation rate is seen to depend clearly on the dimensionality of the electron gas. Most recently, a theoretical explanation for this effect has been given by Shum and Alfano[15] on the basis of an electron degeneracy model. The theory seems to fit the experimental results of Ref's 7 and 14 very well. Other details remain to be confirmed.

Most recent measurements reported by Edlestein, Tang, and Nozick[16] have found for the first time a similar drastic slowing down of the relaxation rate in GaAs/GaP strained super-lattices.

These results show clearly that the recently developed femto-second and pico-second optical techniques offer interesting possibilities of unraveling the dynamics of hot carriers in semiconductors through various nonlinear optical effects.

This work is supported by the National Science Foundation and the Joint Services Electronics Program.

References

1. See, for example, Picosecond Phenomema III, K. Eisenthal, R. Hochstrasser, W. Kaiser, and Lauberau, eds. (springer-Verlag, Berlin, 1982); also, articles in Feature Issue on Femtosecond Optical Interactions, D. Grischkowsky, J. Opt. Soc. of Am. B. (April 1985)

2. J.A. Valdmanis, R.L. Fork, J.P. Gordon: Opt. Lett. Vol. 10, 131 (1985)

3. J-M. Halbout and D. Grischkowsky: App. Phys. Lett. Vol. 45, 1281 (1984)

4. W.H. Knox, R.L. Fork, M.L. Downer, R.H. Stollen, C.V. Shank, and J.A. Valdmanis: App. Phys. Lett. Vol. 46, 1120(1985)

5. A.J. Taylor, D.J. Erskine, and C.L. Tang: J. Opt. Soc. B. Vol. 2, 663-673(1984)

6. D.J. Erskine, A.J. Taylor, and C.L. Tang, App. Phys. Lett. Vol. 45, 1209(1984)

7. Z.Y. Xu and C.L. Tang: App. Phys. Lett. Vol. 44, 692(1984)

8. A.J. Taylor, D.J. Erskine, and C.L. Tang: App. Phys. Lett. Vol. 43, 989(1983); Vol. 45, 54(1984)

9. E.P. Ippen and C.V. Shank, in Ultrashort Light Pulses, S.L. Shapiro, ed. (Springer-Verlag, Berlin, 1977); T.F. Heinz, S.L. Palfrey, and K. Eisenthal: Opt. Lett. Vol. 9, 359(1984); S.L. Palfrey and T. Heinz, J. Opt. Soc. of Am. B. Vol. 2, 674(1985)

10. B.P. Zakharchenya, D.N. Mirlin, V.I. Perel', and I.I. Reshina, Sov. Phys. Usp. Vol. 25, 143(1982)

11. E.O. Kane: J. Phys. Chem. Sol. Vol.1, 249(1957)

12. B.S. Wherrett, A.L. Smirl, and T.F. Bogess: J. Quant. Elec. Vol. QE-19, 680(1983)

13. D. von der Linde, J. Kuhl, and E. Rosengart, J. Lumin. Vol. 24-25, 675(1981)

14. J.F. Ryan, R.A. Taylor, A.J. Turberfield, A. Maciel, J.M. Worlock, A. Gossard, and W. Wiegmann: Phys. Rev. Lett. Vol. 53, 1841(1984)

15. K. Shum and R.A. Alfano: to be published

16. D. Edlestein, C.L. Tang, and A.J. Nozik: to be published

Transient Nonlinear Optical Effects in Semiconductors

J.L. Oudar

CNET, Laboratoire de Bagneux, 196, rue de Paris, F-92220 Bagneux, France

I. Introduction

Nonlinear optical effects offer a wide range of potential applications, among which the possibility of using optics for information processing is considered more and more seriously [1]. To achieve this goal, proper understanding of the nonlinear mechanisms [2] and the development of appropriate materials are essential. This is particularly true when we consider the transient regime of optical nonlinearities, which obviously has implications on the rapidity with which various kinds of optical functions can be performed. On the other hand, the study of transient optical nonlinearities can provide new information about a given material, and add to our knowledge of the fundamental properties of this material [3]. In the present paper, we shall discuss some aspects of the study of transient nonlinear processes, with particular emphasis on those related to the photogeneration of electron-hole pairs in a direct gap semiconductor.

Generally speaking, the transient regime of optical nonlinearities occurs when the materials response to an intense light field is slower than the rate of variation of the light intensity. This is the case when a light pulse causes a material excitation due to an energy-transfer from the light beam to the material. If the degree of excitation is high enough, this can give rise to a sizeable change of the optical properties of the material : transmission or refractive index changes, induced birefringence, etc. The type of material excitation can be of various origins, ranging from the orientation of molecules in liquids, to the accumulation of excited states through light absorption, or may even consist of several distinct processes occurring in cascade, as in the photorefractive effect [4]. In turn, the change of the material optical properties modifies the propagation of light, which in some cases (positive feedback) can lead to beam instabilities [5] or to the phenomenon of optical bistability [6].

At this point of the discussion, it is interesting to compare the class of nonlinear optical processes defined above, sometimes named "dynamic nonlinearities", to another class of nonlinear optical phenomena, called the parametric interactions. In the latter type of processes, the material only acts as an intermediate medium for the interaction between several coherent light beams. In principle, there is no net energy-exchange between light and matter, so the material response is reactive, which has the consequence that energy is conserved among the light beams (Manley-Rowe relations [7]). This reactive response occurs because the light-matter interaction induces only virtual transitions between the energy eigenstates of the materials system. On the contrary, real excitations are involved in the dynamic nonlinearities, giving rise to dissipative processes. It is to be noted that there is not always a clear-cut separation between the two classes of optical nonlinearities, since as soon as resonances are involved in the light-matter interaction, real excitations and parametric processes both contribute to the total optical nonlinearity. This situation is similar to the coexistence of hot luminescence and resonant Raman scattering [8], which in some cases can be distinguished by their temporal evolution, but nevertheless appear as two aspects of the same physical phenomenon [9]. In addition, one can also remark that the context in which parametric processes and dynamic nonlinearities are studied is somewhat

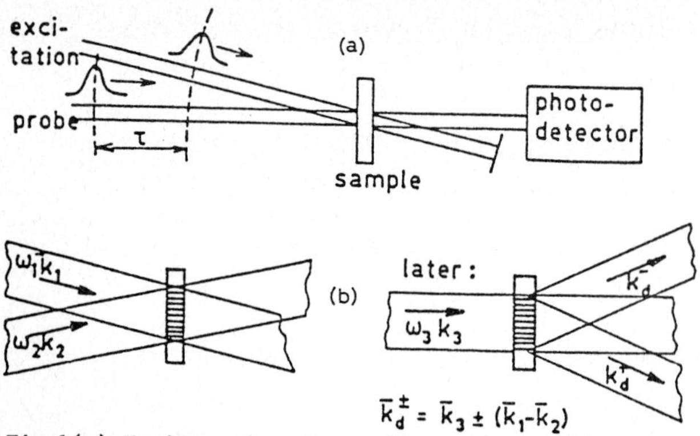

Fig 1(a) Excite-and-probe configuration ; (b) Transient grating configuration

separated, the former giving rise to frequency mixing, harmonic generation, and parametric amplification, while the latter are investigated for applications such as optical gates, phase-conjugation, optical bistability. This is due to the fact that purely parametric processes are in general less efficient in the latter type of phenomena, while the dynamic nonlinearities are generally not effective in the frequency-mixing processes.

Two basic configurations can be used to study experimentally the time response of transient optical nonlinearities, as displayed in Fig 1.
The excite-and-probe configuration (Fig 1a) involves one beam to excite the material out of equilibrium, and another beam to probe the changes of its optical properties. These changes can be observed as transmission variations due to nonlinear absorption phenomena (excited state absorption in three-level systems, absorption saturation in two-level systems). They can also be caused by phase variations due to the nonlinear refractive index, as in the optical Kerr effect, producing changes in the polarization state of the probe beam, or transmission changes in an interferometric arrangement. This arrangement has given rise to much interest in recent years [10], due to the occurrence of optical bistability in Fabry-Perot interferometers containing a nonlinear medium [11]. The other configuration, sketched in Fig 1b, involves two beams to excite the material. This produces an interference pattern in the overlapping region of the beams, and hence a grating of material excitation. A probe pulse sent at a later time on the material can be diffracted by this grating, and the study of diffraction efficiency as a function of probe pulse-delay can reveal the time evolution of the excited state grating [12]. Three limiting cases can be considered, which pertain to distinct physical phenomena. When the two interfering beams have the same frequency and polarization, a stable intensity interference pattern can occur, giving rise to an excited-state population grating. When this is not the case, i.e. when the frequency or the polarization of the interfering beams are not the same, other kinds of excitation gratings are formed. A well-known example is the time-resolved coherent antistokes Raman effect [13], which allows to measure the dephasing time of vibrational excited states. Another example involves the generation of a polarization grating, and is produced by the interference of two orthogonally polarized beams [14]. As will be discussed in section 3, these polarization gratings have a much shorter lifetime than the population gratings, which opens new possibilities for eliminating unwanted slow responses, as in the case of optical sampling techniques. Other configurations, such as the equal-pulse correlation technique [15] are also useful to probe transient nonlinear effects.

In the following, Section 2 will give a brief discussion of the theoretical concepts needed to describe the transient nonlinear response of materials. Section 3 will discuss in more details the particular case of semiconductors, while section 4 will give a short review of recent experimental results obtained on GaAs in the picosecond and femtosecond regime.

2. Theoretical Description of Transient Nonlinear Response

(a) General Considerations

The basic framework appropriate to describing the nonlinear optical response in a phenomenological way is to consider the constitutive relations. These relate the polarization P of the material to the electric field E, which is done through an expansion in power series of E, when a perturbative approach can be adopted. For a complete discussion of this approach, the reader is referred to P.N. BUTCHER [16], of which we shall extract some results pertinent to the discussion of transient nonlinear effects.

Depending on whether the electric field is described in the frequency domain or in the time domain, one can define nonlinear susceptibilities or nonlinear response functions. Although the two descriptions are equivalent in principle (through generalized Fourier transformations), the nonlinear susceptibilities are more convenient to use in the case of monochromatic (steady state) light beams, while the nonlinear response functions are more useful to handle in the case of ultrashort light pulses. This is just a generalization of well-known concepts in the linear response theory. Let $\bar{E}(t)$ be the time dependance of the electric field, and $\bar{E}(\omega)$ its Fourier transform. The two quantities are related by

$$\bar{E}(t) = \int (d\omega/2\pi)\, e^{-i\omega t}\, \bar{E}(\omega) \tag{1}$$

$$\bar{E}(\omega) = \int dt\, e^{i\omega t}\, \bar{E}(t) \tag{2}$$

The most general response function for the polarization $\bar{P}^{(r)}(t)$ which depends on the r-th power of $\bar{E}(t)$ is written as

$$\bar{P}^{(r)}(t) = \int dt_1 \int dt_2 \ldots \int dt_r\, \bar{Q}^{(r)}(t;t_1,t_2,\ldots t_r) \,\vdots\, \bar{E}(t_1)\bar{E}(t_2)\ldots\bar{E}(t_r) \tag{3}$$

where $\bar{Q}^{(r)}$ is a tensor of rank r+1 and the $\vdots$ notation stands for the r-fold dot product of $\bar{Q}^{(r)}$ with the $\bar{E}$ field products. Since $\bar{Q}^{(r)}$ is meant to describe a permanent property of the material, it is invariant with respect to a time translation τ, i.e.

$$\bar{Q}^{(r)}(t+\tau\ ;\ t_1+\tau,\ldots t_r+\tau) = \bar{Q}^{(r)}(t\ ;\ t_1,\ldots t_r) \tag{4}$$

so that one can define a simpler response function $\bar{R}^{(r)}$ by :

$$\bar{Q}^{(r)}\ (t;t_1,\ldots t_r) = \bar{R}^{(r)}(t-t_1,\ldots t-t_r) \tag{5}$$

On the other hand, when the Fourier components $\bar{E}(\omega)$ are adopted to describe the light field, the polarization $\bar{P}^{(r)}(t)$ at order r is expressed by

$$\bar{P}^{(r)}(t) = \int d\omega_1 \ \ldots \int d\omega_r\, \bar{\chi}^{(r)}(\omega_1,\ldots,\omega_r) \,\vdots\, \bar{E}(\omega_1)\bar{E}(\omega_2)\ldots\bar{E}(\omega_r)\ e^{-i(\omega_1+\ldots+\omega_r)t} \tag{6}$$

where the nonlinear susceptibility $\bar{\chi}^{(r)}$ is related to the nonlinear response function $\bar{R}^{(r)}$ through the generalized Fourier transform

$$\bar{\chi}^{(r)}(\omega_1,\ldots,\omega_r) = \int d\tau_1 \ldots \int d\tau_r\ \bar{R}^{(r)}(\tau_1,\ldots,\tau_r) e^{\ i(\omega_1\tau_1+\ldots+\omega_r\tau_r)}. \tag{7}$$

Comparison of eqs (3) and (6) beautifully describes the complete symmetry of the two descriptions. As already mentioned, the time-domain description is well adapted to the case of ultrashort light pulses. In fact it would be ideally-suited to the case of optical pulses of less than one optical period, since $\bar{Q}^{(r)}(t;t_1\ldots t_r)$ describes the material response to a succession of δ-function electric pulses occurring at times $t_1,\ldots t_r$. On the other extreme, the nonlinear interaction of purely monochromatic beams is most easily described using the nonlinear susceptibility concepts as in eq.(6). In that case, the Fourier components $\bar{E}(\omega)$ are in the form of δ-functions, so that the integrals in eq.(6) are straightforwardly evaluated. Since most situations of transient optical nonlinearities lie midway between these two extreme cases, one must consider both descriptions as possible alternatives, and choose which one is more appropriate for each specific case. One is tempted to say that for the excite-and-probe experiments depicted in Fig 1, the time-dependent description of eq.(3) is the more adequate. However, this type of experiment is usually performed to study the time evolution of some excited state of the materials which are in resonance with the excitation or the probe beams. The sole fact that resonance conditions come into play indicates that the frequency domain description cannot be completely forgotten.

Before going further into this problem, let us recall briefly the formal quantum mechanical expression for the nonlinear response function $\bar{Q}^{(r)}$ (see [15] for more details). One considers the total hamiltonian of the material system as the sum of an unperturbed hamiltonian Ho plus a perturbation term $-\bar{p}.\bar{E}(t)$ in electric dipole approximation, i.e.

$$H = Ho - \bar{p}.\bar{E}(t) \tag{8}$$

where the electromagnetic field is treated as a classical variable. For obtaining a perturbative expansion of the induced dipole in power series of $\bar{p}.\bar{E}$, the most direct way is to make the formal expansion using Dyson series [16]. In this approach one transforms the electric dipole operator p_1 into the interaction representation

$$\tilde{p}_1(t) = e^{iHot/\hbar}\, p_1\, e^{-iHot/\hbar} \tag{9}$$

and one obtains directly for the r-th order term, a multiple integral expression in the form of eq.(3), with

$$Q^{(r)}_{ij\ldots n}(t;t_1\ldots t_r)=\left(\frac{i}{\hbar}\right)^r \langle[[\ldots[\tilde{p}_i(t),\tilde{p}_j(t_1)],\ldots],\tilde{p}_n(t_r)]\rangle\Gamma(t>t_1>t_2\ldots>t_r) \tag{10}$$

where $\langle\ \rangle$ designates the ensemble average of the r-fold commutators products ([A,B] means AB-BA) and Γ is a generalized Heaviside function, equal to 1 when the time-ordering is as indicated in the argument of Γ, and equal to 0 otherwise. In eq(10) each of the r-fold nested commutators contributes to a pair of terms, so that one has a total of 2^r different terms. For instance the second-order term is developed as

$[[\tilde{p}_i(t), \tilde{p}_j(t_1)], \tilde{p}_k(t_2)] = [\tilde{p}_i(t)\tilde{p}_j(t_1)-\tilde{p}_j(t_1)\tilde{p}_i(t), \tilde{p}_k(t_2)]$ which gives

$$\begin{aligned}[[\tilde{p}_i(t), \tilde{p}_j(t_1)], \tilde{p}_k(t_2)] &= \tilde{p}_i(t)\tilde{p}_j(t_1)\tilde{p}_k(t_2) - \tilde{p}_k(t_2)\tilde{p}_i(t)\tilde{p}_j(t_1)\\ &\quad -\tilde{p}_j(t_1)\tilde{p}_i(t)\tilde{p}_k(t_2) + \tilde{p}_k(t_2)\tilde{p}_j(t_1)\tilde{p}_i(t)\end{aligned} \tag{11}$$

Finally, in order to make the possible resonances appear with the various excited states, one can develop as usual the operators $\tilde{p}_1(t)$ on the basis of the energy eigenstates of the unperturbed hamiltonian Ho, where eigenstate $|n\rangle$ has an energy E_n. From eq.(9) one obtains for the corresponding matrix element

$$\langle m|\tilde{p}_1(t)|n\rangle = \langle m|p_1|n\rangle e^{\,i(E_m - E_n)t/\hbar} \tag{12}$$

A very convenient way to make the bookkeeping of all the various terms in eq(10), is to use a diagrammatic approach [17], which has the advantage of graphically displaying the various time-ordered interactions. This is particularly helpful for quickly selecting the relevant terms in a given resonance situation[18].

(b) <u>Some Limiting Cases</u>

Coming back to the problem of choosing between a temporal or a spectral development of the nonlinear response function, we need to consider the various characteristic times that come into play for each physical situation. These are summarized in Table I, and it is important to realize that when one of these characteristic times is much smaller or much greater than the others, one is allowed to make different simplifications, which lead to different formulations of the

Light	Pulse duration Coherence time ∝ $1/\Delta\nu$	t_p t_c
Matter	Lifetime Dephasing Time Fluctuation correlation time	T_1 T_2 T_c
Light-Matter Interaction	(offset from resonance)$^{-1}$ (Rabi Frequency)$^{-1}$	$1/\Delta$ $h/\mu E$

Table I : Characteristic times relevant in transient optical nonlinear effects;

general problem discussed in section 2(a). Let us consider briefly some of these limiting cases.

(i) Transparent Materials

When the optical frequencies are all far from any resonance in the material, the linear and quadratic responses are essentially instantaneous with respect to the driving fields. Even with ultrashort light pulses one can adopt the frequency domain formulation of eq(6), since the time-dependent formulation is not useful when no resonance occurs. For third-order processes, it is well known that a resonance behaviour can arise, when the difference of two optical frequencies is equal to some excited state of the material. In that case, the time evolution of atomic motion amplitude must be considered explicitly, which is usually done within the Born-Oppenheimer approximation[19]. This has led to the development of time-resolved CARS experiments [13] to probe the relaxation of Raman-active vibrations.

(ii) Absorbing Materials

Two types of resonant transient behaviour can occur, depending on the relative value of the dephasing time T_2 of the resonant excited state with respect to the light pulse duration t_p. When $T_2 \gg t_p$ the medium can keep memory of light-induced coherence for some time after the extinction of the excitation pulse. This gives rise to a whole variety of transient coherent optical effects[20], such as free induction decay, transient mutation, self-induced transparency, photon echoes, superradiance, etc. On the other limit, when $T_2 \ll t_p$, then much of the light-induced coherence is lost even during the excitation pulse, and the materials excitation essentially consists of an excited state population. We are here in the case of dynamic nonlinearities, which can be described with kinetic equations derived from transition probabilities, e.g. using the Fermi Golden Rule for their quantum mechanical evaluation. In condensed matter, the dephasing times of electronic levels are often very short (in the subpicosecond range), except for the narrow resonances of impurities in solids. As a result, the most frequent situation is that of dynamic nonlinearities due to population build-up. The rapid development of femtosecond lasers will certainly change this state of affairs, making possible the direct probing of coherent transients[21] in an enlarged class of materials excitation.

3. Dynamic Nonlinearities in Semiconductors

The dynamic nonlinearities in semiconductors, and especially direct-gap semiconductors, have been subject to quite a lot of interest in recent years, since in these materials, even a modest level of excitation can lead to important changes in the optical properties. Various physical mechanisms can contribute to the observed optical nonlinearities, among which one can mention the plasma contribution to the nonlinear refractive index[22], the free-carrier screening of excitons[23], two-photon resonances on excitonic molecules[24], bound exciton absorption saturation[25], or the saturation of band-to-band transitions[26]. Many of these effects have been studied in the context of optical bistability, but also in that of

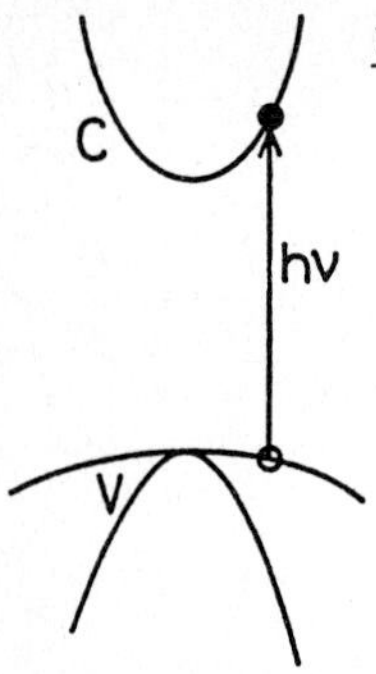

Fig 2 Simplified energy band structure of a direct gap semiconductor

phase conjugation by degenerate four-wave mixing[27]. Since several other lectures in this Course are dedicated to the discussion of some of these effects, we shall restrict the present discussion to the saturation of band-to-band transition in a direct gap semiconductor.

Fig 2 schematizes the energy band structure of such a semiconductor in the vicinity of the band gap. When the material is unexcited, the valence bands are fully occupied by electrons, while the conduction band is empty. The incident light at photon energy $h\nu > E_g$ can promote electrons from a valence band to the conduction band, which has the additional effect of leaving a hole in the corresponding state of the valence band. One thus creates an electron-hole pair, for each absorbed photon of energy $h\nu$. On the other hand, if electrons and holes are injected permanently, as in a semiconductor laser, one can have a medium with "population inversion", which leads to stimulated emission, or gain. In that case an electron-hole pair is destroyed for each emitted photon. In both cases an intense light beam tends to equalize the populations of the optically-coupled states, until the band-to-band absorption or gain becomes saturated. In order to study this mechanism in more detail, we need to describe more precisely the optical coupling between the valence and conduction bands. As suggested in Fig 2, this coupling occurs through vertical transitions between the bands, i.e. the $\bar{k}$ vector labelling the Bloch states in the crystal is the same for the pair of optically-coupled states (actually the selection rule is $\bar{k}_c = \bar{k}_v + \bar{k}_p$, but the magnitude of the photon wave vector $\bar{k}_p$ can usually be neglected). As a result the electronic states of the semiconductor can be described, as far as the light-matter interaction is concerned, as an assembly of two-level systems. For a given optical frequency ω, we need to consider the pair of states which are separated by the energy $\hbar\omega$, and have the same $\bar{k}$ vector. For spherical parabolic bands, this energy separation is

$$E_c - E_v = E_g + \Delta E_c + \Delta E_v = E_g + \frac{\hbar^2 k^2}{2m_c} + \frac{\hbar^2 k^2}{2m_v} \qquad (13)$$

where m_c and m_v are the effective masses of the conduction and valence band respectively. In this approximation, $E_c - E_v$ only depends on the modulus of $\bar{k}$, which is rewritten as

$$E = E_c - E_v = E_g + \frac{\hbar^2 k^2}{2\mu} \qquad (14)$$

where $\mu = m_c m_v / (m_c + m_v)$ is the reduced effective mass of the two-band system. The effective number of two-level systems per unit volume that have an energy-separation comprised in the interval $[E, E + dE]$ is given by $\rho(E)dE$ where $\rho(E)$ is the "joint density of states", expressed by[28] :

$$\rho(E) = (1/2\Pi^2)\,(2\mu/\hbar^2)^{3/2}\,(E - E_g)^{1/2} \qquad (15)$$

In addition to the $\bar{k}$ selection-rule that lead to vertical transitions, an additional selection rule further restricts the number of optically-coupled states in the case of polarized light. This restriction depends on the symmetry properties of the electron wavefunctions in the valence an conduction bands, and leads to preferential coupling with certain direction of the k vectors. In the case of linearly polarized light, the absorption saturation can exhibit some anisotropy, due to the finite lifetime of orientational relaxation for electron and hole wavevectors. This effect will be discussed in Section 4 for the case of GaAs, in view of recent experiments performed on the femtosecond time-scale. If we ignore this effect for the moment, we may assume that the occupation probability only depends on energy, and hence only on the wavevector modulus k. Within this approximation, and assuming that no transient coherent effects are occurring, the linear dielectric susceptibility $\chi(\omega) = \chi_R(\omega)+i\chi_I(\omega)$ can be expressed as :

$$\chi_R(\omega) = A\int d\omega' \rho(h\omega')[1-f_e(\omega')-f_h(\omega')]\ (\omega-\omega')/[(\omega-\omega')^2+\gamma^2] \qquad (16)$$

$$\chi_I(\omega) = A\int d\omega' \rho(h\omega')[1-f_e(\omega')-f_h(\omega')]\ \gamma/[(\omega-\omega')^2+\gamma^2] \qquad (17)$$

where A is a proportionality constant, $f_e(\omega)$ and $f_h(\omega)$ are the electron and hole occupation probabilities in the states optically-coupled with light at frequency ω, i.e. states with wavevectors

$$k = [2\mu(\hbar\omega-E_g)]^{1/2}/\hbar \qquad (18)$$

as deduced from eq.(14). The homogeneous linewidth γ may depend on energy, which is not explicitly mentioned in (16) and (17). Eq.(16) describes the variation of the semiconductor refractive index, and has been used successfully to describe the resonant enhancement of the nonlinear refractive index close to the band gap of III-V compounds[29].

In many cases, f_e and f_h are quasi-equilibrium distribution functions, i.e. they can be written as Fermi-Dirac distribution functions :

$$f_i(\omega) = [1 + \exp(\frac{E_i(\omega) - E_i}{k\,T})]^{-1} \qquad (19)$$

where E_i is the chemical potential of particle i (i=e or h), and $E_i(\omega) = (\mu/m_i)(\hbar\omega-E_g)$ is the excess energy of this particle in the states defined by eq.(18). In most cases $\hbar\gamma << kT$, so that the Lorentzian in eq.(17) may be approximated by a Dirac δ-function. One thus obtains a simplified expression for the absorption coefficient $\alpha(\omega)$ at frequency ω :

$$\alpha(\omega) = \alpha_0(\omega)[1 - f_e(\omega) - f_h(\omega)] \qquad (20)$$

where $\alpha_0(\omega)$ is the absorption coefficient of the unexcited material, i.e. when $f_e=f_h=0$. This equation shows that the measurement of α/α_0 directly provides information on the sum $f_e + f_h$ of the carrier distribution function.

In the case where the e-h pairs are created by ultrashort (subpicosecond) light pulses, f_e and f_h may not be adequately described in terms of quasi-equilibrium distribution functions, since the e-h generation rate can become greater than the collision rate that leads to a thermalized equilibrium distribution. In that case eq.(20) is no longer valid,and one must look for more general equations. This is particularly true when one considers the excitation of e-h pairs by linearly polarized light, since the polarization of the light beam imposes a preferential direction in the orientation of the excited states. In cubic III-V semiconductors such as GaAs, the valence band wavefunctions originate from p-type atomic orbitals quantized along the wavector, and the conduction band from s-type wavefunctions. In the limit of spherical parabolic bands (wavevectors not too far from the centre of the Brillouin zone), it can be shown that the probability of exciting an electron-(heavy) hole pair of wavevectors $\pm \vec{k}$ with linearly polarized light is proportional to

$$|\hat{e}\cdot p_{ehh}|^2 \propto \sin^2\Theta \qquad (21)$$

where Θ is the angle between the wavevector $\vec{k}$ and the electric field vector $\hat{e}$. That is, wavevectors perpendicular to the polarization vector $\hat{e}$ are preferentially excited[30].

The electron and hole distributions $f_{e,h}$ have cylindrical symmetry along the polarization vector $\hat{e}$; therefore each distribution can be expanded in a series of Legendre polynomials of $\cos\Theta$. That is

$$f(\vec{k}) = f_0(k) + f_2(k)\ P_2(\cos\ \Theta) + f_4(k)\ P_4(\cos\ \Theta) + \ldots \qquad (22)$$

where k is the modulus of $\vec{k}$ and P_i is the ith order Legendre polynomial and the odd orders are discarded by symmetry. Since the transition probability is proportional to

$$\sin^2\Theta = \frac{2}{3}\left[1-P_2(\cos\Theta)\right] \qquad (23)$$

the interaction of light with the semiconductor excites mainly the angular distributions f_0 and f_2, because of the orthogonality of the Legendre polynomials. More precisely, the probability of exciting a pair with wavevectors in the direction Θ, is given by

$$\frac{d}{dt}\ f(\Theta) \propto \frac{3}{2}\ \alpha_0\ I\ \sin^2\Theta\ \left[1 - f(\Theta)\right] \qquad (24)$$

where $f(\Theta) = f_e(\Theta) + f_h(\Theta)$. Substituting eq.(22) and performing the angular averages, we obtain

$$\frac{d}{dt}\ f_0 \propto \alpha_0\ I\ \left[1 - f_0 + (1/5)\ f_2\right] \qquad (25a)$$

$$\frac{d}{dt}\ f_2 \propto -\ \alpha_0 I\ \left[1 - f_0 + (5/7)f_2 - (2/7)f_4\right] \qquad (25b)$$

and so on.

In an excite-and-probe configuration, an e-h pair excited by the pump beam contributes to the bleaching measured by the probe beam, since its contribution to the absorption coefficient of the probe beam is :

$$\alpha = \alpha_0\ (1 - f(\Theta))\ \sin^2\Theta' \qquad (26)$$

Where Θ and Θ' are the angles between the electron wavevector $\vec{k}$ and the polarization vectors of the excitation and probe beams respectively. Summing the contributions of all pairs and using the properties of the Legendre polynomials, we may obtain the absorption coefficient for a probe beam whose polarization forms an angle χ with the excitation polarization [31,32] as

$$\alpha(\chi) = \alpha_0\ \left[1-f_0 + (1/5)f_2\ P_2\ (\cos\ \chi)\right] \qquad (27)$$

which constitutes a generalization of eq.(20), keeping in mind that in eq.(27) f_0 and f_2 are the sum of the electron and hole contributions of appropriate symmetry.

The above discussion did not take explicitly into account the many-body effects that come from the interactions between a high density of electrons and holes. Among them, the screening of excitons is obviously not taken into account in eqs(16) and (17), which directly refer to the band-to-band transitions. Another manybody effect is the band-gap renormalization, which can be accounted for by introducing in eqs(15) and (18) a carrier density dependent energy- gap in the form[33]

$$E_g = E_g^0 + (a+br_s)/(c+dr_s+r_s^2) \qquad (28)$$

where $r_s = (3/4\pi n a_x^3)^{1/3}$, a_x is the excitonic Bohr radius, n is the density of e-h pairs, a= -4.8316, b=-5.0879, c=0.0152 and d=3.042. This correction alone is not satisfactory, since it does not account for the modifications of transition oscillator strength due to the electron-hole interaction. This interaction leads to an enhancement of the oscillator strength for $\hbar\omega$ E_g as first described by Elliott[34]. However, when a large number of photoexcited carriers are present, the electron-hole interaction is partly screened, reducing the magnitude of this enhancement. The theoretical treatment of this problem lead to very complex calculations[35] even in the quasi-equilibrium case, although with some approximations the nonequilibrium case can be treated [36].

4. Picosecond and femtosecond dynamics of absorption saturation in GaAs.

As discussed above, only part of the electronic states are directly involved in the optical transitions of such a direct-gap semiconductor as GaAs. In the limit of dynamic nonlinearities, the absorption saturation depends on the population of these states only, which leads to the description of the valence and conduction bands as an ensemble of two-level systems. The frequent collisions experienced by electrons and holes lead to a very rapid cross-relaxation between the two-level systems, leading to an internal thermal equilibrium on a very short time scale. Therefore, if the pulse duration is long enough, an electron-hole plasma is formed, which fills the band states up to the energy-level of the optically-coupled states (band-filling or dynamic Burstein shift). If the pulse duration is shorter than the cross-relaxation within the bands (intraband relaxation) the photoexcited carriers could remain in the optically coupled states during a significant part of the pulse duration ; this effect has been termed state-filling[37], as opposed to the band-filling associated with thermalized distributions.

In the picosecond regime, the bleaching induced on a thin sample of GaAs by an intense beam was observed[38] by measuring the transmission of that beam as a function of its intensity. The incident beam consisted of pulses obtained from an optical parametric generator driven by the frequency-doubled output of a mode-locked Nd^{3+} : YAG laser. The output of the parametric generator was tunable between 0.7 and 2μm and consisted of single pulses of 7ps duration and up to 10μJ pulse energy at 3 Hz repetition rate. This tunability allowed to study the wavelength dependance of the absorption saturation, which was done in the range 0.7 to 0.82μm, i.e. at photon energies slightly above the GaAs band gap (1.519 eV or 816 nm at liquid helium temperature). The sample was 1.5μm thick GaAs layer sandwiched between two $Al_xGa_{1-x}As$ layers (x=0.35) grown by molecular beam epitaxy. The resulting curves (see Fig 3) are typical of the saturation of absorption through the accumulation of optically excited electron-hole pairs. Note that under such short pulse excitation, the carrier density can be connected to the incident light intensity, neglecting any recombination or diffusion process.

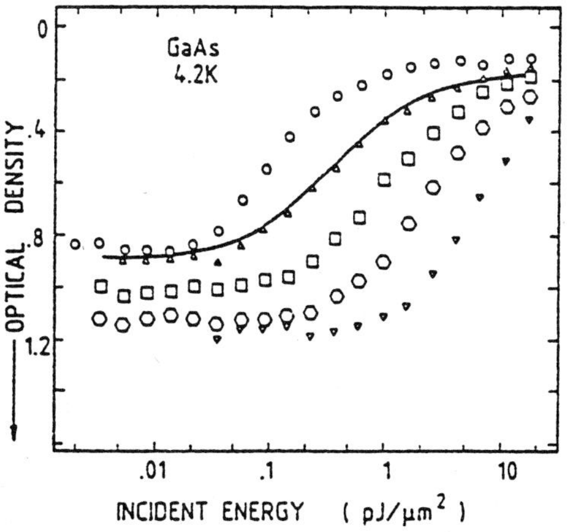

Fig.3. Absorption saturation in GaAs at different wavelengths. Excess above the bandgap is : (O)1 meV ; (Δ)10 meV ; (□)40 meV ; (○)100 meV ; (∇)270 meV. The curve at 10 meV was fitted using a thermalized distribution at a temperature T=90K. (From [38]).

It was found that a band-filling model, based on eqs(19)-(20) gives a fairly good account of the experimental data[38]. It should be noted however, that such single beam experiments are not sensitive enough to evaluate the state-filling contribution. Further experiments[39] have been done on the same sample, using two independently tunable optical parametric generators, both pumped simultaneously and in parallel by the same frequency-doubled, mode-locked YAG laser pulse. This allowed to use the excite-and-probe configuration, thus yielding information on the spectral and temporal variations of the bleaching induced by the intense excitation pulse. Fig 4 presents the transmission of an excited sample as a function of the probe beam frequency. The pump beam was tuned at 1.53 eV, i.e. approximately 18 meV above the band gap at 77K, (the sample temperature for these experiments). The pump and probe pulses arrived simultaneously on the sample. We note that the spectral variation of the bleaching is fairly smooth, except for some regular oscillations in the transmission curves, due to the Fabry-Perot interference effect on the sample surfaces. This spectral variation can be fitted with theoretical curves derived from eq.(19)-(20), assuming a thermal distribution at a temperature of 120 K for the electron and hole energy distributions. As evident from these data, the excitonic effects which produce the transmission dip at 1.508 eV for $I_{pump=0}$ are completely screened at the electron densities of these experiments ($N = 10^{17} - 10^{18} cm^{-3}$).

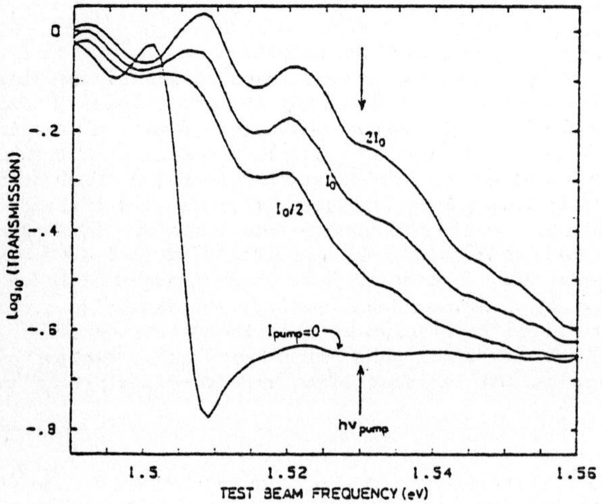

Fig.4 Bleached transmission spectrum of GaAs. For the pump I_o=100μJ/cm^2 (from Ref.32)

The observation of a quasi-thermalized distribution indicates that within the pulse duration of 7 ps there is an extensive energy redistribution of the photoexcited electrons and holes, which closely resembles the thermalization process. In spite of the small excess energy of 18 meV above the band gap, an effective temperature higher than the lattice temperature is obtained. Subsequent cooling of the electronic temperature, as well as the radiative recombination of e-h pairs contribute to the subsequent evolution of the absorption saturation. Fig. 5 presents the results of experiments in which the relative time-delay between the excite and probe pulses (both at the same frequency) is scanned. The absorption saturation recovery is much faster at higher excess energy (1.6 eV) than at zero excess energy (1.51 eV). This is because carrier cooling is much more effective in decreasing the population of higher energy states than that of the lower energy states. In addition the larger carrier density necessary to achieve an important bleaching at the higher photon energies are such that the radiative (bimolecular) recombination occurs on a shorter time-scale.

These experiments give information about the carrier dynamics in the quasi-equilibrium regime where an electronic temperature can be defined. However their time resolution is not enough to give information on the earlier steps of

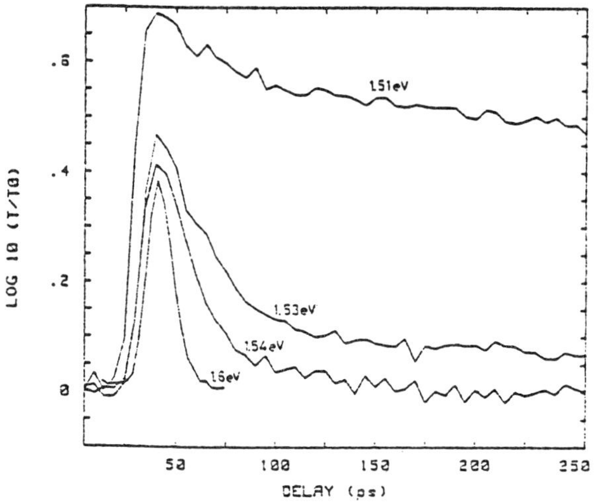

Fig.5 Bleached transmission of GaAs vs. probe delay, at different wavelengths. (from Ref.32)

carrier dynamics by which the quasi-thermalized distribution is attained. Four-wave mixing or transient grating experiments (Fig.1b) performed with orthogonally interfering beams have revealed the existence of very short-lived orientational gratings due to anisotropic state-filling. Their observation was first done by Smirl and coworkers[14] on Germanium crystals at ~ 1 GW/cm^2 intensity levels. Subsequent experiments on GaAs were performed in our laboratory, and four-wave mixing due to orientational gratings could be observed[40] at intensities down to 5 MW/cm^2, suggesting that the lifetime of such orientational gratings was at least one order of magnitude longer in GaAs than in Ge, i.e. $\simeq 10^{-13}$s rather than $\simeq 10^{-14}$s. This relaxation time could not be time-resolved however, and it appeared that femtosecond temporal resolution would be necessary. To this end, an experiment was performed[31],in which two subpicosecond pulses were used in an excite-and-probe configuration, exploiting the possibility of inducing a transient linear dichroism, as described by eq.(27). The difference of absorption coefficients parallel and perpendicular to the excite beam polarization is calculated from eq.(27) as

$$\alpha_{/\!/} - \alpha_{\perp} = (3/10)\ \alpha_o\ f_2 \qquad (29)$$

hence the absorption anisotropy is directly proportional to the second-order angular distribution f_2. Due to this anisotropy, the probe beam propagating through the excited material undergoes a rotation of its polarization in addition to the attenuation of its intensity. For a small rotation angle ε, we have

$$\varepsilon = \frac{1}{4} \sin 2\chi \int_0^d (\alpha_{/\!/} - \alpha_{\perp})\ dz \qquad (30)$$

where the integral is over the sample thickness d. Thus, a time-resolved measurement of the polarization rotation monitors the time-evolution of f_2, which in turn gives information on the relaxation processes that lead to a randomization of the electron and hole wavevectors.

This measurement was performed with subpicosecond pulses whose cross-correlation, represented on Fig 6a, could be fitted to a Gaussian $G(\delta) = \exp(-\delta^2/\delta_o^2)$ with $\delta_o = 0.25$ps, which represents the instrument response. The excitation pulse had an estimated energy of ~100$\mu J/cm^2$, and was centered at 806nm (~1.54eV),that is about 25 meV above the GaAs band gap, with a spectral width of about 12nm (~25 meV). The probe pulse had a broad spectrum, but its detection was done at the pump wavelength. The polarization of the probe pulse, initially set at $\chi \simeq 30°$, was analyzed, after propagation through the sample, into two mutually perpendicular components. The ratio of their intensities, which is a measure of the angle of polarization rotation, was plotted as a function of the relative time delay δ between the excitation and probe pulses. The curve obtained, represented on Fig 6b, is

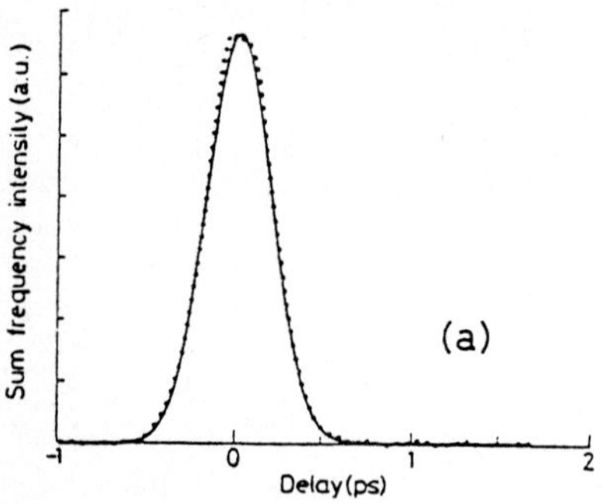

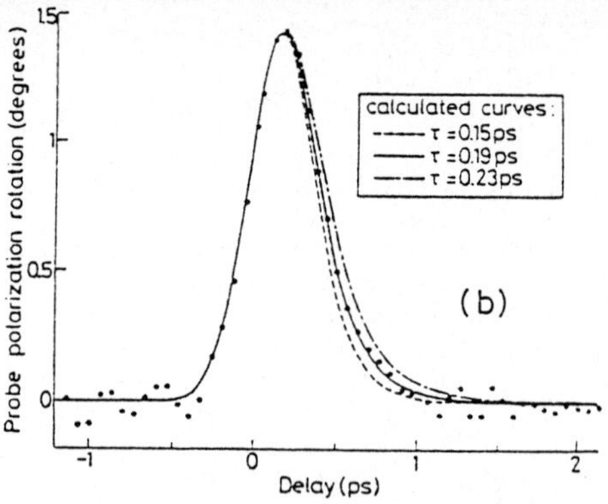

Fig.6 Measurement of orientational relaxation of photoexcited carriers in GaAs. (a) pump-probe intensity cross-correlation curve ; (b) Probe polarization rotation E vs probe delay (from [31]).

slightly wider than the cross-correlation curve of the two pulses and is asymmetric towards the positive delay times (i.e. test after pump), indicating that the anisotropy decays at time scales comparable to the duration of the pulses. An estimate of this decay time was obtained by calculating the convolution of the Gaussian G (δ) of Fig6a with decaying exponentials of characteristic time-constant τ. A fit to the experimental points gives a value of τ = 0.19ps, which provides a direct determination of the orientational relaxation of the photogenerated electron-hole pairs. Due to the small excess energy of these carriers, their dynamics is probably dominated by the rate of intercarrier collisions (electron-electron, electron-hole and hole-hole), a quantity on which very few experimental data were available up to now.

5. Conclusion

When considering the theoretical description of transient nonlinear optical effects, we have seen that in many situations of practical interest, their dynamic behaviour is governed by the population kinetics of excited states, through a mechanism that has been termed "dynamic optical nonlinearity". In contrast with the parametric interactions, it is a dissipative process that is usually predominant when the incident light is significantly absorbed by the sample, leading to a resonant enhancement of the optical nonlinearity, but also to a slowing down of the materials response.

These concepts have been applied to the dynamics of absorption saturation in direct-gap semiconductors. With pulses longer than a few picoseconds, this saturation essentially comes from the dynamic Burstein shift, described as band-filling by thermalized distributions of electrons and holes. In that case, recombination and cooling of the hot electron-hole plasma determine the dynamics of the absorption saturation. With subpicosecond pulses, however, the electron-hole pairs cannot reach a quasi-equilibrium during the excitation pulse, which leads to ultrafast transients in the absorption saturation. The observation of a transient anisotropy in the absorption saturation is one of the experimental tools available for investigating the primary steps of carrier relaxation as described here for GaAs. Another possibility is to measure the spectral dependence of the bleaching induced by intense subpicosecond pulses, since in favorable conditions, a dynamic spectral hole-burning effet [41] is observable.

Acknowledgement

The work briefly described here has been performed in close collaboration with I. Abram, R. Raj, J. Dubard and C. Minot at Centre National d'Etudes des Télécommunications in Bagneux, and with D. Hulin, J. Etchepare and A. Antonetti at ENSTA - Ecole Polytechnique in Palaiseau.

References

1. P.W. Swith, Bell Syst. Techn. Journal 61, 1975 (1982)
   A.M. Glass, Science, 226, 657 (1984)
2. Y.R. Shen, The Principles of Nonlinear Optics, Wiley NY (1984)
3. D.H. Auston in Ultrashort Light Pulses, ed. by S.L. Shapiro, Topics in Appl. Phys., Vol.18 (Springer, Berlin, Heidelberg, New York 1977)
4. J.P. Huignard, same proceedings, p.
5. C. Flytzanis, same proceedings, p.
6. B.S. Wherret, same proceedings, p.
7. J.M. Manley and H.E. Rowe, Proc. IRE 47, 2115 (1959)
8. Y.R. Shen, Phys. Rev. 9, 622 (1974)
9. P.F. Williams, D.L. Rousseau, S.H. Dworetsky, Phys. Rev. Lett. 32, 196 (1974)
10. Special Issue on Optical Bistability, IEEE J. Quant. Electron. QE-17 (1981)
11. H.M. Gibbs, S.L. McCall, and T.N.C. Vankatessan, Phys. Rev. Lett. 36, 1135 (1976) ; H.M. Gibbs, S.L. McCall, T.N.C. Vankatessan, A.C. Gossard, A. Passner and W. Wiegmann, Appl. Phys. Lett. 35, 451 (1979)
12. H.J. Eichler, Advances in Solid State Physics, Vol.XVIII, p 241 Ed. by J. Treusch (Vieweg, Braunschweig, 1978)
13. A. Laubereau and W. Kaiser, Rev. Modern. Phys. 50, 607 (1978)
14. A.L. Smirl, T.F. Boggess, B.S. Wherrett, G.P. Perryman and A. Miller, Phys. Rev. Lett. 49, 933 (1982)
15. C.L. Tang, same proceedings, p.
16. P.N. Butcher ; Nonlinear Optical Phenomena (Columbus 1965)
17. T.K. Yee and T.K. Gustafson, Phys. Rev. A18, 1597 (1978)
18. J.L. Oudar and Y.R. Shen, Phys. Rev. A22, 1141 (1980)
19. R.W. Hellwarth, Prog. in Quant. Electron. 5, 1 (1977)
20. L. Allen and J.H. Eberly, Optical Resonance and Two-level Atoms (Wiley, New York, 1975)
21. A.M. Weiner and E.P. Ippen, Opt. Lett. 9, 53 (1984)
22. R.K. Jain and M.B. Klein in Optical Phase Conjugation, ed. by R.A. Fisher, Academic Press, NY (1982)
23. J. Shah, R.F. Leheny and W. Wiegmann, Phys. Rev. B16, 1577 (1977)
24. A. Maruani, J.L. Oudar, E. Batifol and D.S. Chemla, Phys. Rev. Lett. 41, 1372 (1978)
25. M. Dagenais and W.F. Sharfin, Appl. Phys. Lett. 46, 230 (1985)
26. D.A.B. Miller, S.D. Smith and B.S. Wherett, Opt. Commun. 35, 221 (1980)
27. R.K. Jain, Optical Engineering, 21, 199 (1982)
28. G. Harbeke in Optical Properties of Solids ed. by F. Abelès North-Holland (1972)
29. D.A.B. Miller, C.T. Seaton, M.E. Prise, and S.D. Smith, Phys. Rev. Lett 47, 197 (1981)
30. Related phenomena have been observed in polarization studies of hot luminescence. For a review, see B.P. Zakharchenya, D.N. Mirlin, V.I. Perel' and I.I. Reshina, Usp. Fiz. Nauk. 136, 459 (1982) [Sov. Phys. Usp. 25, 143 (1982)]
31. J.L. Oudar, A. Migus, D. Hulin, G. Grillon, J. Etchepare and A. Antonetti, Phys. Rev. Lett. 53, 384 (1984)
32. J.L. Oudar, I. Abram, A. Migus, D. Hulin and J. Etchepare, J. of Luminescence 30, 340 (1985)
33. P. Vashista and R.K. Kalia, Phys. Rev. B25, 6492 (1982)
34. R.J. Elliott, Phys. Rev. 108, 1384 (1957)
35. C. Klingshirn and H. Haug, Phys. Rep. 70, 315 (1981)
36. H. Haug, J. of Luminescence 30, 171 (1985)
37. D.K. Ferry, Phys. Rev. B18, 7033 (1978)
38. C. Minot, J. Chavignon, H. Le Person and J.L. Oudar, Sol.State Commun 49, 141 (1984)
39. I. Abram, R. Raj, J. Dubard and J.L. Oudar (unpublished)
40. J.L. Oudar, I. Abram and C. Minot, Appl. Phys. Lett. 44, 689 (1984)
41. J.L. Oudar, D. Hulin, A. Miguo, A. Antonetti and F. Alexandre (to be published)

Picosecond Luminescence Studies of Electron-Hole Dynamics in Semiconductors

E.O. Göbel

Philipps-Universität, Fachbereich Physik, Renthof 5, D-3550 Marburg, F. R. G.

This lecture will deal with the experimental investigation of electron-hole dynamics in semiconductors by means of picosecond luminescence spectroscopy. The basic experimental techniques will be described briefly. We then shall present picosecond luminescence measurements to investigate the energy relaxation of photoexcited electrons and holes in GaAs and CdS, and the effects of free carrier screening, nonequilibrium phonons and localization of electronic states will be discussed. The second topic will deal with the dynamics of excitons, exciton molecules and the electron-hole plasma in CdS. Finally, some recent results of picosecond luminescence studies of the carrier dynamics in GaAs/AlGaAs quantum well structures will be presented.

1. Introduction

The rapid progress in ultrafast laser spectroscopy has made it feasible to study the very initial relaxation processes of nonequilibrium carriers in semiconductors [1,2]. Coherent or incoherent excite-and-probe techniques have to be employed in order to fully utilize the time resolution provided by femtosecond laser systems, like the colliding pulse mode-locked dye ring laser. The time resolution of luminescence experiments presently is limited to about 1 ps and thus does not allow the study of the very early relaxation processes taking place on a subpicosecond time-scale. Nevertheless, picosecond luminescence studies can provide useful and interesting information on e.g. energy relaxation and recombination dynamics of nonequilibrium carriers, as will be discussed by some examples in this paper.

We will first briefly describe the experimental techniques. Next, experimental studies of the energy relaxation in GaAs and CdS will be discussed, including the particular case of heavily doped GaAs. We will then describe some picosecond luminescence experiments on excitons, exciton molecules and the EHP in CdS and finally report some recent results of time-resolved luminescence in GaAs/AlGaAs quantum well structures.

2. Experimental

The time resolution of standard photoluminescence techniques using photomultipliers (conventional or channel plate PMT) and photon counting techniques is limited by the speed of the PMTs to some 100 ps. Higher time resolution can be achieved by applying streak cameras as fast photon detectors, or by employing nonlinear techniques like up conversion in nonlinear optical crystals [3,4] or a fast optical Kerr gate [5,6].

An example for experimental set-up, based on an active-passive mode-locked Nd:YAG laser system employing the optical Kerr gate, is shown schematically in Fig. 1. The active-passive mode-locked Nd:YAG laser produces a pulse train with about 25 ps width of the individual pulses and a typical pulse energy in the order of several mJ. The optical beam is divided into two parts, one of them is amplified and single pulses are selected out of the pulse train by optically triggered Pockels cells. The beam used for the excitation passes an optical delay line, can be frequency doubled (SHG) or tripled, and finally is imaged onto the sample. The

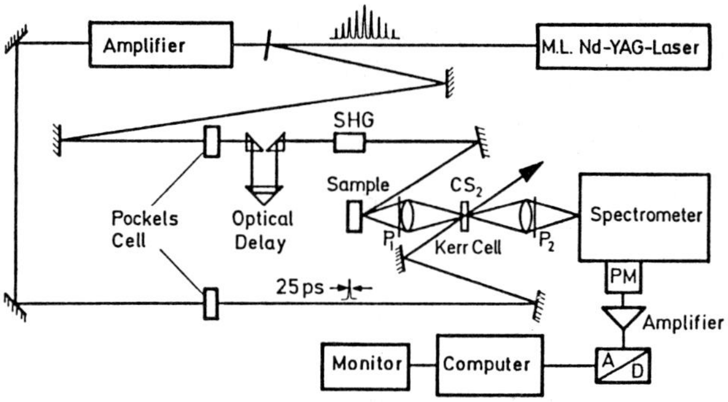

Fig. 1. Experimental set-up for picosecond luminescence measurements using a mode-locked Nd:YAG laser and an optical Kerr gate

luminescence light emitted from the sample, collected by a lens and linearly polarized by a polarizer (P1) is focused into a Kerr cell containing liquid CS_2 and subsequently imaged onto the entrance slit of a spectrometer. A second polarizer (P2) tilted by 90° with respect to P1 blocks the luminescence if the Kerr cell is not activated. Activation of the Kerr cell is achieved by focusing the second beam at its fundamental wavelength at 1.064 µm into the Kerr cell in spatial overlap with the luminescence beam. The linearly polarized optical field introduces an optical anisotropy in the CS_2 liquid proportional to the square of the electric field amplitude, which results in a phase-retardation of the luminescence light passing thru the cell,and hence the luminescence partly passes P2 and can be detected behind the spectrometer. The time-resolution is limited by the molecular orientation relaxation times of the Kerr liquid, which has been determined to 2.1 ps for CS_2 [6]. A response even slightly below 1 ps has been reported for benzene [7]. Time-resolved spectra can be measured by varying the time delay between the excitation pulse and the pulse, which activates the Kerr cell in a similar fashion,as performed in a conventional electronic box-car integrator. The peak intensities required to obtain sufficient rotation of the plane of polarization of the luminescence are in the order of 100 MW/cm^2, and thus the optical Kerr shutter is generally employed with high power, low repetition rate picosecond laser systems.

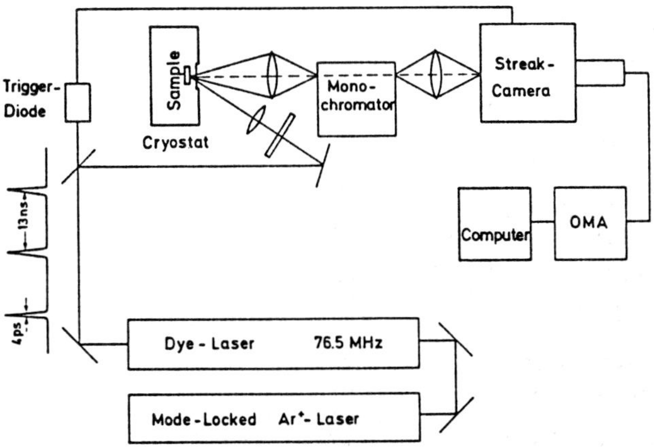

Fig. 2. Experimental set up for picosecond luminescence measurements with a synchronously pumped dye laser and a synchroscan streak camera

Synchronously mode-locked cw dye laser systems provide repetition rates in the order of 80 MHz with pulse duration in the order of 1 ps and pulse energies up to 1 nJ. The high repetition rate can be fully utilized in time-resolved luminescence experiments by using a synchroscan streak camera for detection. A respective experimental setup is shown in Fig. 2. A synchronously mode-locked dye laser pumped by an actively mode-locked Ar^+ ion laser is used for photoluminescence excitation. The average power of the Ar^+ ion laser typically is 600 mW at 80 MHz repetition rate and a pulse width of 90 ps. Different dyes covering the spectral range of about 560 nm to 870 nm are used for synchronous mode-locking and typically 50 to 100 mW average output power at 1 to 5 ps pulse width can be achieved. The luminescence light passes a spectrometer (spectral resolution typically about 1 nm) and is imaged onto the entrance slit of the streak camera. The streak camera is synchronized with the excitation pulse train by a fast Si-APD. The time resolution of the synchroscan streak camera is limited by the jitter of the mode-locked pulse train and the jitter between the sweep voltage and the actual optical pulse train to about 8 ps for our system (Hadland, Imacon 500).

3. Experimental Results

3.1. Carrier Thermalization in GaAs and CdS

Nonequilibrium (hot) carriers in semiconductors relax their energy by interaction with the lattice. The coupling strength of the carriers to the respective phonons determines the energy-relaxation rate. The coupling of charge carriers to longitudinal optical phonons via Coulomb interaction (Fröhlich coupling) dominates in polar semiconductors like GaAs and CdS [8]. This Fröhlich interaction, however, can be screened and a carrier density-dependence of the energy relaxation rates thus is expected. In addition, for both polar and nonpolar materials, it has to be considered that the phonon occupation numbers of the modes involved in the energy relaxation are changed with respect to their equilibrium values during the relaxation process, and nonequilibrium phonon effects will be important depending on the phonon lifetimes [9,10,11].

An experimental measure of the energy-relaxation rate yields the phonon coupling, which e.g. enters directly into the mobility. Convenient methods for the experimental determination of the energy-relaxation rates are stationary, and time-resolved luminescence or absorption spectroscopy [12,13,14]. In Figure 3 time-resolved luminescence spectra of GaAs (3a) and CdS (3b) are depicted for one-photon excitation with 25 ps pulses of the Nd:YAG laser at 0.53 µm (second harmonic of the Nd:YAG laser) and 0.355 µm (third harmonic), respectively. The luminescence is due to free carrier band-to-band recombination within the electron hole plasma. The spectral shape, in particular the width and the slope of the high-energy side of the emission spectra, is determined by the carrier density and the effective free carrier temperature T_c. The respective values for the density and effective temperature can be obtained by a lineshape analysis, including many-particle interactions [15] and carrier diffusion [16]. The effective temperature can be approximately determined from the high-energy side of the spectrum (well above the quasi-Fermi level spacing), which varies according to $I(\hbar\omega) \sim \omega^2 e^{-\hbar\omega/kT_c}$ if reabsorption is neglected. Typical experimental results for T_c versus time for GaAs and CdS are depicted in Fig. 4a and 4b, respectively. The full lines in Figs. 4a and 4b represent calculations, which are described in detail elsewhere [17]. The basic conclusions can be summarized as follows: The energy relaxation of the electrons and holes by longitudinal optical phonons via the Fröhlich interaction is strongly reduced at these excitation intensities, basically due to a high nonequilibrium population of the respective phonon modes, resulting in appreciable phonon reabsorption. Cooling then is overtaken by TO-phonons, which couple to the holes via the optical deformation potential. Finally, at later times, where the mean energy of the electrons and holes is well below the optical phonon energies, further cooling proceeds via acoustic phonon interaction, with a much lower energy-relaxation rate. An equilibrium of the electronic system and the lattice is actually not achieved during the carrier lifetime for both GaAs and CdS. Screening of the Fröhlich interaction, though present, does not affect the cooling rates appreci-

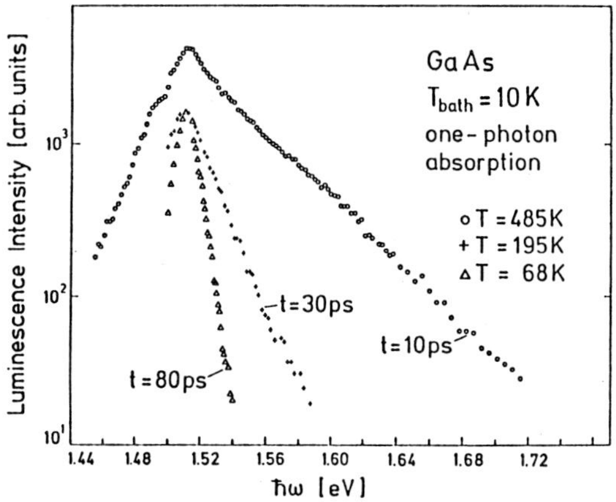

(a)

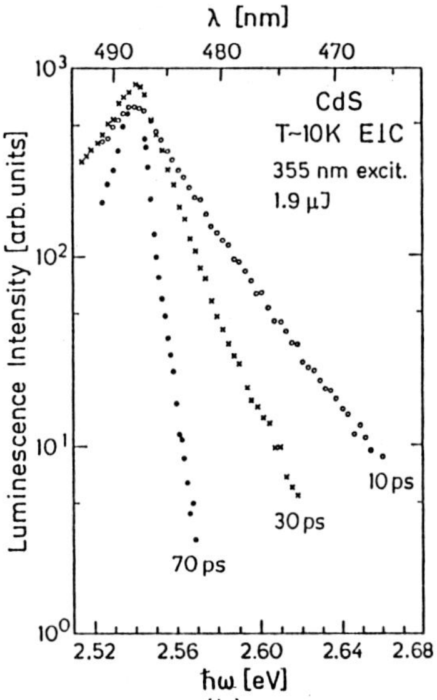

(b)

Fig. 3. Time-resolved spontaneous luminescence spectra of GaAs (a) and CdS (b) for high intensity one photon excitation with pulses of the passively mode-locked Nd:YAG laser system

ably, due to the much stronger effects of the nonequilibrium phonon population. The presence of these nonequilibrium phonons and their lifetime has been measured directly by a picosecond Raman experiment [18]. The same effective temperature for the electrons and holes has been always assumed in the above calculations. The cooling rates will be modified if this condition is not fulfilled, as discussed recently by ASCHE et al. [19].

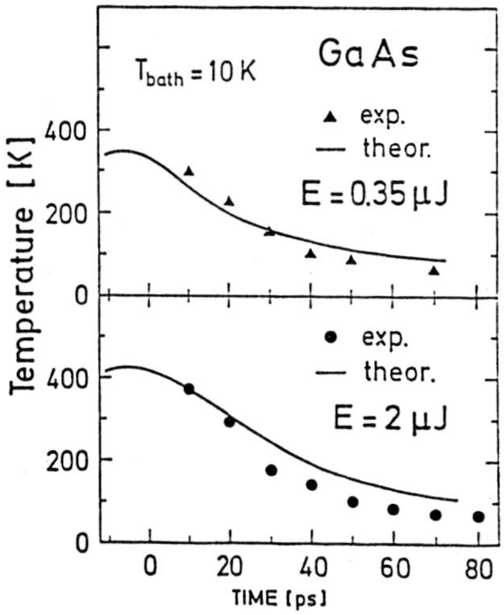

(a)

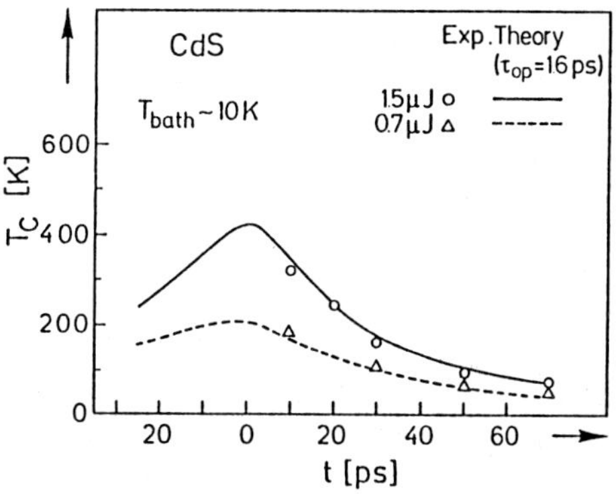

(b)

Fig. 4. Effective carrier temperature T_C versus time for GaAs (a) and CdS (b) as obtained from the time resolved luminescence spectra (c.f. Fig. 3). The full lines represent calculations by P. Kocevar as described in ref. 11 and 14.

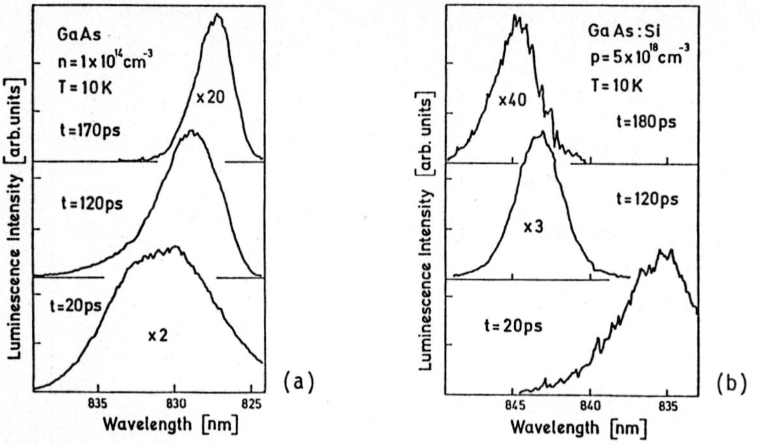

Fig. 5. Time-resolved emission spectra of an undoped (a) and heavily Si-doped (b) GaAs epitaxial layer for two-phonon excitation with pulses of the passively mode-locked Nd-YAG laser system

Thermalization took place within the continuum states of the conduction and valence band in the experiments discussed so far. In heavily doped semiconductors the continuum states extend below the actual band gap; however, these lower-lying states (tail states) are partly localized, whereas the "true" band states are fully delocalized. The localization of the band tail states has a pronounced influence on carrier thermalization [20],as illustrated in Fig. 5, where low-temperature time-resolved luminescence spectra of an undoped (5a) and a heavily Si-doped GaAs epitaxial layer (5b) are compared for the same experimental conditions. The samples are excited by two-photon absorption of the mode-locked pulses of the Nd:YAG laser at 1.064 µm to provide homogeneous excitation of the sample. Optical amplification due to stimulated emission affects the spectra under these conditions, however, because of the large excited volume. The striking difference in the time-resolved spectra of the undoped and heavily doped sample is the opposite shift of the emission band with time. The high energy shift for the undoped sample is due to the reduction of the band gap renormalization [21] because of carrier recombination, consistent with the decrease of the spectral width. The luminescence of the Si-doped sample occurs at lower energies, i.e. corresponds to transitions involving the quasi-localized tail states. The shift to lower energies with time reflects directly the relaxation of the excited electrons towards lower energies (note that the residual doping is p-type, so the valence band tail states are occupied by holes). Energy relaxation within the tail state region at low temperatures thus is appreciably slower than within true band states. The relaxation within the tail states becomes fast at higher temperatures, too. These experimental findings can be consistently explained by a thermally activated multiple trapping process for the energy relaxation [20,22]. In this process, energy relaxation involving electronic states at different lattice sides occurs via thermal reemission of a localized carrier into a delocalized state and subsequent retrapping. This mechanism results in a continuous low energy-shift of the carrier distribution with time, because the high-energy states are more frequently reemitted. The low-energy shift continuous till the energetically lowest tail states are completely filled (saturation). The multiple trapping relaxation, which has been originally suggested to explain the dispersive transport in amorphous semiconductors [20,21], thus seems to be a quite general process in materials with some degree of disorder.

3.2. Excitons, Exciton Molecules and the Electron Hole Plasma in CdS

The optical properties of highly excited CdS are described in detail in a review article by KLINGSHIRN and HAUG [25]. Free excitons, exciton molecules and an electron hole plasma can be observed depending on temperature, excitation intensity and crystal quality. The electron-hole plasma can only be excited at very high excitation intensities corresponding to carrier densities in the $10^{18}cm^{-3}$ range. The exciton molecule and free exciton polariton are stable at densities below $10^{18}cm^{-3}$ and sufficiently low temperatures. The transformation of the electron hole plasma into exciton molecules and excitons can be observed directly by picosecond luminescence experiments [26]. Figure 6 depicts time-resolved spontaneous emission spectra of CdS for one photon band-to-band excitation with the third harmonic of the mode-locked Nd:YAG laser. Broad luminescence spectra corresponding to recombination within the electron-hole plasma are observed just after excitation. The broad spectra transform into a relatively narrow line after about 200 ps, which corresponds to exciton molecule recombination. This interpretation of a transition of the electron-hole plasma into excitons and exciton molecules is further supported by data for the decay of the spectrally integrated luminescence intensity as shown in Fig. 7. The open circles correspond to an excitation intensity, where only the narrow line at about 2.543 eV is observed. An exponential decay with a time constant of about 300 ps is found, corresponding to the lifetime of the exciton molecules in equilibrium with the free excitons. A faster initial component begins to appear as the excitation intensity is increased (full points) and simultaneously the broad emission band is observed in the spectrally resolved data. The same behaviour is found for still higher excitation with a slight increase of the temporal region where the fast component due to electron-hole plasma recombination prevails. The electron-hole plasma thus transforms smoothly into excitons and exciton molecules as its density decreases due to carrier recombination, as further supported by picosecond excite and probe as well as transient grating experiments [26]. The decay of the exciton molecule can be characterized by a time-constant of about 300 ps.

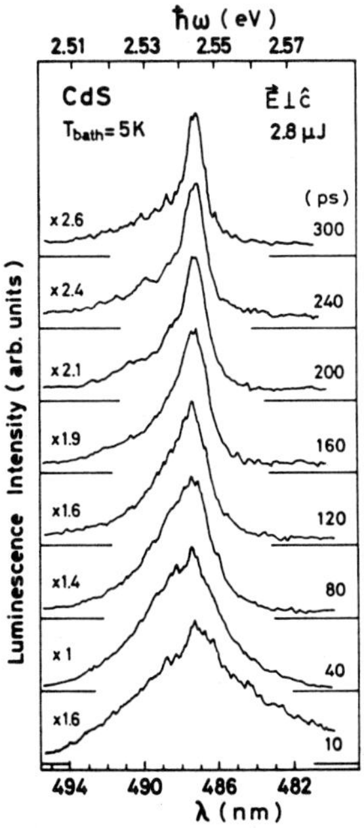

Fig. 6. Time-resolved spontaneous luminescence spectra of CdS for excitation with the third harmonic pulses of the passively mode-locked Nd:YAG laser system

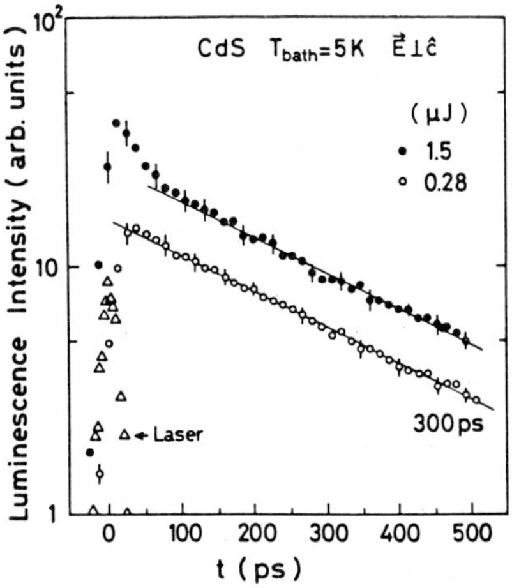

Fig. 7. Spectrally integrated luminescence intensity versus time of CdS for two different excitation intensities

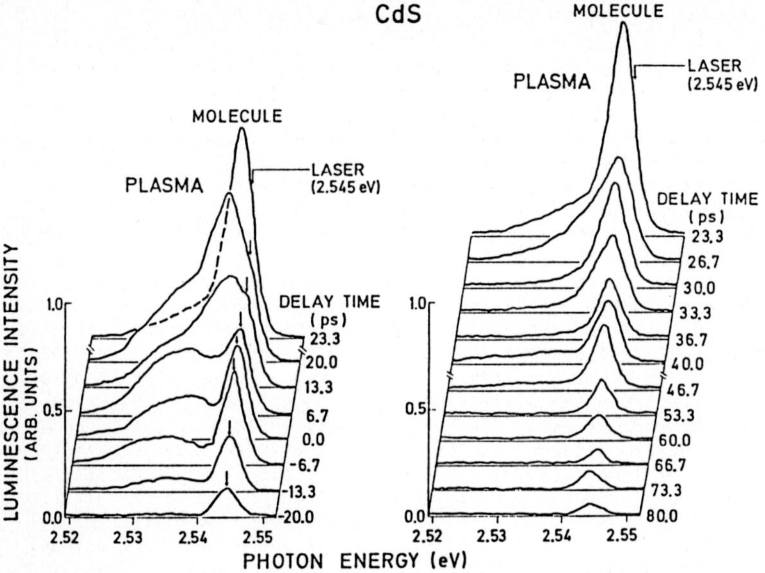

Fig. 8. Time-resolved luminescence spectra of CdS for resonant excitation of the exciton molecule with a mode-locked dye laser synchronously pumped by the third harmonic pulses of the passively mode-locked Nd:YAG laser system

Appreciably shorter decay times are observed if the exciton molecules are resonantly excited [27]. Figure 8 depicts time-resolved luminescence spectra of CdS for resonant excitation of the exciton molecules. A dye laser synchronously pumped by the third harmonic of the passively mode-locked Nd:YAG laser is used for excitation, and the optical Kerr shutter for gating of the luminescence. The half-width (FWHM) of the excitation pulses amounts to about 25 ps. The luminescence at the end of the excitation pulse (delay time of about 20 ps) is dominated by recombination of resonantly created exciton molecules with some contribution of electron-hole plasma recombination at lower photon energies. The very fast decay of the exciton molecules is obvious by comparing the intensity of the uppermost (23.3 ps) and the lowest (80.0 ps) spectrum on the right-hand side. The position of the molecule luminescence shifts slightly towards higher photon energies with time, due to some cooling of the biexciton gas. The fact that the energy position of the molecule luminescence at 80 ps coincides exactly with the laser photon energy demonstrates the resonant excitation. An exciton molecule lifetime of 20 ps is determined from the data, which is more than one order of magnitude shorter than for nonresonant excitation. The short lifetime of 20 ps is consistent with the large oscillator strength of the exciton molecule [28], which, however, can be determined by a time-resolved luminescence experiment only for resonant excitation. For nonresonant excitation, where the molecule is formed by fusion of two excitons , the luminescence decay is determined by the much longer exciton-polariton lifetime [27].

3.3. Picosecond Luminescence of Quantum Wells

The results of picosecond luminescence studies of GaAs/AlGaAs quantum well structures have been reviewed in several recent papers [29,30,31]. We therefore present only two examples in the present paper, which shall illustrate the potential of picosecond luminescence spectroscopy to study carrier relaxation and recombination in quantum wells.

Quantum wells can be excited either directly by photon absorption within the quantum well layers only or indirectly by trapping of carriers excited in the barrier layers. The photoexcited carriers have to diffuse or, in the presence of internal and/or external electric fields, to drift towards the quantum well layers in the case of indirect excitation, where they have to lose their energy in order to get trapped. The population of the quantum well layers with electrons and holes thus will be delayed depending on the time of flight of the carriers to reach the quantum well. This is clearly demonstrated by the experimental results depicted in Figure 9, where the time behaviour of the lowest subband exciton (n = 1, heavy hole exciton) is compared for indirect (I) and direct excitation (II). The sample was a GaAs/AlGaAs double quantum well with thickness of the two GaAs layers of 9 nm (separated by a 4.5 nm thick AlGaAs (x = 0.2)) and the AlGaAs cladding layers of 1 μm, respectively. A synchronously pumped dye laser with photon energy of the mode-locked pulses slightly above and below the band gap of the AlGaAs is used for the indirect and direct excitation, respectively. The rise of the exciton luminescence is appreciably slower for indirect excitation (I) than for the direct excitation (II) due to the delayed trapping. An average velocity of the carriers in the AlGaAs towards the quantum well layers of about 10^5cm/s can be estimated from these data for this particular sample and excitation conditions. The rise time of the luminescence in the case of direct excitation, with photon energies higher than the lowest subband edge, is still slower than the actual excitation pulse, which reflects the thermalization of the carriers within the quantum wells.

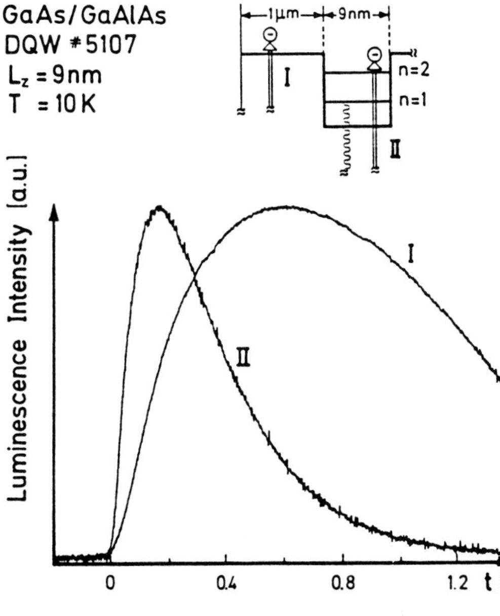

Fig. 9. Comparison of the time behaviour of the n = 1 heavy hole exciton luminescence of a GaAs/GaAlAs double quantum well (L_Z = 9 nm) for indirect (I) and direct excitation taken with the experimental set up shown in Fig. 2

Different subband transitions can contribute to the absorption for direct excitation of the quantum wells depending on the photon energy. Energy relaxation of photoexcited electrons and holes consequently has to proceed via intra- and intersubband scattering if more than the lowest subband are involved, whereas only intrasubband scattering is important otherwise. This difference in the relaxation processes, however, does not affect the time behaviour of the lowest exciton recombination as demonstrated in Fig. 10, where the Rayleigh scattered light of the excitation pulse is plotted together with the time response of the n = 1 heavy hole exciton luminescence. The result shown in the upper part corresponds to excitation via n = 1 and n = 2 subband transitions, i.e. the photon energy of the laser pulses is higher than the n = 2 subband edge, whereas absorption takes place only via n = 1 transitions for the data depicted in the lower part. The time response of the exciton luminescence is almost identical in these two cases,

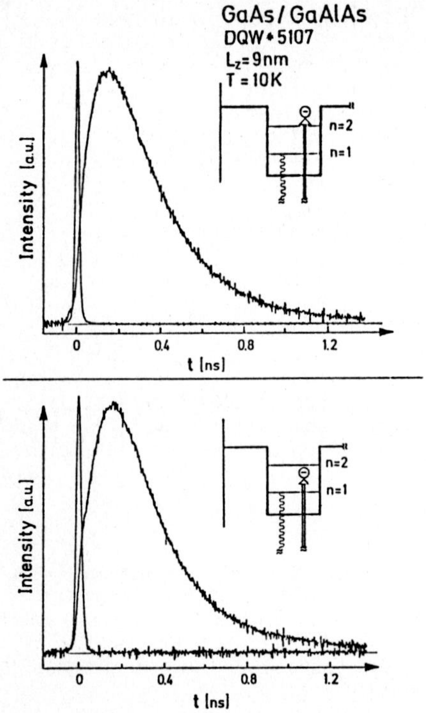

Fig. 10. Comparison of the time response of the n = 1 heavy hole exciton luminescence of a GaAs/GaAlAs double quantum well for direct excitation of the n = 1 subband only (lower part) and simultaneous excitation of the n = 1 and n = 2 subband (upper part)

which leads to the conclusion that energy relaxation is not appreciably delayed if intersubband scattering has to be involved, at least on the time-scale accessible to present time resolved luminescence experiments.

The decay of the exciton luminescence at low temperatures is faster in quantum wells compared to bulk GaAs. This reduction of the exciton lifetime has been attributed to an increase of the exciton oscillator strength due to the shrinkage of the exciton volume in quantum wells [32,33], which reflects the transition from a three-dimensional to a two-dimensional electronic system.

4. Conclusions

Though the time resolution of luminescence experiments could not keep up with the rapid recent progress in short optical pulse generation,it still can provide new and interesting information on the dynamics of electrons and holes in semiconductors. The application of synchroscan streak cameras in conjunction with synchronously mode-locked dye laser systems had made it feasible to study also low-intensity luminescence signals with a time resolution still in the order of 10 ps. Three examples have been briefly discussed in this paper to illustrate the potential of picosecond luminescence spectroscopy to investigate semiconductor materials, namely energy-relaxation in undoped and doped GaAs and undoped CdS, the kinetics of the electron-hole plasma, exciton molecules and excitons in CdS and finally a few recent results obtained on GaAs/AlGaAs quantum wells have been shown. In particular, the latter subject offers new and intriguing possibilities for picosecond luminescence experiments. e.g. by inclusion of electric fields to affect the transport, trapping and recombination behaviour of carriers in quantum well structures.

Acknowledgement

The subjects briefly discussed here represent parts of our research activities during my most enjoyable time at the Max-Planck-Institute für Festkörperforschung in Stuttgart. I am very much indebted to all my former colleagues and coworkers for the numerous contributions to this work. In particular, I would like to thank Dr. J. Kuhl for the very pleasant cooperation over the years and Prof. H.J. Queisser for his continuous interest, encouragement and support.

References

[1] C.L. Tang: same proceedings, p.

[2] J.L. Oudar: same proceedings, p.

[3] T. Daly, H. Mahr: Solid State Comm. 25, 323 (1978)

[4] K. Kash, J. Shah: Appl.Phys.Lett. 45, 401 (1984)

[5] M.A. Duguay, J.W. Hansen: Appl.Phys.Lett. 15, 192 (1969)

[6] E.P. Ippen, C.V. Shank: Appl.Phys.Lett. 26, 92 (1975)

[7] D. Hulin, J. Etchepare, A. Antoniette, L.L. Chase, G. Grillon, A. Migus, A. Mysyrowicz: Appl.Phys.Lett. 45, 993 (1984)

[8] E.M. Conwell: Solid State Physics, Suppl. 9 (ed. F. Seitz, D. Turnbull, H. Ehrenreich) Academic Press, N.Y. 1967

[9] E.J. Yoffa: Phys.Rev. B 21, 2415 (1980)

[10] J. Collet, A. Cornet, M. Pugnet, T. Amand: Solid State Comm. 42, 883 (1982)

[11] W. Pötz, P. Kocevar: Phys.Rev. B 28, 7040 (1983)

[12] R. Ulbrich: Solid State Electr. 21, 51 (1978)

[13] J. Shah: J. de Physique 42, C7-445 (1981)

[14] J. Shah, R. Leheny: "Hot Carriers in Semiconductors Probed by Picosecond Techniques" in Semiconductors Probed by Ultrafast Spectroscopy, R.A. Alfano (ed.), Academic Press, 1984, p. 45

[15] H. Haug, D.B. Tran Thoai: phys.stat.sol. (b) 98, 581 (1980)

[16] A. Forchel, H. Schweizer, G. Mahler: Phys.Rev.Lett. 51, 501 (1983)

[17] P. Kocevar: "Hot Phonon Dynamics" in Proc. 4th Int. Conf. on Hot Electrons in Semiconductors, Physica B, to be published

[18] D. von der Linde, J. Kuhl, H. Klingenberg: Phys.Rev.Lett. 44, 1505 (1980)

[19] M. Asche, O.G. Sarbey: phys.stat.sol. (b) 126, 607 (1984)

[20] E.O. Göbel, W. Graudszus: Phys.Rev.Lett. 48, 1277 (1982)

[21] for a discussion of the many body effects see e.g. E.O. Göbel, G. Mahler "Extended Phase Diagrams of Excited Semiconductors" in Festkörperprobleme (Advances in Solid State Physics) XIX, J. Treusch (Ed.), Vieweg, Braunschweig 1979, p. 105

[22] see e.g. T. Tiedje, A. Rose: Solid State Comm. 37, 49 (1980)

[23] H. Scher, E.W. Montroll: Phys.Rev. B 12, 2455 (1975)

[24] J. Orenstein, M. Kastner: Phys.Rev.Lett. 46, 1421 (1981)

[25] C. Klingshirn, H. Haug: Physics Reports 70, 316 (1981)

[26] H. Saito, E.O. Göbel: Phys.Rev. B 31, 2360 (1985)

[27] R. Baumert: unpublished

[28] E. Hanamura: Solid State Comm. 12, 951 (1973)

[29] I.F. Ryan: Physica 127 B, 343 (1984)

[30] E.O. Göbel, J. Kuhl, R. Höger: J. Luminesc. 30, 541 (1984)

[31] E.O. Göbel, R. Höger, J. Kuhl: J. de Physique, in press

[32] E.O. Göbel, H. Jung, J. Kuhl, K. Ploog: Phys.Rev.Lett. 51, 1588 (1983); R. Höger, E.O. Göbel, J. Kuhl, K. Ploog, G. Weimann: Proc. 17th Int. Conf. Phys. Semicond., in press

[33] J. Christen, D. Bimberg, A. Steckenborn, G. Weimann: Appl.Phys.Lett. 44, 84 (1984); D. Bimberg, J. Christen, A. Steckenborn, G. Weimann, W. Schlapp, J. Luminesc. 30, 562 (1985)

Part III

Nonlinear Optical Materials

Prospect of New Nonlinear Organic Materials

G.R. Meredith

Central Research and Development Department,
E.I. du Pont de Nemours and Company, Experimental Station,
Wilmington, DE 19898, USA

I. Introduction

Interest in the field of nonlinear optical properties of organic materials has grown substantially since the demonstrations of enhancement of second- and third-order nonlinear polarizabilities a decade ago. In other lectures of this school a greater amount of detail will be presented about specific materials or structures. In this lecture a broad view of organics in a subset of nonlinear optical effects (predominantly electronic nonresonant second- and third-order effects) will be given. Some reviews of research in this area,which contain far more detail and more complete references than are presented herein,are referenced for readers' convenience[1-8].

A prerequisite to intelligible discussions of "organic materials in relation to nonlinear optics" must be the specification and differentiation of nonlinear processes, including wavelength and power regimes, for which materials are to be considered. There is a double risk of being overly vague and unfocussed in a discussion of this type, since both nonlinear optics and organic materials are richly diverse fields. It is hoped that the Basics section appearing below will provide a conveniently intuitive view which is useful for nonexperts in this regard. If inventive chemists and materials scientists can gain a conceptual appreciation of the physics involved without becoming discouraged by the mire of formal and experimental optical detail, the prospect of many new organic nonlinear optical materials being developed and reported in the upcoming years is very high. Whereas a decade ago one could only point to a few small research groups whose pioneering scientific studies began to outline the potential attributes and/or enhancements of these nonlinearities in organic compounds, there is currently a growing number of chemical or materials-oriented academic, government and industrial researchers entering this field, worldwide.

The multiplicity of molecular and polymeric compounds and structures, due to the maturity of chemical science, is essentially infinite. While this is often cited as foundation for "molecular engineering" approaches, it is essential to create a unified set of structure-property rules if such an approach is to be taken seriously. Unfortunately, these are still too complex, and, being couched in the mathematical and symbolic language of physics, are not particularly useful to the inventive materials researchers who might achieve the promise of organics. While delineating such rules is an important goal, it is also wise for the physically trained to utilize information from the related and mature fields of molecular photophysics or dye and color chemistry[9]. Consider,for example, how the advancements in tunable and short-pulse lasers has occurred in large part for this reason. In this lecture, rather than a thorough review of research results, an attempt to contribute to both these goals will be made.

II. Basics

The simplest view of nonlinear optical phenomena useful in this context is the semiclassical approach (since, despite the potential of larger but more dispersive and lower time-scale nonlinearity, resonant optical processes are excluded from consideration). A material has certain nonlinear constitutive relations which relate, in the dipole approximation, the induced polarization density ($\underline{P} - \underline{P}_0$) to the Maxwellian electric field $\underline{E}$,

$$\underline{P} - \underline{P}_{(0)} = \underline{\chi}^{(1)} \cdot E + \underline{\chi}^{(2)} \cdot\cdot \underline{EE} + \underline{\chi}^{(3)} \cdot\cdot\cdot \underline{EEE} + \ldots \quad (1)$$

The nth-order nonlinear susceptibility tensors $\underline{\chi}^{(n)}$ are intrinsic material properties which we seek to understand and control. For molecular or polymeric materials, a corresponding microscopic picture relates the total induced dipole moment ($\underline{p} - \underline{\mu}$) of a basic unit via the hyperpolarizability tensors to the local electric field $\underline{F}$,

$$\underline{p} - \underline{\mu} = \underline{\alpha} \cdot \underline{F} + \underline{\beta} \cdot\cdot \underline{FF} + \underline{\gamma} \cdot\cdot\cdot \underline{FFF} + \ldots \quad (2)$$

Consider for a moment the physical implications of such a polarization expansion. The assumption of simple linear polarizability is equivalent to assuming the molecular charge displacement, which creates the dipole moment in response to the local field, is governed by a harmonic or quadratic potential. This is physically unreasonable, since all charge separations must approach a finite potential energy limit at large displacements (i.e. ionization potentials and dissociation energies). Hyperpolarizability must, therefore, be a general property of molecules. This simple argument also suggests that polarizability increases with increased charge displacement, and therefore γ will be positive, as is generally observed. β is not necessarily involved, since it can be shown to be the first correction for the asymmetry of polarizability (or potential energy of charge displacement) on reversing the field polarity. It therefore vanishes in centrosymmetric molecules. Obviously, a crude formula for the enhancement of β is to arrange very different electronegativities at the extremes of a molecule (but also requiring substantial orbital delocalization and overlap, otherwise the electron motion necessary for polarization could not be efficiently optically driven). Considerations of electronegativities and orbital shape, amplitudes and perturbations are second nature to synthetic, physical organic and dye or color chemists in considerations of stability, reactivity, acidity, etc. and optical spectra. Undoubtedly, with a common language, in short time optical physicists and such experts could jointly identify and produce numerous new and interesting molecular routes to large β.

When the driving fields (assuming that $\underline{E}$ and $\underline{F}$ may be considered to be made up of a number of monochromatic waves, which for mathematical convenience occur in pairs of positive and negative frequency) or certain of their combinations approach molecular resonance frequencies, dynamics must be considered. We intend to avoid this topic, except to note that the presence of molecule specific resonances gives rise to dispersion in the hyperpolarizability tensors. It is important to recognize that different molecular motions, associated with these resonances, contribute to hyperpolarizability, and that as the driving frequency increases from below, resonance enhancement occurs, followed by anomolous dispersion as resonance is passed through, and the contribution diminishes as the zero-valued high-frequency asymptote is approached. For example, β can involve both electronic and nuclear displacements. However, for purely optical processes such as frequency doubling or mixing, where all light is in the near infrared or bluer, nuclear motions cannot contribute. Nuclear contributions to a process such as the linear electro-optic effect (change of refractive index which is functionally linear in an applied low-frequency or dc field) can be the major contribution. It is essential then to specify by parameters the driving frequencies, and often the resultant is included, (e.g. $\beta(-2\omega,\omega,\omega)$ or $\chi^{(2)}(-2\omega,\omega,\omega)$ for second harmonic generation or $\beta(-\omega,\omega,0)$ or $\chi^{(2)}(-\omega,\omega,0)$ for linear electro-optic effect, etc.)

It is assumed that in some manner we can make a connection between the molecular polarization properties of (2) and the ensemble susceptibility properties of (1). This is by no means straightforward, because of the mutual polarization effects of molecules in dense phase. Many uncertainties having to do with molecular size, shape, polarization distribution and correlations, quantum interactions and cooperativity occur. But there are various levels of approximation which allow conceptualization and discussion. If one naively assumes point polarizabilities, and limits the intermolecular interactions (after perhaps adopting an averaged condensed-phase-molecule description) to those of mutual polarization by virtue of additions to the local field, oriented (or frozen) molecular ensemble relationships may be found. It turns out, using the normal approach wherein the linear polarizability dominates the mutual polarization and leads to tensors which function as local field tensors (i.e. convert Maxwellian to local fields), that infinite expressions for the susceptibility tensors are obtained. There are the obvious dominant terms where molecular nth-order hyperpolarizability contributes directly to the nth-order susceptibility. But because both the molecular dipole moment and the nonlinear polarization contribute to local fields, the former "bring down" higher-order hyperpolarizabilities into lower-order susceptibilities and the latter "cascade" lower-order hyperpolarizabilities into the higher-order susceptibilities. This problem has received little attention and because of space limitations little more can be said here (but see below). Nevertheless, adopting the existence of such fixed configuration ensemble susceptibilities, they describe the entirely molecular-level nonlinearities.

Whenever driving frequencies (fields or combinations) approach zero, one must take account of alteration of the ensemble distribution because of time-averaged interactions. One could then write

$$P_0 = \{P_0\}_{E=0} \tag{3}$$

$$\chi^{(1)} = \{\chi^{(1)} + dP_0/dE\}_{E=0} \tag{4}$$

$$\chi^{(2)} = \{\chi^{(2)} + d\chi^{(1)}/dE + d^2P_0/dE^2\}_{E=0} \tag{5}$$

$$\chi^{(3)} = \{\chi^{(3)} + d\chi^{(2)}/dE + d^2\chi^{(1)}/dE^2 + d^3P_0/dE^3\}_{E=0} \tag{6}$$

where the only field-dependence of the material parameters is the implicit variation of the molecular ensemble spatial distribution whose derivative is taken here at zero-field strength. One can associate specific motions with various terms, thus recognizing when they might contribute to specific nonlinear susceptibility tensors. As illustration, if E is monochromatic in the optical range, only the first term of (4), the intramolecular electronic susceptibility, is involved in linear dielectric behavior. At infrared or lower frequencies, infrared-active vibrational modes enhance this term. At radio frequencies or lower, permanent molecular dipoles μ can follow the field. The second term is identified to be the well known $\mu^2/3kT$ orientational contribution, etc. In (5) the first term contains intramolecular contributions as dispersion dictates. The second term describes a change of the intramolecular linear susceptibility by an applied field (e.g. an orientational contribution which might occur for molecules in an orientational electret having highly anisotropic α - see below), and the last term contains an orientational contribution to optical rectification (in addition to contributions necessarily present from the earlier terms). In (6) the second term contains the important orientational term in electric field-induced second harmonic generation described below. The other terms can be easily thought through. While this may seem transparent, just the enhancement of this second term in fluid p-nitroaniline was misunderstood to be a contributor to n_2 in a recent survey [10], and this error has been propagated into further papers[11]!

Returning to the semiclassical approximation, it is assumed that Maxwell's equations, subject to the polarization constraint of (1), describe the classical EM fields. Numerous nonlinear optical effects arise in this model, due to (1). There are two important generalized types: 1) color conversion and 2) modification of propagation. These are different in an important way. The first type requires a flow of energy from one light wave into one or more others; there are very severe restrictions on the relative propagation of these waves, and therefore not only on the correlations between linear and nonlinear susceptibility anisotropies, but also on the quality of material required (e.g. for harmonic generators and parametric oscillators). The second type has substantially relaxed simultaneous requirements on linear and nonlinear susceptibility tensors, and consists of very material-forgiving processes: propagation constant (e.g. refractive index, birefringence, absorption, etc.) alteration due to variation of a single beam's intensity, a second beam's intensity or due to the presence of a dc or low-frequency electric field. This class of processes (which are automatically phase-matched) while less exciting and useful in laboratories, have much greater importance in the developing optical communications, computing and signal processing fields.

To complete this section, consider the tensorial nature of the material dielectric properties. Tensorial analysis of the vector equations (1) and (2) shows each nth-order nonlinear tensor to be n+1 rank in cartesian form. Thus, even in low-order effects, the tensors are quite large and complex to envision. Fortunately, decomposition of cartesian tensors into irreducible sherical tensors aids conceptualization. Irreducible tensors have components which mix only among themselves when symmetry transformations of the coordinate system occur, and for which no further reduction in the number of mixing components is possible. For the group of all possible rotations, the spherical tensors of angular momentum theory (i.e. integral L with 2L+1 components) are appropriate. For example, the L=0 tensor has one component, a scalar invariant, such as the trace of α which is associated with the refractive index of a fluid. Jerphagnon, Chemla and Bonneville [12] have extensively treated this topic, showing how odd (even) rank cartesians give rise to odd (even) L sphericals and even (odd) L spherical pseudotensors. Pseudotensors are generally unimportant, and arise only as certain dispersion contributions cause violation of global symmetries such as Kleinmann's. The beauty of the method is that it tells directly the number of experimentally determinable parameters in fluids and how they relate to molecular tensors. For example, in third harmonic generation (THG) a single scalar component arises. Furthermore, if the molecule is made up of portions to which we wish to assign additivity contributions, (if no redirecting of local field is included) the linear relationship among the tensors means there is an additivity of scalar components. In contrast, for β only a pseudoscalar component survives orientational averaging in a fluid. To observe some significant aspect of β a perturbation of the distribution function must be created. Application of a dc electric field is one way. It has a vectorial L=1 interaction with the molecular dipole. Therefore, a vectorial L=1 projection of β onto μ can be induced along the field. This is the basis of the extremely important electric field-induced harmonic generation (EFISH) technique, which allowed the first systematic study of structure-property relationships over a range of molecules. And as for THG, if bond or group additivity is assumed, it is the L=1 or vector portions of β which can be summed. Unfortunately, with these techniques little is learned of hyperpolarizability anisotropy. With the increasing number of molecular computations being done, it is likely that increased confidence in their results will allow them to serve as trusted models.

III. Properties of β

It is important to recognize that although (the vector component projected onto μ of) β values are often quoted for molecules, only if the determination was done in the vapor phase can one have substantial confidence in its validity. The uncertainty in condensed phase measurements derives from the fact that a model of liquid structure and condensed phase dielectric behavior must be adopted to extract β from the EFISH susceptibility. In actuality, it was the EFISH $\chi^{(3)}$ which was experimentally determined. Solvent effects, intermolecular correlations, local field uncertainties, even invalid analysis, etc. are complicating factors facing the experimentalist and theoreticians who extract values from the literature.

Nevertheless, it is worthwhile to reveal the nature of β by quotation of condensed phase EFISH determined values. Table 1 contains a set of EFISH liquid results (except difluoromethane).

The large range of values in Table 1 is striking. This is fortunate, since, despite the uncertainties whether these numbers reflect true molecular values, there are very plain structure-property relationships. Ideas of bond additivity, induction, mesomerism, charge-transfer and resonance theory have been useful empirical concepts.

Bond Additivity. A small set of saturated compounds have been characterized. It has been concluded that vectoral addition of bond hyperpolarizabilities yields reasonable correlation with determined molecular betas. It's been further suggested that these bond values are appropriate to account for the sigma-bonding framework contribution to β in unsaturated compounds if a sigma-pi separability is assumed. In the overall range of β's these contributions are very small, so that serious examination of these approximations is mainly of academic interest.

Inductive Effect. Many of the simple pi-orbital systems, such as benzene, are centrosymmetric, and therefore have vanishing β (but do possess respectable γ tensors). A simple hydrogen substitution yields noncentrosymmetry. The resulting β may be thought to contain a bond term, β(C-X) - β(C-H), plus a term due to inductive

Table 1. Second-Order Properties. $\chi^{(2)}$ values are largest components.

Compound	$\beta\,[10^{-30}$ esu]	$\chi^{(2)}\,[10^{-9}$ esu]
difluoromethane	-0.2	
polyvinylidene fluoride		2
chlorobenzene	-0.3	
fluorobenzene	-0.7	
aniline	1.1	
nitrobenzene	2.2	
m-nitroaniline (mNA)	6	80
o-nitroaniline	10	
p-nitroaniline	35	
2-methyl-4-nitroaniline (MNA)		1000
4-nitropyridine-1-oxide	35	
3-methyl-4-nitropyridine-1-oxide (POM)		46
2,4-dinitroaniline	21	
methyl-(2,4-dinitrophenyl)-aminopropanoate (MAP)	88	
2-(4-dicyanomethylenecyclohexa-2,5-dienylidine)-imidazolidine	-230	
4-dimethylamino-4'-nitrostilbene (DANS)	450	
2 wt. % DANS in liquid crystalline polymer poled at 3KV/mm		3
4'-dimethylamino-N-methyl-4-stilbazolium methylsulfate (DMSM)		~9000
CH3-NC5H4-CH=CH-C6H4=O (merocyanine)	~1000	

perturbation of the pi-system. One may calculate the latter from spectroscopically-derived orbital perturbation parameters,or may view the effect as arising through an equivalent internal electric field (EIEF) acting on γ to create a pi-system β. Table 1 shows this effect to be larger than sigma bond contributions. Such an approach has been used for multiply-substituted benzenes as well.
Mesomeric Effect. For many substituents there are orbitals available which overlap the base pi-system. Mixing causes extended molecular orbitals. It's known that delocalization enhances μ in such molecules, requiring identification of mesomeric moments. Table 1 shows beta to be even further enhanced by mesomerism. The EIEF model has been applied here as well.

Charge Transfer. While the above concepts work well on many compounds, it was seen that for donor and acceptor substituted benzenes a large unaccounted β appeared in ortho and para isomers. Table 1 shows this to be a drastic enhancement. (Reasonable) Charge transfer resonance structures can be drawn for those two, but not the meta isomer. These structures are charge separated forms which enhance the asymmetry of polarizability mentioned earlier.

Resonance Theory. There are more complex molecular types, such as the merocyanine dye in Table 1, which have no simple donor and acceptor groups, but which display very large β. One mode of understanding is resonance theory, in which different bonding configurations are drawn and an attempt is made to deduce their contribution to the normal ground state and their potential for contributing asymmetric polarizability. This is a very imprecise approach which can be replaced with straightforward molecular orbital considerations.[9]

Two-Level Model (TLM). A number of investigations have supported the idea that the primary source of β enhancement is the occurrence of a dominant, low-energy, intense, charge-transfer transition in the optical absorption spectra of these compounds [13-15]. This author is aware of errors, and has objections to assumptions, points of view or limitations of several of these, but the model is still convenient in the absence of detailed and reliable quantum calculations. Table 2 contains the predictions of β and γ when only the two states of that transition are retained.

Table 2. Predictions of the Two-Level Model

Quantity	Expression
$\beta_{yyy}(-2\omega,\omega,\omega)$	$3(\mu_e-\mu_g)\mu_{ge}{}^2\omega_e{}^2/h^2(\omega_e{}^2-4\omega^2)(\omega_e{}^2-\omega^2)$
$\beta_{yyy}(-3\omega,2\omega,\omega)$	$\{(\omega_e{}^2-7\omega^2/3)/(\omega_e{}^2-9\omega^2)\}\beta_{yyy}(-2\omega,\omega,\omega)$
$\gamma_{yyyy}(-3\omega,\omega,\omega,\omega)$	$4\mu_{ge}{}^2\{(\mu_e-\mu_g)^2F-\mu_{ge}{}^2F'\}$

where $F = \omega_e(\omega_e{}^2+\omega^2)/h^3(\omega_e{}^2-9\omega^2)(\omega_e{}^2-4\omega^2)(\omega_e{}^2-\omega^2)$

$F' = \omega_e/h^3(\omega_e{}^2-9\omega^2)(\omega_e{}^2-\omega^2)$

The TLM allows one to utilize knowledge from the areas of dye and color chemistry and photophysics. This is natural, since the large oscillator strength transitions required in Table 2 are the subject of interest in those fields. Maximizing β, and to some extent γ, requires maximization of $(\mu_e - \mu_g)\mu_{ge}{}^2$. There is a wealth of information available for donor-simple acceptor, donor-complex acceptor, asymmetrical cyanine, merocyanine, oxonol, etc. dyes which can aid such an effort[9].

The other aspect of the TLM which deserves mention is the energy denominators. Obviously, minimization of these factors, without actually entering the regime of substantial imaginary value, enhances hyperpolarizability. Similar enhancement of linear polarizability occurs, so that a play-off of reduced electric field strengths in the light waves (recall that intensity is proportional to n times E squared) occurs, as does increased wavelength and birefringence or orientation sensitivity.

Quantum Mechanical Calculations. Many varieties of quantum mechanical calculations of β tensors have been undertaken[14-17]. Despite a high degree of sophistication in methodology, it's been shown recently that small molecule calculations are inaccurate[18]. Large molecule calculations appear superior, and have functioned as predictive tools.

IV. Utilizing β at the Macroscopic Level

As seen in the last section, there is a wide range of molecular types which will possess large beta tensors. The impact of these large values is illustrated by the fact that a nonenhanced, saturated compound such as cholesterol or the crystalline polymer polyvinylidene fluoride have $\chi^{(2)} \sim 2 \times 10^{-9}$ esu, whereas a crystal of the merocyanine dye of Table 1, if in a parallel polar alignment, would exhibit $\chi^{(2)} > 10^{-5}$ esu. The challenge in second-order nonlinear organics is achieving polar molecular alignment in a suitable, stable, high optical quality, conveniently produced structure. Much of the work to date has emphasized the alignment problem. The other equally important quality factors have not received nearly as much attention, since the field is young and these studies occur as specific materials are chosen for development for specific functions. Here we will also deal only with alignment.

There are three obvious polar alignment methods: crystallization (in polar space groups, if one happens to occur for a compound), electric field poling of a microscopically orientable medium, and orientation at surfaces or interfaces.

Noncentrosymmetric Crystals. The factors relating molecular structure and crystal structure are not refined to the point that an engineering approach can be adopted. Rather, there are various strategies which can be adopted, which increase the likelihood of achieving favorably aligned molecular crystals. Alternatively, as has been done recently, one can survey the crystallographic literature, identifying appropriately coincident molecular and crystal structures. These strategies include use of low molecular symmetry to prevent inversion pairing or high symmetry packing, use of chiral compounds to force noncentrosymmetry (a mathematical argument which does not assure alignment of the β-enhancing molecular segment, as seen in Table 1 for MAP, and which is similar to the occurrence of polar liquid crystals such as smectic C phase, where the ensemble-preserved dipole is only that associated with the mandatory chiral center, which is usually positioned at the end of the mesogen), reduction of molecular dipoles (large values of which logically accompany enhanced beta) to reduce probability of inversion-pairing as in POM, resort to molecular salts to overpower dipole-dipole interactions with those of monopoles as in DMSM, the use of molecules for which hydrogen bonding will cause parallel stacking, and the use of orienting matrices such as cyclodextrins, where nonlinearly polarizable molecules are oriented by virtue of inclusion complexation. There are examples of all of these in the literature [1,19].

The cyclodextrin inclusion complexation is an exciting development [19]. A naive picture attributes the crystal structure to the cyclodextrin host, with the guest being surrounded and shielded by it. Guests are oriented by specific interactions (e.g. hydrogen bonding to internal hydroxyl groups). Here is a method wherein true engineering of properties might be achieved for such purposes as SHG; the necessary linear properties might be fine-tuned by small changes of guest shape or functionality, with little impact on the crystal nonlinear properties.

Recently, analysis of molecular orientation in the various crystal classes assuming effectively one- or two-dimensional hyperpolarizability tensors (the appropriate form for enhanced-β molecules) has been published [20]. It shows how some classes cause large unavoidable symmetry-related reductions of molecular hyperpolarizability when projected into the unit cell. One obvious example is that nonpolar noncentrosymmetric classes suffer substantial reduction, as in POM.

It is obvious, but warrants comment, that molecular crystals are substantially different than inorganic dielectrics or semiconductors in their physical properties. Their use technologically will probably demand clever exploitation of their properties, as in the encapsulated growth described by Nayar in this school, to avoid their fragility and the expense of single crystal preparation and fabrication for use.

Microscopically Orientable Media. By "microscopically poled media" we mean orientational electrets, not the process of rectifying the multidomain structure of polar crystals (e.g. lithium niobate). One example, polyvinyledene fluoride, is a crystalline polymer which undergoes crystal structure change to a polar form and whose crystallites can be similarly aligned by suitable perturbations. There being no enhancement of β in this system, $\chi^{(2)}$ measured by SHG is comparable to that of the lesser inorganics such as KDP. Molecules can be doped into polymers, poled (as in the EFISH technique) and frozen in this nonequilibrium alignment. Such an approach was used with DANS (see Table 1) in a liquid crystalline polymer [1]. The latter provides an enhanced alignability and solubility. There are unanswered questions about the physics and stability of this medium, and there are fairly obvious tacts to be taken to increase the achievable macro-nonlinearity. A fundamental limitation is that the alignment is described statistical mechanically by the interaction energy of molecular dipoles, with the poling field relative to the thermal kT energy unit. The ratio of energies is small, even for such dipolar species as DANS ($\mu \sim 7.5$ D). Incorporation of enhanced β species into rigid dipolar macromolecules would allow total alignability, and would probably increase the stability of the alignment. Helices are a natural structure for such grafting or inclusion. It is not difficult to imagine the fabrication advantages if instead of single crystals, one could use a polymeric medium for second-order nonlinearities.

Surface Alignment. It is obvious that an interface is locally noncentrosymmetric. At a planar interface one can, with sufficent sensitivity, detect nonlinear effects (e.g. SHG) arising from the asymmetry of polarizability of the materials involved. In recent years it's been shown that the asymmetry can be impressed onto an ensemble of adsorbed molecules as well, for instance, as seen by SHG from a fused silica plate to which a rhodamine dye had adsorbed. Proof of the rhodamine participation is the occurrence of resonance enhancement as the harmonic is tuned through the $S_0 \rightarrow S_2$ electronic resonance. An impressive enhacement is achieved when both fundamental and harmonic are made nearly resonant for an adsorbed dye such as nile blue A [21]. The maximum harmonic conversion was reported to be 20% that from a phase-matched 1mm thick KDP crystal. If general, nonresonantly-enhanced-β molecules can be made to adopt a polar alignment at an interface, very exciting possibilities arise (even though the spectacular double resonance enhancement doesn't occur for the more important linear electro-optic effect). Two obvious approaches to this task are the use of the Langmuir-Blodgett methodology of polar alignment at an air-water interface, followed by transfer to a substrate [22], or the use of specific chemical or physical interactions to cause preferential orientation (i.e. "self-ordering assemblies").

IV. Properties of γ

Far less is known about the structure-property relationships of γ hyperpolarizability in molecules. It was straightforward to derive a crude but useful description of the meaning of β and how it might be enhanced. No such picture seems to be available for γ. Furthermore, dispersion behavior is far more complex than β. Since there are up to three driving frequencies, the possibility of resonance enhancements are high. Most commonly, these are two-photon electronic and Raman vibrational terms. As described earlier, if a zero-frequency single or combination occurs, ensemble alterations can be the dominant effects in susceptibilities. There have, in fact, been several enhanced response media reported, which utilize this fact (e.g. colloidal suspensions, liquid crystals), but they suffer from very slow relaxation times. There is a hope that because electron-delocalization in molecules can enhance hyperpolarizability, and because the response is on the time-scale of electron motion and relies on no specific resonance condition, very broadband and rapidly responding media might come from study of organics. It is this electronic hyperpolarizability which will be discussed below.

There are four categories into which it is convenient to divide molecular substances: mononuclear, small (one to about six atoms), intermediate and larger, and linear chain-conjugated molecules. The first lends itself to very precise calculations and/or experimentation, where the narrow electronic resonances can be exploited. The second has allowed precise characterizations in vapor and liquid phases, with little need for materials efforts. For these systems the physics is the primary interest. Because of the small sizes detailed, testing of calculations or models of polarizability are possible. Enhanced nonlinearity in these systems requires use of resonances, and therefore suffers extreme wavelength specificity. Broadband enhancement has been associated with delocalization in larger molecules. However, in comparison to the first two types, very little reliable experimentation has been reported. In fact, at this time there are far more molecules and classes of large molecules for which γ hyperpolarizabilities have been calculated by various methods [23-26]. We will not review these calculations, since their general validity really needs empirical testing. It is anticipated that work in progress at various laboratories, particularly using the most straightforward technique, third harmonic generation, will increase the data base against which they can be judged. Remembering that only one scalar component of the gamma tensor is determinable, it must be obvious that understanding of significant structure-property correlations in geometrically complex molecules will require reliance on calculated tensor properties.

The linear chain-conjugated molecules, though falling in the set of large molecules, have had their third-order nonlinear polarizabilities studied so much more thoroughly that they form a separate group. This has occurred because even fairly naive models (e.g. free electron) were able to correlate with the enhancements characterized by Ducuing and coworkers [8]. Unfortunately, some subtle experimental errors occurred. However, their works have shown that the potential for enhancement is real, and have led others to consider more rigorous calculations (which, for instance, detail the consequences of bond alteration, superalteration and chain-pairing, and which provide a view of the consequences of effective limitations of delocalization [6,25]) or to make prediction of enhancement of γ as a consequence of the need both to achieve large pi-electron polarizability and to acquire anharmonicity [26]. The latter point is very important. In second-order hyperpolarizability the arrangement of electronegativity difference and considerations of charge localization and overlap are essential aspects beyond simple electron delocalization, which allow the engineering of enhancement so large that there is little question that useful materials will be forthcoming. The limits of enhancement of nonresonant γ have not yet been determined, and it is probably the appreciation of some similarly formulated concepts that will lead us there.

V. Macroscopic Third-Order Nonlinearities

As represented in (6) there are multiple types of molecular motion which can contribute to $\chi^{(3)}$. If the intent is to use an organic for its n_2 behavior, the ensemble perturbation terms will need to be clearly characterized in addition to understanding the internal contributions. It was not the intent of this presentation to cover those effects, since even a cursory treatment requires substantially more time and space. However, there are two mechanisms which are very fast, can enhance $\chi^{(3)}$ to the level of the best delocalized electron systems, but increase due to the enhancement of β rather than γ.

First, there is a mechanism involving second-order cascaded interactions through the local fields. If one uses the TLM and an Onsager local field model, one can show how this contribution to the $\chi^{(3)}$ of THG, say, can be larger than the direct term [27]. Table 3 contains some calculated values for enhanced-β compounds using formulas from Table 2.

Table 3. Predictions of the Two-Level Model for Solutions in Low Polarizability Solvents (Values in 10-36 esu)

compound	direct	cascaded	observed
p-nitroaniline	22	18	35
DANS	390	250	1000

The second mechanism involves second-order cascading through some intermediate macroscopic fields. This topic is mathematically detailed [28] and is not reproduced. In general, an intermediate wave resulting from second-order interaction of fields associated with a pair of the arguments of the third-order effect can interact with the remaining field to yield an effective, apparently third-order result, which looks to have been generated from a $\chi^{(3)}$ as approximated in (7).

$$\chi^{(3)} \approx \chi^{(2)'} \{4\pi/O(\varepsilon)\} \chi^{(2)} \qquad (7)$$

$O(\varepsilon)$ denotes a value on the order of the dielectric constant or the difference of optical dielectric constants if a transverse optical wave serves as intermediate. Since $\chi^{(2)}$ can in principle be made as large as 10^{-5} esu, effective $\chi^{(3)}$ can be as large as or greater than 10^{-9} esu, which is larger than values quoted for the popular polydiacetylene materials.

VI. Conclusion

It is well established that organic molecules can be made to have very highly enhanced beta. This, together with the diversity of properties of organics, presents opportunity for creative materials and fabrication invention. On the other hand, knowledge of factors affecting and limiting third-order nonresonant hyperpolarizability is sketchy. It may be that organics, as with inorganics, will be most useful for third-order effects under resonant conditions. The disparity between "enhanced organics" with $\chi^{(3)}$ in the range of 10^{-10} esu and semiconductor media where $\chi^{(3)}$ can approach unity is tremendous, making the case for third-order organics much weaker than appeared a decade ago.

References

1. D. J. Williams: Nonlinear Optical Properties of Organic and Polymeric Materials (American Chemical Society, Washington, DC 1983)
2. D. J. Williams: Angew. Chem. Int. Ed. Eng. 23, 690 (1984)
3. S. Basu: Ind. Eng. Chem. Prod. Res. Dev. 23, 183 (1984)
4. J. Zyss: J. Non-Cryst. Sol. 47, 211 (1982)
5. D. S. Chemla: Rep. Prog. Phys. 43, 1191 (1980)
6. C. Flytzanis: Nonlinear Behavior of Molecules, Atoms and Ions in Electric, Magnetic or Electromagnetic Fields (Elsevier, New York 1979) p. 185
7. B. Levine: Dielec. Rel. Mol. Proc. 3, 73 (1977)
8. J. Ducuing: Nonlinear Optics, ed. P. G. Harper and B. S. Wherrett (Academic, New York 1977) p. 11
9. J. Griffiths: Colour and Constitution of Organic Molecules (Academic, New York 1976)
10. T. Y. Chang: Opt. Eng. 20, 220 (1981)
11. A. M. Glass: Science 226, 657 (1984)
12. J. Jerphagnon, D. S. Chemla and R. Bonneville; Adv. Phys. 27, 609 (1978)
13. J. L. Oudar and D. S. Chemla: J. Chem. Phys. 66, 2664 (1977); J. L. Oudar: J. Chem. Phys. 67, 446 (1977)
14. S. J. Lalama and A. F. Garito: Phys. Rev. A20, 1179 (1979); S. Lalama, K. D. Singer, A. F. Garito and K. N. Desai: Appl. Phys. Lett. 39, 940 (1981); C. C. Teng and A. F. Garito: Phys. Rev. B28, 6766 (1983)
15. J. Morrell and A. C. Albrecht: Chem. Phys. Lett. 64, 46 (1979)
16. J. Waite and M. G. Papadopoulos: J. Comp. Chem. 4, 578 (1983); J. Mol. Struct. 108, 247 (1984)
17. J. Zyss: J. Chem. Phys. 70, 3333, 3341 (1979); 71, 909 (1979); J. Zyss and G. Berthier: J. Chem. Phys. 77, 3635 (1982)
18. J. W. Duddley, II and J. F. Ward: J. Chem. Phys. 82, 4673 (1985)
19. Y. Wang and D. F. Eaton: submitted for publication
20. J. Zyss and J. L. Oudar: Phys. Rev. A26, 2028 (1982)
21. G. Marowsky, A. Gierulski and B. Dick: Opt. Comm. 52, 339 (1985)
22. L. M. Blinov, N. V. Dubinin, L. V. Mikhnev and S. G. Yudin: Thin Sol. Film 120, 161 (1984)
23. J. Waite and M. G. Papadopoulos: J. Chem. Phys. 82, 1427, 1435 (1985); Chem. Phys. Lett. 114, 539 (1985); J. Chem. Soc. Far. Trans, 81, 433 (1985); and C. A. Nicolaides: J. Chem. Phys. 77, 2527, 2536 (1982)
24. O. Zamani-Khamiri and H. F. Hameka: J. Chem. Phys. 71, 1607 (1979); 73, 5693 (1980); and E. F. McIntyre: J. Chem. Phys. 72, 1280, 5906 (1980)
25. G. P. Agarwal, C. Cojan and C. Flytzanis: Phys. Rev. B17, 776 (1978)
26. S. C. Mehendale and K. C. Rustagi: Opt. Comm. 28, 359 (1979)
27. G. R. Meredith and B. Buchalter: J. Chem. Phys. 78, 1938 (1983)
28. G. R. Meredith: J. Chem. Phys. 77, 5863 (1982)

Photorefractive Materials for Optical Processing

J.P. Huignard

Thomson-CSF LCR, B.P. 10, F-91401 Orsay, France

G. Roosen

Institut d'Optique, Lab. Associé CNRS, BP 43, F-91406 Orsay Cédex, France

1. Introduction

Over the past several years, a new field of interest has emerged in coherent and nonlinear optics. This field includes the possibility of performing operations in real time and with low-power lasers on the phase and on the amplitude of optical wavefronts. The use of such nonlinear interactions has resulted in numerous new applications, such as phase-conjugate wavefront generation, image transmission and amplification, optical resonators.

The first part of this paper reviews the basic properties of the nonlinear photorefractive crystals involved in the mixing of optical beams, while the second part is devoted to the application of photorefractive crystals such as $Bi_{12}SiO_{20}$, $Bi_{12}GeO_{20}$, $BaTiO_3$, $KNbO_3$... This includes the demonstration of phase conjugation with gain, dynamic holographic interferometry and self-induced optical cavities.

2. General presentation of the photorefractive effect

The photoinduced change in the refractive index of electrooptic crystals was first discovered at Bell laboratories [1] when an intense blue-green laser beam was focused into a $LiNbO_3$ or $LiTaO_3$ crystal (Fig. 1). This index inhomogeneity leads to distortions of the incident wavefronts and therefore limits the use of these materials for frequency doubling or high-speed modulation. It was later recognized that the crystal could be returned to its original state by heating or by uniform illumination, and that we can take advantage of the effect for recording and erasing in real-time holographic fringe patterns (photorefractive effect).

Photorefraction has been later found in a variety of electro-optic crystals including $BaTiO_3$, $KNbO_3$, KTN, SBN, $Bi_{12}SiO_{20}$, $Bi_{12}GeO_{20}$... and also in PLZT ceramics [2, 3, 4]. Photorefractive crystals were firstly investigated by different laboratories in view of their application to high-density holographic storage (phase volume holograms) but their main attractive capabilities are

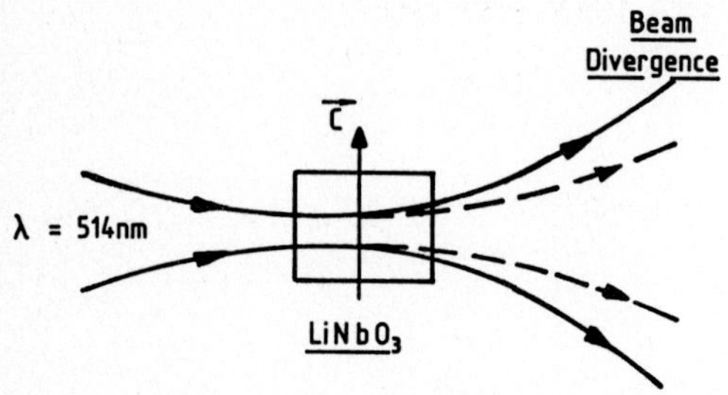

Fig. 1. Index-change induced by a laser beam focused in an electro-optic $LiNbO_3$ crystal.

now in the field of parallel optical signal processing and optical phase conjugation.

Indeed,some of these crystals have the required optical quality,and show a large photoinduced index change with a low incident visible or near infrared laser beam. The photorefractive effect in electro-optic crystals results from a polarization change due to a photoinduced charge-transfer. Compared to materials having a nonlinear polarizability and where the index-change, having an electronic origin occurs instantaneously, photorefractive crystals have inertia. Indeed, in these materials, the refractive index-change results from electro-optic effects driven by space-charge fields. Therefore the time-constant to build-up a phase volume grating will firstly depend on the efficiency of the charge transport process and will secondly be inversely proportional to the incident beam intensities. As a consequence of this remark, photorefractive crystals exhibit a large spectrum of time-constants for a given laser beam intensity. For example, it varies from milliseconds in BSO-BGO to few seconds in $Ba\,Ti\,O_3$ or $Li\,Nb\,O_3$ for an incident recording beam intensity of 100 [mW cm^{-2}] at the green line of an argon laser. In the same manner, holograms can be stored in the dark for time-scales of milliseconds to hours or years,depending on the crystal dark conductivity [4].

The volume nature of thick phase gratings photoinduced in dynamic electro-optic crystals permits the interference of the incident recording beam with its own diffracted beam. This creates a new grating,which can add or substract to the initial one, depending on the position of the photoinduced index modulation with respect to the incident intensity fringe pattern. This, in a two-wave mixing experiment, gives rise to an energy redistribution between the pump and the probe beams interfering in the crystal. In particular, a low-intensity probe beam can be efficiently amplified through energy-transfer from the pump.and the optimum of the effect occurs when the index modulation is spatially phase-shifted by $\pi/2$ [3].

Similarly, in a four wave mixing set-up having two anti-parallel pump beams, the phase conjugate beam generated by this interaction can be much more intense than the incident probe beam (phase conjugation with gain), when an optimized phase shift exists between the index and the intensity patterns.

In conclusion, the refractive index-change induced by the recording beams gives rise to a redistribution of the intensity and phase of the interference field. The resulting beam coupling, via self diffraction process, will be useful for coherent light amplification of stationary or time-varying light beams [5, 6]. Moreover since the grating phase-shift depends on the recording mechanism, the measurement of the energy-transfer, either in two wave or four wave mixing experiments, will give additonal information on the physical mechanism responsible for the photorefractive effect (sign of the photocarriers, density of traps, drift and diffusion length of the photocarriers).

After the presentation of the physical effects involved in the photorefractive effect, the amplitude of the photo-induced space charge-field will be derived from KUKHTAREV'S equations [7] as well as the time-constant to build-up the grating. The last part of the paper deals with the applications of the two important configurations : two wave and four wave mixing respectively applied to wave front amplification and to wave conjugation for phase-distortion correction of laser beams.

3. Physical mechanism of the photorefractive effect in electro-optic crystals

3.1. General description

In order to account for the observed results, detailed descriptions of the photorefractive effect have been proposed. They all assume the existence of charges in a crystalline material. Although the origin of these charges is often uncertain, it is considered that they are located in low-lying traps formed by impurities or defect sites in the crystal. (Fig. 2)

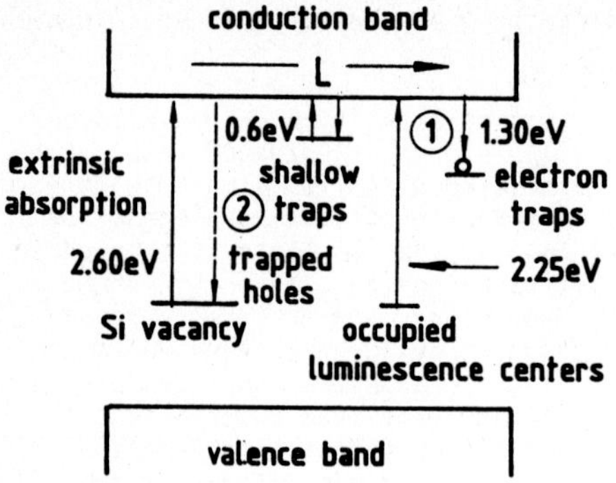

Fig. 2. Energy band diagram of $Bi_{12}SiO_{20}$ at room temperature.

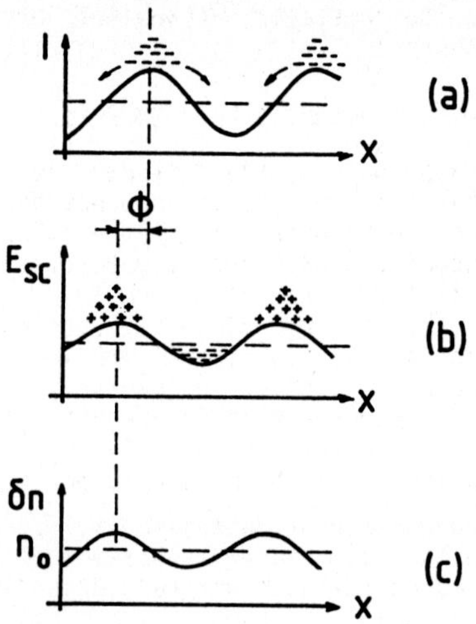

Fig. 3. Basic principle for grating recording mechanism in photorefractive crystals.
a) illumination-carrier photoexcitation and charge transport.
b) phase-shifted photoinduced space-charge field.
c) related refractive index modulation.

In absence of light the charges are frozen in place by the small dark conductivity of the material. However, under illumination the trapped charges can migrate between trapping sites. The hopping model [8] assumes that carrier transport occurs through hoppings from a filled donor site to a nearly located empty trap. The problem is then discussed in terms of hopping probability, the hopping rate being proportional to the local optical intensity.

The band transport model [7] considers that carriers, say electrons, are optically excited from filled donor sites to the conduction band. These carriers then migrate before recombining into an empty trap.

When the illumination is not uniform, such as a spatially periodic interference pattern, Fig. 3 , photoionization causes a grating to appear in both the charge carriers and the ionized donors on the time-scale for ionization.

If the recombination time is not too short, the carriers migrate by diffusion or drift in the local electric field and thus recombine in a different location from where they were ionized. They consequently leave a space-charge grating in the ions. Associated with this grating is a space-charge electrostatic field that produces a change in the refractive index of the crystal by the linear electro-optic effect, the Pockel's effect, provided the crystal lacks inversion symmetry.

The photoinduced refractive index-change corresponds to a phase volume hologram that persists after illumination ends, for seconds to months, depending on the dark conductivity of the material.

It can be read out in a quasi non-destructive manner with a light beam whose wavelength is out of the crystal sensitivity domain.

The phase grating is erasable under uniform illumination. Photocarriers are thus reexcited and uniformly redistributed, giving rise to a relaxation of the index modulation.

The time-constant for both writing or erasing the grating can be controlled by monitoring the incident light power.

3.2. Mathematical description

As in the analytical description proposed by KUKHTAREV et al. [7], we use the band transport model and consider a single charge carrier and one species of ions that contribute to the photorefractive effect. Despite the limitations, this model explains correctly the experiments presented in the following chapters.

The set of equations that describes the charge generation, migration and trapping is the following :

- rate equation : $$\partial N_D^{\,i}/\partial t = (sI + \beta)(N_D - N_D^{\,i}) - \gamma n\, N_D^{\,i} \qquad (3\text{-}1)$$

in which N_D is the total number density of dopants or impurities and $N_D^{\,i}$ the number density of ionized dopants that act like traps ; s, β and γ are respectively the photoionization cross-section, the thermal generation rate and the electron trap recombination rate ; n is the number density of electrons in the conduction band and I the total illumination.

- charge conservation equation : $$\partial n/\partial t - \partial N_D^{\,i}/\partial t = \mathrm{div}\, \vec{j}/e \qquad (3\text{-}2)$$

where $\vec{j}$ is the current density and e the charge of the carrier.

- Ohm's law : $$\vec{j} = n\, e\, \mu\, \vec{E} + \mu\, kT\, \overrightarrow{\mathrm{grad}}\, n \qquad (3\text{-}3)$$

where μ is the mobility of the charge carrier . In this equation, the first term corresponds to migration by drift in the total electrostatic field and the second one to diffusion. This charge-migration produces an electric field.

- Poisson's equation: $$\mathrm{div}\, \vec{E} = - (n + N_A - N_D^{\,i})\, e/\varepsilon_s \qquad (3.4)$$

where ε_s is the static dielectric constant and N_A is the number density of negative ions that compensate for the charge of the traps existing in the dark. N_A is assumed to be a constant and not to take part in the photorefractive process. All other quantities are time and space dependent.

The illumination produced by interference of two light beams of intensities I_1 and I_2 is :

$$I = I_0(1 + m \cos Kx)$$

where $m = 2(I_1\, I_2)^{1/2}/I_0$ is the modulation index, $I_0 = I_1 + I_2$ and K is the grating wave vector ($K = 2\pi/\Lambda$, Λ grating spacing).

By definition of the linear electro-optic effect, the electric field variation δE, induced by the modulated part of the illumination, produces a change in the optical permittivity of the material given by :

$$\delta\varepsilon^{\,\omega} = n_0^{\,4}\, r\, \delta E \qquad (3\text{-}5)$$

where r is an effective electrooptic coefficient depending on material properties and experimental conditions and n_0 the material refractive index.

To determine the optical field diffracted by the hologram recorded in the photorefractive crystal, one has to solve the wave equation using the coupled mode analysis [9] :

$$\nabla\vec{\xi} + \omega^2\, n^2\, \vec{\xi}/c^2 = - \omega^2\, n^4\, r\, \delta E\, \vec{\xi}/c^2 \qquad (3\text{-}6)$$

In this equation, ξ and ω are the amplitude and frequency of the optical field

The right-hand term contains the polarization field induced in the crystal through the Pockel's effect by the electrostatic field. Its value is obtained through equations (3-1) to (3-4). However, the solution of this set of equations is difficult to obtain because they are non-linear. If one wants to solve them analytically, one has to consider that they can be linearized in the grating modulation index, the amplitude of the spatially varying part being much smaller than the uniform one.

As in the following part of this article, we consider continuous illumination with low-power laser beams; we thus remain far below saturation, and the characteristic time of the photorefractive effect is large compared to the time-constant of the unmodulated phenomenon. Therefore we can use the steady-state approximation in the zeroth-order electron number density, and obtain an analytical expression of the photoinduced space-charge electrostatic field [7, 10].

During the recording process one gets :

$$\delta E(x,t) = E_{sc}\left[\cos(Kx + \omega t + \psi) \times \exp(-t/\tau_W) - \cos(Kx + \psi)\right] \qquad (3\text{-}7)$$

The space-charge electrostatic field presents an oscillatory behavior with time-constant τ_W and frequency ω given by :

$$\tau_W = \tau_i(\mu\tau\, kT/e)^2\,(E_q/E_D)\,\frac{[K^2 + (e/\mu\tau\, kT)^2]^2 + (Ke\,E_0/kT)^2}{(1+E_q/E_D)\,[K^2+(e/\mu\tau kT)^2] + (KeE_0/kT)^2}$$

$$\omega = (E_D/E_q\,\tau_i)(e/\mu\tau\, kT)^2\,(e\,E_0/K\,kT)\,\frac{(e/\mu\tau\, kT)^2 - K^2\,E_q/E_D}{[K^2 + (e/\mu TkT)^2]^2 + (KeE_0/kT)^2}$$

with :

E_D : diffusion field, $E_D = kT\,K/e$

E_q : maximum space charge field amplitude that can be photoinduced in the crystal $E_q = e\,N_A(N_D - N_A)/\varepsilon_s\,K\,N_D$

τ_i : dielectric relaxation time under incident illumination I_0

τ : electron-trap recombination time

E_0 : applied electric field.

The amplitude of the photoinduced space charge field is related to material and experimental recording parameters. In the steady state it is given by :

$$Esc = m\left[\frac{E_0^{\,2} + E_D^{\,2}}{(1 + E_D/E_q)^2 + (E_0/E_q)^2}\right]^{1/2} \qquad (3\text{-}8)$$

Equations (3-8) and (3-5) demonstrate that the steady-state refractive index change is independent on the total light intensity but depends on the relative intensity of the incident beams. This is in contrast to other physical mechanisms for which the refractive index-change increases with any increase in the total light intensity. However, the speed of the photorefractive effect increases with the light intensity.

Another important peculiarity of the photorefractive effect is its nonlocal character. A maximum index change does not necessarily occur where the illumination is the largest, as shown by (3-5) and (3-7). The value of the phase-shift ψ between the incident illumination and the photoinduced space-charge grating is given by :

$$\tan \psi = (E_D/E_0)(1 + E_D/E_q + E_0^2/E_D E_q) \qquad (3\text{-}9)$$

This specificity of the photorefractive effect will permit to achieve the coupling between beams having the same frequency, as it will be presented in Chap. 4.

For erasure under uniform illumination, the field evolution is given by :

$$\delta E(x,t) = E_{sc}^{i} \cos(Kx).\ \exp(-t/\tau_W)$$

where E_{sc}^{i} represents the initial amplitude of the space-charge electric field in the erasure process.

From the wave equation (3-6), one determines the scattering efficiency of the photorefractive grating :

$$\eta = \exp(-\alpha\ell)\ \sin^2(\pi\ell\Delta n_s/\lambda \cos\Theta)$$

where Δn_s is the steady-state amplitude of the photoinduced refractive index modulation given by :

$$\Delta n_s = n_0^3\ r\ E_{sc}/2 \qquad (3\text{-}10)$$

and ℓ the crystal length, α the absorption coefficient, λ the wavelength of the read-out beam and Θ the Bragg angle inside the crystal.

4. Stationary energy-transfer between writing beams in two-wave mixing

Grating formation in a photorefractive crystal is accompanied by an intensity redistribution between the two interfering light beams, i.e. the pump and the low intensity probe. This intensity transfer (beam coupling) initially observed in $LiNbO_3$ by STAEBLER [3] is due to a permanent phase mismatch between the holographic grating and the fringes, "Fig. 4.a", as demonstrated in Chap. 3.

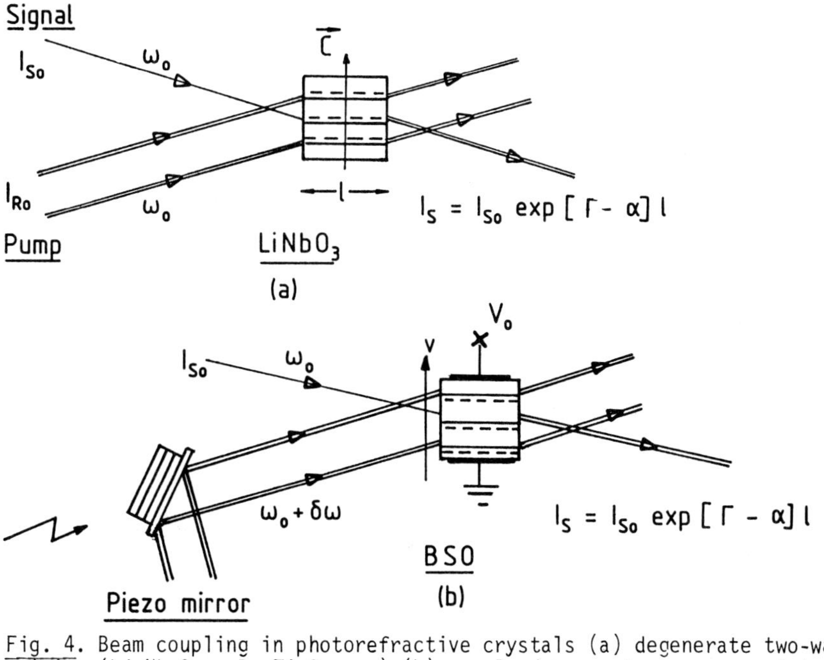

Fig. 4. Beam coupling in photorefractive crystals (a) degenerate two-wave mixing ($LiNbO_3$, $BaTiO_3$ ...) (b) nearly degenerate two-wave mixing in BSO - moving grating recording ;—— : Intensity pattern ; --- : Index modulation.

The qualitative interpretation of this energy-exchange is the following : in a dynamic material, the self interference of the incident beam with the diffracted beam creates a new holographic grating which can add to the initial one. Since the diffracted wave is phase-delayed by $\pi/2$ with respect to the reading beam, the maximum of energy transfer is obtained when the incident fringe pattern and the photoinduced index change are shifted by $\pi/2$.

In photorefractive crystals a $\pi/2$ phase shift exists when the photoinduced index modulation occurs by diffusion of the photocarriers (no external applied electric field).

As a consequence of this phase-shift, a permanent and efficient amplification of a probe beam has been obtained in $Ba\,Ti\,O_3$, $Li\,Nb\,O_3$ or $K\,Nb\,O_3$ crystals. However, the same experiment done with highly photoconductive BSO (or BGO) crystals leads to a very limited energy-transfer for the following reasons [11] :

- By "diffusion" ($E_0 = 0$) the required phase-shift $\psi = \pi/2$ is established (see (3-9)) but the index modulation at saturation is low in the high spatial frequency domain ($\Lambda^{-1} > 1000$ [mm^{-1}]).

- By "drift" ($E_0 \neq 0$) the index modulation is much higher for $\Lambda^{-1} < 300$ [mm^{-1}] but the corresponding phase shift is small.

Nevertheless, an efficient beam coupling can be obtained if the fringe pattern is moved at a constant velocity [12, 13]. The speed is adjusted so that at any time the index modulation is recorded but with a spatial phase shift with respect to the incident fringe pattern, (Fig. 4.b).

This moving grating results from the interference of the signal beam with a frequency-shifted pump beam (nearly degenerate two-wave mixing).

From the coupled wave theory [9], the equations that express the coherent interaction of the two waves R and S ($R \ll S$) are the following (undepleted pump beam approximation) :

$$dS/dz = \Gamma\, S/2 - \alpha\, S/2$$

$$dR/dz = -\,\Gamma\, S^2/2R - \alpha\, R/2$$

The transmitted probe beam intensity resulting from the two beam coupling is given by :

$$I_S = I_{So}\, \exp[(\Gamma - \alpha)\ell]$$

where α is the absorption coefficient, ℓ the crystal interaction length and Γ the exponential gain of the interaction related to the maximum amplitude of the photoinduced modulation Δn_s given by (3-10), through the following relation :

$$\Gamma = (4\pi\, \Delta n_s/\lambda\, \cos\Theta)\, \sin\psi$$

ψ being the spatial phase-shift of the grating given by (3-9). Γ is maximum for $\psi = \pi/2$.

When the beam coupling is induced by the moving grating, Γ is related to the fringe velocity v by :

$$\Gamma = (4\pi\, \Delta n_s/\lambda\, \cos\Theta)\, \left[Kv\, \tau_W/(1 + K^2\, v^2\, \tau_W^{\,2})\right] \qquad (4\text{-}1)$$

where $K = 2\pi/\Lambda$, Λ grating spacing and τ_W the time-constant required to build-up the index modulation via the photorefractive effect. τ_ω and Δn_s depend on the recording conditions (Λ, E_0,...) and on the physical parameters of the crystal such as electrooptic coefficient, trap density, drift and diffusion length of the photocarriers [10].

From (4-1), in the moving grating recording mode, the maximum of Γ is obtained for $v_0 = \Lambda/2\pi\ \tau_W$ which corresponds to a frequency detuning $\delta\omega = 1/\tau_W$ on the pump beam.

However, [14] shows that in photorefractive BSO (or BGO) the validity domain of (4-1) is limited. The general formalism starting from KUKHTAREV'S equations will result in a resonance of the gain Γ versus both fringe velocity and grating spacing. In BSO the gain is optimum for a grating spatial frequency $\Lambda^{-1} \# 50\,|mm^{-1}|$ and for very low probe beam intensities ($\beta = \frac{I_{R_0}}{I_{S_0}} > 10^3$) [14]. In such conditions, efficient amplification of a probe beam is achieved (10^3x), the related value of the exponential gain coefficient being $\Gamma \#$ 8 to 12 $[cm^{-1}]$.

Similar values of gain are obtained with ferroelectric $Ba\ Ti\ O_3$, SBN, $LiNbO_3$ and $K\ Nb\ O_3$ at zero-applied field but with an optimum around the spatial frequency $\Lambda^{-1} \# 1000\ [mm^{-1}]$.

5. Optical phase conjugation by four-wave mixing

In the four-wave mixing configuration shown in Fig. 5.a , two counterpropagating pump beams R_1 and R_2 and a low-intensity signal (or probe) beam interact in the volume of the photorefractive crystal. Using the holographic approach firstly proposed in [15], the generation of the conjugate beam arises from diffraction of one of the pump beams on the index grating recorded by the two other interfering beams. If the transmission type grating is dominant, R_1 and S_1 are the recording beams while R_2 is the readout beam of the photoinduced grating. The term of interest in the holographic recording is S^*R_1 which, after read-out by R_2, generates the wave-field :

$$A_c \propto S^*\ R_1\ R_2$$

This wavefront is the phase conjugate of the incident wavefront. The efficiency of the interaction is defined by the intensity ratio :

$$\rho = I_c/I_s$$

ρ is the phase conjugate mirror reflectivity.

As in two-wave mixing, the self-diffraction process plays an important role for obtaining an efficient interaction between the recording beams. The intensity changes arising from the two dominant transmission-type gratings (R_1 - S, R_2 - S_c) can be derived by solving the wave equation. For a $\pi/2$ phase-shifted grating (diffusion recording or moving grating under optimum conditions) the wavefront reflectivity is given by :

$$\rho = (I_2/I_1)\ (\exp \Gamma\ell - 1)^2$$

Depending on the crystal orientation ($\Gamma > 0$) the conjugate mirror reflectivity can be much larger than unity, thus allowing wavefront conjugation with gain. In a first approximation, ρ depends linearly on the pump beam ratio $r = I_2/I_1$ but, in a more general formalism derived by FISHER et al. |16|, an optimum of the pump beam ratio also exists, as well as an optimum phase-shift between the index modulation and the fringe pattern.

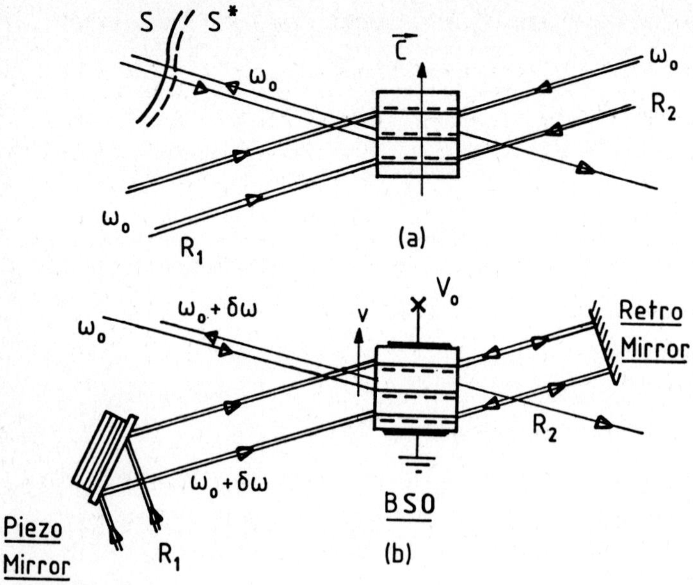

Fig. 5. Optical phase conjugation via degenerate four-wave mixing (a) and nearly degenerate four-wave mixing (b).

In the photorefractive BSO crystals,this phase-shift, like in two-wave mixing, is induced by a correct frequency detuning of the pump beam R_1, "Fig. 5.b." (nearly degenerate four-wave mixing).

Photorefractive crystals show attractive capabilities for efficient phase conjugation,since reflectivities between 1 and 30 have already been obtained in $Ba\,Ti\,O_3$, SBN, $K\,NbO_3$ and Bi_{12} (Si, Ge, Ti) O_{20}. For photorefractive $BaTiO_3$ and SBN, configurations are found where no external pump beams are required for phase conjugate wavefront generation (self-pumped conjugate mirror).

In this case, the phase conjugate of the incident beam is due to a photoinduced grating between the incident beam and the scattered noise [6, 17] . After efficient two-beam coupling, reflection of a conjugate beam is obtained. The corresponding experimental arrangement shown in "Fig. 6" is now extremely simple and resembles that used for phase conjugation by stimulated Brillouin backscattering.

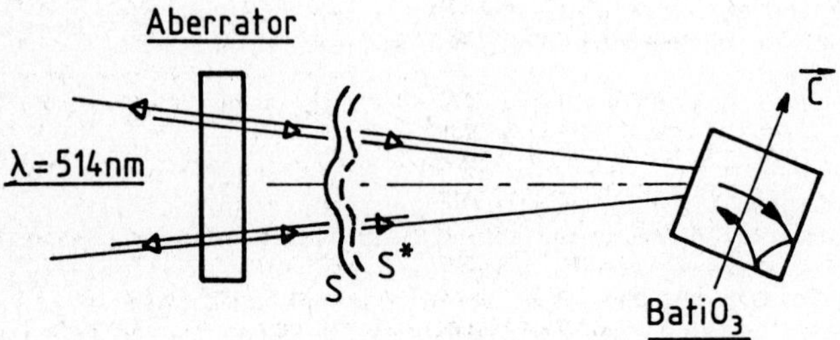

Fig. 6. Phase conjugate wavefront generation by a self-pumped $BaTiO_3$ photorefractive crystal.

6. Summary of the experimental results obtained with different types of photorefractive crystals

The following tables summarize some of the specific peculiarities of different photorefractive crystals. The time-constants τ_W for grating formation are related to the same incident intensity of 100 [mW.cm^{-2}] at the green line of argon laser, λ = 514 [nm] (excepted for Ga As and InP which work in the near IR).

Ferroelectric	τ_W	Γ	ρ
$LiNbO_3$, $BaTiO_3$ SBN, $KNbO_3$, KTN	seconds	10-20 [cm^{-1}]	1-50

Comments

- single domain crystals
- No applied field $E_0 = 0$
- optimum grating spacing $\Lambda \# 1$ [µm]
- self-pumped capabilities

Non Ferroelectric	τ_W	Γ	ρ
$Bi_{12}(Si,Ge,Ti)O_{20}$	10 [ms]	8-15 [cm^{-1}]	1-30
Ga As ; In P-Fe	20 [µs]	0.4 [cm^{-1}]	10^{-3}
(near IR-λ = 1.06 µm)	4 [W cm^{-2}]	[18-19]	

Comments

- Applied field E_0 = 10 [kV cm^{-1}] and moving grating recording
- Optimum grating spacing (for BSO) Λ = 20 [µm], critical
- ms speed.

7. Applications of the photorefractive effect

Photorefractive crystals have a great potential for many applications in coherent optical devices [20-5]. Indeed, materials exist with the required optical quality, and in some of them, fast response time and high index modulation can be achieved, thus allowing efficient interaction in wave mixing experiments. Moreover, when the optical signal is fedback to a photorefractive amplifier, different types of oscillators may be realized. The aim of this chapter is to review several of these experiments, and to discuss applications to image amplification, dynamic holographic interferometry, phase conjugation and photorefractive oscillators.

7.1. Image amplification based on the two-wave mixing gain

The large values of the gain coefficient permit the amplification of a low-intensity signal wavefront containing spatial information. As shown in Fig. 7, a binary photographic transparency is inserted on the signal beam path, and an amplified image due to the energy-transfer from the reference beam is projected onto a screen. The fringe velocity and reference signal beam angles are adjusted such that the maximum gain corresponds to the hologram spatial

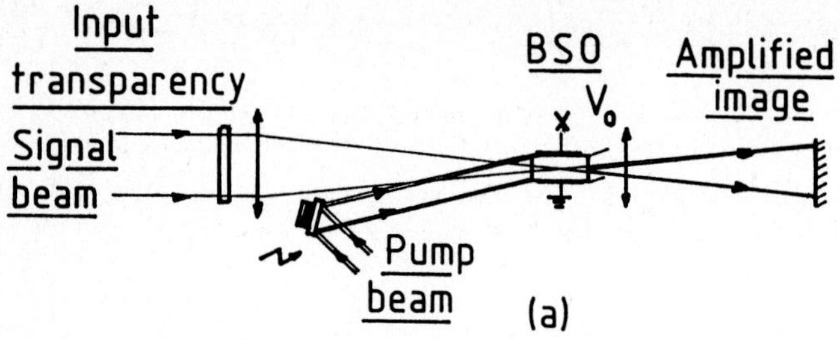

Fig. 7. Image amplification by two-wave mixing in BSO.

carrier frequency $\Lambda^{-1} \# [45\ mm^{-1}]$; but due to the bandpass response of the photorefractive amplifier the difference in gain for the various spatial frequencies may be noticeable and limit the size of the image to be amplified. For an applied field $E_0 = 6\,[kV\ cm^{-1}]$ the image is amplified 20 x [14]. A higher value of the gain is possible for $E_0 = 10\,[kV.cm^{-1}]$, but it would result in a loss in image uniformity and quality. Similar experiments have also been reported in photorefractive $BaTiO_3$; $KNbO_3$, using the diffusion recording mode (no applied field ; no moving grating).

7.2. Phase-conjugate wavefront generation using four-wave mixing

Phase conjugation by degenerate (or nearly degenerate) four-wave mixing can be used for restoration of phase-distorted images or wavefronts [15]. The implications of this concept are therefore important for coherent light transmission in the atmosphere, optical fibers, and in laser cavities. Before giving experimental results with photorefractive crystals, we explain the difference between a classical and a phase-conjugate mirror. A classical mirror doubles the phase distortions induced by an aberrator, while a conjugate mirror cancels it after double passing. Practical implementation of the experiment is shown in figure 8

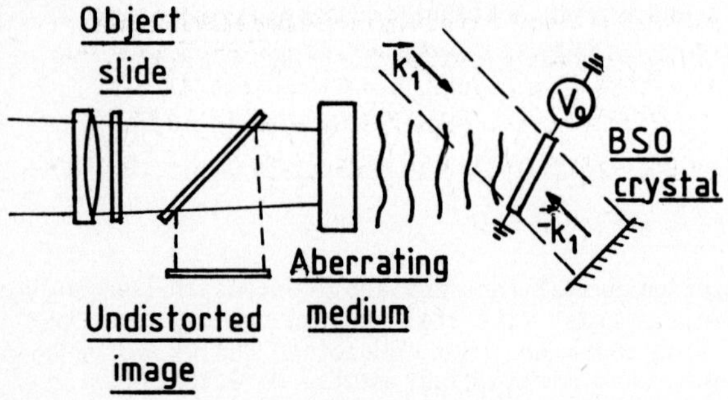

Fig. 8. Phase conjugation applied to imaging through a phase-disturbing media.

using a photorefractive BSO crystal as conjugator. The same set-up can be used for high-resolution photolithography, and recently, a photorefractive $BaTiO_3$ crystal was used as an intracavity distortion correction device in an Argon laser |6|. In conclusion, with optimum recording conditions which are specific to each crystal, conjugate beam reflectivities largely exceeding unity have been obtained on elementary probe beams or signal beams containing spatial information. These attractive results, demonstrated with CW single-mode laser of moderate power (< 1 W) brings up possibilities of laser resonators using phase-conjugate mirrors.

7.3. Ring oscillator

Self-starting optical resonators are obtained by adding an optical feedback to the photorefractive amplifier. These coherent oscillators have been reported in $BaTiO_3$ and $LiNbO_3$ [21, 22, 23] crystals due to the high gain resulting from their high electro-optic coefficients, and they are now obtained with BSO because of the gain enhancement due to self-induced moving gratings when an electric field is applied to the crystal.

The experimental setup for obtaining a unidirectional ring oscillator from a photorefractive crystal amplifier is shown in "Fig. 9" . The BSO crystal is introduced into the beam path defined by three plane mirrors M_1, M_2, and M_3, and the angle 2Θ between the pump beam I_{R_0} and the M_1-M_2 direction is chosen so as to correspond to the optimum ring spacing for the energy-transfer in the two-wave-mixing interaction (Λ = 23 µm, 2Θ = 1.5°) [24]. The condition for oscillation is given by :

$$(1 - R_{BS})\, R^3 \exp(\Gamma - \alpha_t) \geqq 1$$

$$L = k\lambda$$

where R and R_{BS} are respectively the dielectric mirror and beam splitter reflectivities, L is the cavity length, Γ is the exponential gain of the two-wave interaction, α_t represents the total losses. The values of Γ obtained ($\Gamma > 8$ cm^{-1}) largely exceed the cavity losses ($\alpha_t < 1$ cm^{-1} at λ = 568 nm, R = 0.98, $R_{BS} = 8.10^{-2}$), and therefore oscillation can occur. The oscillation in the cavity is self-starting : the optical noise due to the pump beam is sufficient to generate a weak probe beam that is then amplified after each round trip in the cavity. The required detuning $\delta\omega$ between the pump beam and the cavity beam in the ring oscillator is also self-induced. In other words, the crystals choose from the optical noise spectrum the frequency component shifted by $\delta\omega$ that will be optimally amplified in the cavity.

7.4. Oscillator with a phase conjugate mirror

Since phase conjugation with gain has been demonstrated in several photorefractive crystals, it is possible to induce oscillations between a plane mirror M and a photorefractive conjugate mirror, and TEM_{00} mode emission of the oscillator is observed in spite of any phase-distorting element placed inside the resonator.

The first demonstration was obtained in reference [21], using a $BaTiO_3$ crystal. and more recently in photorefractive BSO due to the gain enhancement with the

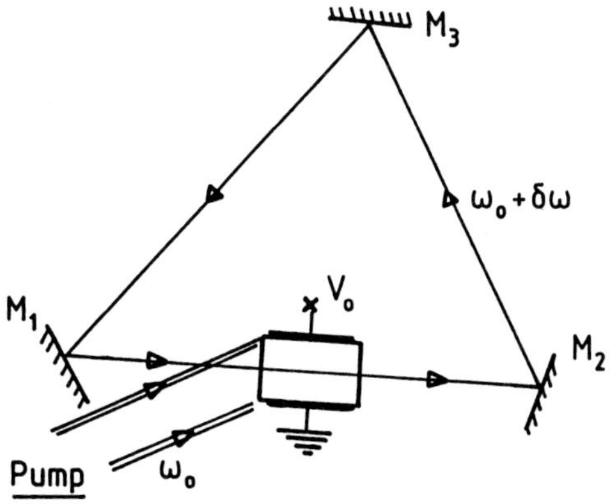

Fig. 9. Ring oscillator with a photorefractive crystal

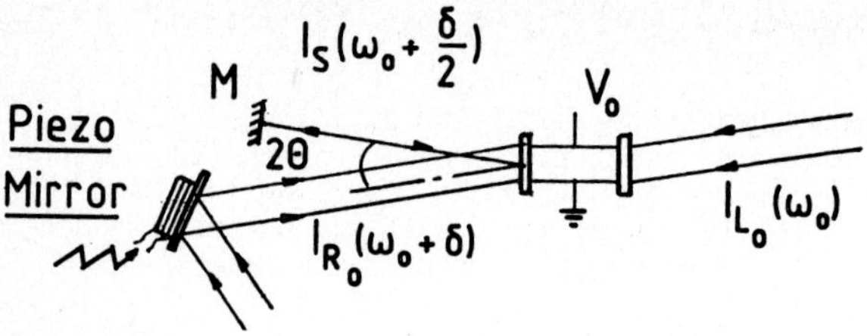

Fig. 10. Oscillator with a photorefractive BSO phase-conjugate mirror.

moving grating [24]. The experiment is shown in Fig. 10 . The oscillation starts in the cavity when a frequency-shift δ is introduced via the piezomirror on the pump beam. This generates in the cavity a beam oscillating at frequency $\omega + \delta/2$. Recently a photorefractive $BaTiO_3$ crystal was used as self-pumped conjugate mirror in a dye laser, thus permitting a better mode control, a reduction of laser bandwith, as well as a self-scanning of the laser frequency [25]. The unique properties araising from these new types of oscillators incorporating phase-conjugate mirrors are still the subject of important theoretical and experimental investigations.

7.5. Other applications of the photorefractive effect

In recent years several other applications of photorefractive materials in coherent image-processing and dynamic holography have emerged. It is not the object of this chapter to review all of these applications, which are described in detail in several papers, but it appears that there is an increasing interest in the following optical processing operations :

- image edging and processing [26];
- image convolution and correlation [27-28] ;
- parallel optical logic ; [33]
- image division [29] ;
- light beam deflection [30-31]
- interferometry [32].

All of these operations have already been demonstrated with dynamic photorefractive crystals, and are in progress in different laboratories. In parallel, further studies on the charge-generation and transport processes should be pursued, as well as the identification of donor and trapping centers. This should lead to the optimization of the photorefractive effect for the applications discussed in this paper.

References

(1) A. Ashkin, G.D. Boyd, J.M. Dziedzik, R.G. Smith, A.A. Ballman, K. Nassau Appl. Phys. Lett., 9, 72 (1966).

(2) A.M. Glass - Opt. Eng., 17, 470 (1978).

(3) D.L. Staebler - In holographic recording materials. Ed. H.M. Smith - Springer Verlag - 101 - (1977).

(4) P. Günter - Phys. Report., 93, 199 (1982).

(5) J.P. Huignard, H. Rajbenbach, Ph. Refregier, L. Solymar, Opt. Eng., 24 586 (1985).

(6) M. Cronin-Golomb, B. Fischer, J.O. White, A. Yariv - IEEE - J.Q.E. 20, 12 (1984).

(7) N.V. Kukhtarev, V.B. Markov, S.G. Odulov, M.S. Soskin, V.L. Vinetskii Ferroelect. 22, 949 (1979).

(8) J. Feinberg, D. Heiman, A.R. Tanguay, R.W. Hellwarth - J. Appl. Phys., 51, 1297, 1980.

(9) H. Kogelnik - Bell Syst. Tech. Journ., 48, 2909 (1969).

(10) G.C. Valley and M.B. Klein - Opt. Eng., 22, 704 (1983).

(11) J.P. Huignard, J.P. Herriau, G. Rivet, P. Günter - Opt. Lett. 5, 102, (1980).

(12) J.P. Huignard, A. Marrakchi - Opt. Comm., 38, 249, (1981).

(13) S.I. Stepanov, V. Kulikov, M. Petrov - Opt. Comm. 44, 19 (1982).

(14) H. Rajbenbach, J.P. Huignard, B. Loiseaux - Opt. Comm. 48, 247 (1983).

(15) A. Yariv - IEEE - J.Q.E., QE 14, 650 (1978).

(16) B. Fisher, M. Cronin-Golomb, J.O. White, A. Yariv - Opt. Lett. 6, 519 (1981).

(17) J. Feinberg - Opt. Lett. 7, 486, (1982).

(18) A.M. Glass, A.M. Johnson, D.H. Olson, W. Simpson, A.A. Ballman Appl. Phys. Lett., 44, 948 (1984).

(19) M.B. Klein - Opt. Lett. 9, 350 (1984).

(20) S.I. Stepanov, M.P. Petrov - Optica Acta, 31, 1335 (1984).

(21) J. Feinberg, R.W. Hellwarth - Opt. Lett., 5, 519 (1980) - 6, 257 (1981).

(22) J.O. White, M. Cronin-Golomb, B. Fischer, A. Yariv - Appl. Phys. Lett., 40 450 (1982).

(23) S. Odulov, M. Soskin - JETP Lett., 37, 289 (1983).

(24) H. Rajbenbach, J.P. Huignard - Opt. Lett., 10, 137 (1985).

(25) J. Feinberg, G.D. Bacher - Opt. Lett. 9, 420 (1984).

(26) Y.H. Ja - Opt. Comm., 42, 377 (1982).

(27) J.O. White, A. Yariv - Appl. Phys. Lett., 5, 37 (1980).

(28) L. Pichon, J.P. Huignard - Opt. Comm. 36, 277 (1981).

(29) E. Ochoa, L. Hesselink, J. Goodman - Appl. Opt. 24, 1826 (1985).

(30) G. Pauliat, J.P. Herriau, A. Delboulbé, G. Roosen, J.P. Huignard submitted to J.O.S.A..

(31) R. Rak, I. Ledoux, J.P. Huignard, Opt. Comm., 49, 302 (1984).

(32) J.P. Huignard, A. Marrakchi, Opt. Lett., 6, 22 (1981).

(33) T.K. Gaylord, M.M. Mirsalehi, C.C. Guest, Opt. Eng., 24-48 (1985).

Optical Fibres with Organic Crystalline Cores

B.K. Nayar

British Telecom Research Laboratories, Martlesham Heath,
Ipswich, Suffolk IP5 7RE, U.K.

1. Introduction

Research on monomode fibre systems has resulted in the development of fibres with very large bandwidths and transmission losses close to the limit set by Rayleigh scatter and IR absorption, as well as semiconductor lasers emitting high powers (> 10 mW). Work on coherent transmission systems has shown that, using lasers with extremely narrow linewidths (< 1 MHz), near quantum noise limited detection is possible. These advantages can only be fully exploited by using all optical systems which do not have the speed and bandwidth limitations of electronic components.

As a precursor to all optical signal processing, research effort is underway to realise a number of discrete active optical devices. These devices make use of non-linear processes which generally occur in materials at high optical intensities. This non-linear behaviour can be described by expressing the induced polarisation, $\underline{P}$, in the material as function of the applied field, $\underline{E}$:

$$\underline{P} = \varepsilon_0 [\hat{\chi}^{(1)} . \underline{E} + \hat{\chi}^{(2)} : \underline{E}\,\underline{E} + \hat{\chi}^{(3)} \vdots \underline{E}\,\underline{E}\,\underline{E} + \ldots] \qquad (1)$$

$\hat{\chi}^{(1)}$, $\hat{\chi}^{(2)}$, $\hat{\chi}^{(3)}$,... are the first, second, third, ... order susceptibilities. These are tensorial quantities and give a measure of the polarising effect of the optical fields on outer valence electrons. The magnitude of the susceptibilities is dependent upon the chemical bonding of the materials. The first term in the expansion is responsible for linear optical properties while the second and third terms give rise to non-linear three and four wave mixing processes respectively.

Three wave mixing processes include the linear electro-optic effect, second harmonic generation (SHG), sum and difference frequency mixing, and parametric amplification. These effects can be used to fabricate modulators, frequency convertors, parametric amplifiers and oscillators. In this paper three wave mixing and in particular optical SHG is considered. The $\hat{\chi}^{(2)}$ is non-zero only in non-centrosymmetric materials. For SHG it is customary to use $\hat{d}$ instead of $\hat{\chi}$ and the two are related by $\hat{\chi}^{(2)} = 2\hat{d}$. SH tensor, $\hat{d}$, is a third-rank tensor and has twenty seven independent terms. In practice there are only a few non-zero elements and they are dependent upon the crystal point group symmetry[1].

A number of devices exploiting three wave mixing have been fabricated in both bulk and guided wave form using inorganic materials eg $LiNbO_3$ and III-V semiconductors. Presently used inorganic materials tend to have either small non-linearity or low optical damage threshold thereby limiting their application. In recent years several organic materials have been developed using 'molecular engineering' with large second-order susceptibility and high optical damage threshold[2-4]. As an example the d_{12} and d_{11} coefficients of MNA[5] are 5.8x and 40x respectively larger than the d_{31} coefficient of $LiNbO_3$. Also, MNA has a damage threshold of > 200 MW/cm^2 at 1064nm wavelength. These materials generally have molecules which include benzene rings or conjugated bonds with acceptor and donor groups. Lately, there have been reports of powder SHG signals obtained using organics up to two orders of magnitude greater than that obtained using $LiNbO_3$

powder[6-7]. These developments have generated interest in the use of organic materials for the fabrication of non-linear devices.

Bulk devices require high optical powers for non-linear behaviour, and this renders them unsuitable for signal processing applications using semiconductor laser sources. An alternative strategy is to use waveguiding structures. Waveguides provide enhanced power density due to the small transverse dimensions of the guiding region and maintain this over long interaction lengths.

Optical waveguides have been fabricated in organic materials in the form of planar guides by vapour growth on a substrate[8], stripe guide by growth from the melt in a channel in glass/silicon substrate[9-10], and by growth from the melt in a hollow glass capillary. The last method is becoming increasingly popular, as it affords protection to the organic crystal core, and the crystalline cored fibres are compatible with the optical fibres. In this chapter fabrication, propagation and three wave mixing in these fibres is described in some detail.

2. Crystal Cored Fibres

The growth of organic crystal-cored fibres was pioneered by STEVENSON and DYOTT[11]. They fabricated a fibre polariser by growing a m-nitroaniline crystal in a Schott LaF N7 glass capillary. This work was followed up by a number of researchers[12-15] who studied crystal growth mechanism in small bore capillaries and investigated the fabrication of non-linear optical devices.

Organic crystal cored fibres are fabricated by the growth of a single crystal from the melt in glass capillaries using a modified vertical Bridgman technique. In this method, the crystal melt is progressively crystallized as it moves through a temperature gradient in a furnace. The method of crystal growth is only applicable to materials which are stable on melting, having melting points less than the glass transformation temperature (typically 350°C to 700°C). Also, for optical guidance it is necessary to ensure that the capillary glass refractive index is less than that of the organic material.

The cladding glass, for a given core material, can be selected from a range of commercially available glasses[16-17]. These have refractive indices in the range 1.44 to 1.95 measured at 588nm wavelength. The cladding glass is machined to give a suitable preform, which is then drawn into long lengths of uniform bore capillaries using a fibre drawing furnace.

The capillaries, typically 10cm in length, are filled with the crystal melt by capillary action in a furnace maintained at a temperature few degrees higher than the crystal melting point. In practice, capillary filling and crystal growth is carried out in the same furnace. The filling time is of the order of minutes. Prior to filling, it is important to ensure that there are no bubbles present in the melt because they can be 'frozen in' as voids on crystal growth.

The furnace for single crystal growth from the melt incorporates 'hot' and 'cold' zones, which are above and below the crystal melting point respectively, with a sharp temperature gradient. The sharp temperature gradient tends to prevent supercooling and spurious nucleations after the initial nucleation. On progressively moving the melt through the temperature gradient the melt crystallises. The crystallisation is accompanied by a reduction in volume and, for void-free crystal growth, it is necessary to optimise the growth-rate to enable the flow of melt to the crystal interface. Slow crystal growth-rate is also necessary to allow the latent heat of crystallisation to be conducted away and help to maintain a planar melt-crystal interface for prevention of void or dislocation formation. BALLENTYNE and AL-SHUKRI[18] have modelled growth kinetics, for single crystal growth in small bore capillaries, and this model can be used to determine optimum growth conditions for a given material.

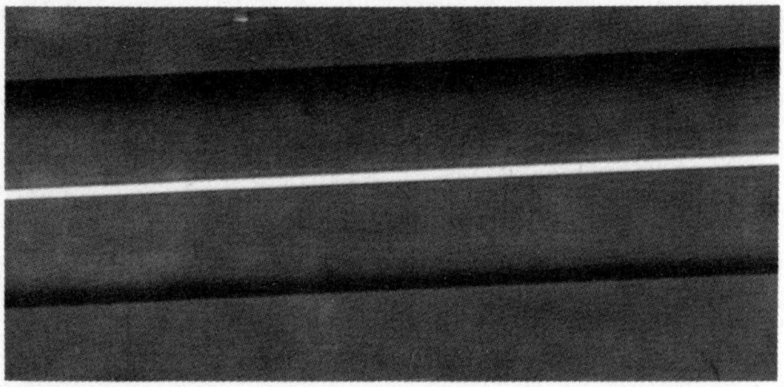

Fig 1. Monomode benzil crystal-cored fibre between crossed polarisers

Single crystals of acetamide, benzil, 2-bromo-4-nitroaniline, formyl-nitrophenylhydrazine, m-dinitrobenzene and meta-nitroaniline have been successfully grown in glass capillaries with 15 μm to 80 μm bore diameter and up to 50mm long. Recently, similar lengths of void-free crystals of benzil[19-20] and m-dinitrobenzene[15] have been successfully grown in capillaries with bore diameters in the 2 μm to 10 μm range for single-mode operation. In our furnace a pulling speed of 18mm/hr and a temperature gradient of 5°C/mm were found to give optimum conditions for void-free growth of single-mode benzil crystal-cored fibres having lengths up to 5cm. The crystal length was limited by the furnace design. Fig 1 shows a photograph of a monomode benzil crystal-cored fibre (core diameter = 3.8 μm) between partially crossed polarisers.

For three wave mixing, the crystal orientation in the capillary is important, as it determines both the pertinent tensor coefficient and the required field polarisations. In the case of m-dinitrobenzene and m-nitroaniline they grow in capillaries with their crystal c-axis along the growth direction. For both of these materials, their largest tensor co-efficients are along the c-axis. In 'weakly guiding' fibres (section 3) the fields are very nearly transverse, so these fibres are not useful for three wave mixing. A possible approach to change the crystal orientation is to have also a transverse temperature gradient and use a very slow growth-rate to prevent nucleation along the preferred growth direction. At present 2-methyl-4-nitroaniline and N-(4-nitrophenyl)-(L)-prolinol[10] are being investigated for growth in capillaries, as they afford correct orientation for three wave mixing.

3. Wave Propagation

A crystal-cored fibre is a cylindrical waveguiding structure and consists of a core with high refractive index, n_1, and diameter, 2a, surrounded by a cladding of a lower refractive index, n_2, fig 2. The mode propagation behaviour is found by solving the wave equation with certain boundary conditions. The core material

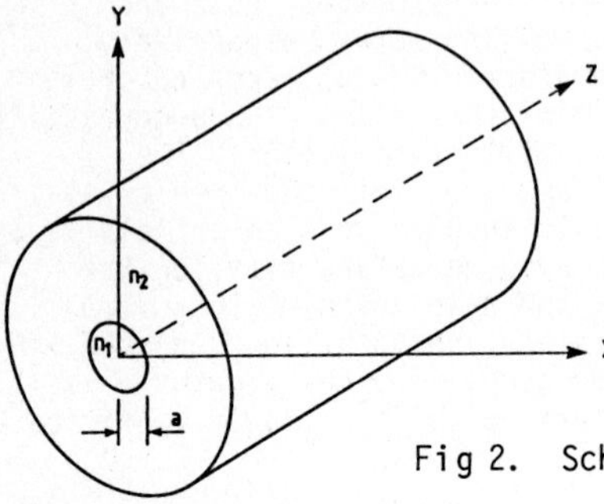

Fig 2. Schematic of a crystal-cored fibre

can, in general, be biaxial, in which case the wave equation has to be solved numerically. Analytical solutions to the wave equation can be obtained for an isotropic core in an infinitely large cladding. The solutions for such a structure are complicated and are given in a number of texts[21-22]. Analytical solutions can also be obtained for a fibre with uniaxial crystal-core having its optical axis along the fibre axis[23]. Here we consider the salient features of wave propagation in isotropic fibres.

The fibre can support a set of discrete guided modes and a continuum of radiation modes. The guided modes propagate with an effective mode index, $n_e = \beta/k_0$ where β and k_0 are the mode and free-space propagation constants respectively, such that $n_2<n_e<n_1$. The number of modes a fibre can support is determined by its 'normalised frequency' or 'V-value' which is given by

$$V = k_0 a n_1 \sqrt{\Delta} \tag{2}$$

where,

$$\Delta = 1 - \left(\frac{n_2}{n_1}\right)^2 \tag{3}$$

The fibre modes are characterised as HE_{nm}, EH_{nm}, TE_{nm} and TM_{nm} modes, where n and m are integer mode indices. The HE_{nm} and EH_{nm} modes have non-zero longitudinal field components. A great simplification can be made by assuming the fibres to be 'weakly guiding' ie $\Delta << 1$.

GLOGE[24] has shown that for small Δ the ratio of the magnitude of the longitudinal and transverse field components is of the order of $\sqrt{\Delta}$, so $E_z << E_x, E_y$. In this case, the fields can be considered as linearly polarised and are labelled as LP_{nm} modes. However, it should be borne in mind that LP_{nm} modes are not the true fibre modes but are combinations of TE_{nm}, TM_{nm}, HE_{nm} and EH_{nm} modes. As the fields are largely transverse, for three wave mixing it is necessary that the crystal core has a non-zero transverse SH tensor coefficient. The transverse field distribution in a fibre supporting N guided modes and a continuum of radiation modes can be expressed by

$$\underline{E}_t = \frac{1}{2}\left[\sum_{\mu=1}^{N} A_\mu(z)\, \underline{\mathcal{E}}_{\mu t}\, e^{i(\omega t-\beta_\mu z)} + c.c + \sum \int_0^\infty A_\rho(z,\rho)\, \underline{\mathcal{E}}_{\rho t}\, e^{i(\omega t-\beta z)}\, d\rho\right] \tag{4}$$

The expansion coefficients, A, are taken to be only z-dependent. The summation in front of the integral indicates summation over all types of radiation modes. The parameter ρ describes the complete set of radiation modes and is given by

$$\rho = (k_0^2 n_2^2 - \beta^2)^{\frac{1}{2}} \quad ; 0 < \rho < \infty \tag{5}$$

In the above representation of the mode field distribution, the power carried by the μth mode is given by $|A_\mu|^2$. Thus $\underline{\mathcal{E}}_{\mu t}$ must satisfy the following orthogonality relation which is the integral of Poynting's vector for unit power in mode μ.

$$\frac{1}{2}\int_{-\infty}^{\infty}\int_{-\infty}^{\infty} \hat{e}_z \cdot (\underline{\mathcal{E}}_{\mu t} \times \underline{\mathcal{E}}^*_{\nu t})\, dxdy = \frac{1}{2}\left(\frac{\beta}{\omega\mu_0}\right)\int_{-\infty}^{\infty}\int_{-\infty}^{\infty} \underline{\mathcal{E}}_{\mu t}\cdot\underline{\mathcal{E}}^*_{\nu t}\, dxdy = \delta_{\mu\nu} \tag{6}$$

where,

$$\underline{\mathcal{H}}_t = \sqrt{\frac{\varepsilon_o \varepsilon_r}{\mu_o}} \left(\hat{e}_z \frac{\partial}{\partial z} \times \underline{\mathcal{E}}_t \right)$$ has been assumed as $|\mathcal{E}_z/\mathcal{E}_t| << 1$.

For guided modes, $\delta_{\mu\nu}$, is Kronecker's delta and is unity for $\mu = \nu$ and is zero for $\mu \neq \nu$.

For radiation modes, $\delta_{\mu\nu}$, is Dirac-delta function. It is infinitely large for $\mu=\nu$, and is zero for $\mu\neq\nu$.

The field must satisfy the wave equation and be finite in both core and cladding. Thus, field distribution in the core region is taken as J-Bessel function to represent the oscillatory behaviour, while in the cladding as K-Bessel function to represent the field decay. The electric field components (for one polarisation) of nth mode are then given by[21].

$$\left.\begin{aligned} \mathcal{E}_x &= C\, J_n \left(U \frac{r}{a}\right) \begin{Bmatrix} \text{Cos } n\,\phi \\ \text{Sin } n\,\phi \end{Bmatrix} \\ \mathcal{H}_y &= n_e \sqrt{\frac{\varepsilon_o}{\mu_o}}\, \mathcal{E}_x \end{aligned}\right\} \text{ for } r < a \qquad (7)$$

$$\left.\begin{aligned} \mathcal{E}_x &= C\, \frac{J_n(U)}{K_n(W)}\, K_n \left(W \frac{r}{a}\right) \begin{Bmatrix} \text{Cos } n\,\phi \\ \text{Sin } n\,\phi \end{Bmatrix} \\ \mathcal{H}_y &= n_e \sqrt{\frac{\varepsilon_o}{\mu_o}}\, \mathcal{E}_x \end{aligned}\right\} \text{ for } r > a \qquad (8)$$

$$\mathcal{E}_y = \mathcal{H}_x = 0 \qquad (9)$$

where,

$$U = k_o a\, (n_1^2 - n_e^2)^{\frac{1}{2}} \qquad (10)$$

$$W = k_o a\, (n_e^2 - n_2^2)^{\frac{1}{2}} \qquad (11)$$

U, W and V are related by, $V^2 = (U^2+W^2)$ (12)

The choice of cos $n\phi$ and sin $n\phi$ is arbitrary, as two degenerate sets of modes rotated by $\pi/2$ from each other can exist.

C is a constant and is evaluated by substituting the field expressions (7) and (8), into the orthogonality relation (6) and is given by

$$C = \left[\frac{4\, (\mu_o/\varepsilon_o)^{\frac{1}{2}}}{e_n\, n_e \pi\, a^2}\right]^{\frac{1}{2}} \frac{W}{V|J_{n-1}(U) \cdot J_{n+1}(U)|^{\frac{1}{2}}} \qquad (13)$$

where,

$$e_n = \begin{cases} 2 \text{ for } n = 0 \\ 1 \text{ for } n \neq 0 \end{cases}$$

The solution to the wave equation is obtained in both core and cladding. Matching the tangential and transverse field components at the interface gives the characteristic equation which relates U and W.

$$\frac{U\, J_{n\pm 1}(U)}{J_n(U)} = \mp \frac{W\, K_{n\pm 1}(W)}{K_n(W)} \qquad (14)$$

This is used to compute effective mode indices and hence obtain field distributions. SNYDER[25] has shown that the errors in the effective mode indices obtained using this equation in comparison with the exact solution are <1% for $\Delta<0.1$ and <10% for $\Delta<0.25$. The use of the 'weakly guiding' approximation is justifiable in pratice, as monomode operation with $\Delta = 0.25$ requires the core diameter to be less than the optical wavelength. In fig 3, the effective mode index is plotted as a function of k_0a for various modes. From the plot, it can be seen that at a given wavelength and hence V-value the optical signal can have different effective indices, depending upon the mode number. This effect is referred to as 'modal dispersion' and can be used for phase-matching in non-linear processes. For V<2.405 only the lowest order propagates and higher order modes are cut-off. The fibres supporting higher order modes are referred to as multimode fibres.

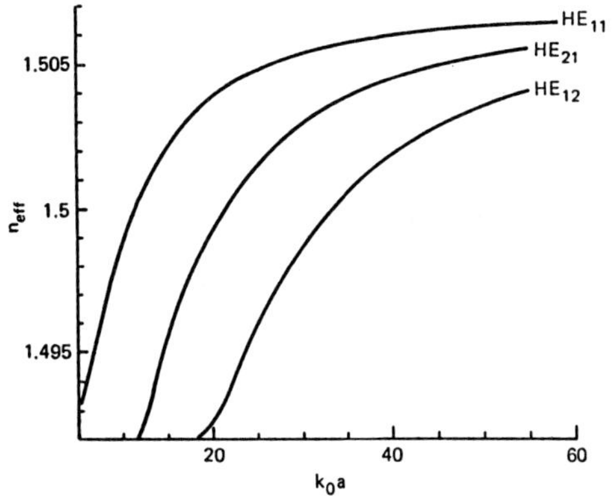

Fig 3 Mode dispersion diagram for a fibre having n_1 = 1.507 and Δ = 0.01

4. Three Wave Mixing

The process of three wave mixing in crystal-cored fibres can be treated as a coupled mode interaction in which the second order susceptibility acts as the perturbation, giving rise to frequency mixing. The coupling equation for such an interaction can be derived from the wave equation[26]. It was shown in the previous section that crystal-cored fibres are weakly guiding structures, and as a consequence mode field distributions are very nearly transverse. The transverse wave equation for propagation through a charge-free, isotropic and uniform, non-megnetic, and loss-less material is given by

$$\nabla^2 \underline{E}_t - \mu_0 \varepsilon_0 \varepsilon_r \frac{\partial^2 \underline{E}_t}{\partial t^2} = 0 \qquad (15)$$

where,

$\underline{E}_t$ is the transverse electric field strength.
μ_0 and ε_0 are the free space permeability and permittivity respectively.
ε_r is the relative permittivity.

For high electric field strengths, the electric polarization also has non-linear term, $\underline{P}_{nl}$, and the wave equation becomes

$$\nabla^2 \underline{E}_t - \mu_0\varepsilon_0\varepsilon_r \frac{\partial \underline{E}_t}{\partial t^2} = \mu_0 \frac{\partial^2 \underline{P}_{nl}}{\partial t^2} \qquad (16)$$

This follows directly from Maxwell's equations on using the electric displacement, $\underline{D} = \varepsilon_0\varepsilon_r \underline{E} + \underline{P}_{nl}$. Here, for simplicity, it has been assumed that $\underline{P}_{nl}$ is also transverse but in general this may not be the case. To obtain the coupling equation the fibre transverse field distribution (4) is substituted into (16) and with the assumptions that

i. the field distribution (4), (7) and (8) also satisfies the unperturbed wave equation (16).

ii. the slowly-varying amplitude approximation can be applied, ie the mode amplitude, A_μ, changes slowly in a distance of wavelength,

$$\left|\frac{d^2A_\mu}{dz^2}\right| \ll \beta \left|\frac{dA_\mu}{dz}\right|.$$

Thus

$$\sum_\mu - i\beta_\mu \frac{dA_\mu}{dz} \underline{\mathcal{E}}_{\mu t}\, e^{i(\omega t-\beta_\mu z)} + \text{c.c} + \sum \int_0^\infty -i\beta \frac{dA_\rho}{dz} e^{i(\omega t-\beta z)}\, d\rho = \mu_0 \frac{\partial^2 \underline{P}_{nl}}{\partial t^2} \qquad (17)$$

The coupling equation for guided modes can then be obtained by multiplying both sides of the above equation by $\underline{\mathcal{E}}^*_{\mu t}$ and integrating over the guide cross-section. This gives, using the orthogonality relation (6)

$$\frac{dA_\mu}{dz} e^{i(\omega t-\beta_\mu z)} + \text{c.c.} = \frac{i}{2\omega} \frac{\partial^2}{\partial t^2} \int_{-\infty}^{\infty}\int_{\infty}^{\infty} \underline{P}_{nl}\cdot\underline{\mathcal{E}}^*_{\mu t}\, dxdy \qquad (18)$$

The non-linear polarisation for three wave mixing is given by

$$\underline{P}_{nl} = \underline{P}_i = \varepsilon_0\, \chi^{(2)}_{ijk} : \underline{E}_j\, \underline{E}_k = \frac{1}{2}\left[p_i\, e^{i(\omega t-\beta_i z)} + \text{c.c.}\right] \qquad (19)$$

where, the complex polarisation amplitude, p_i, is

$$p_i = \frac{1}{2}\varepsilon_0\, \chi^{(2)}_{ij} : A_j^{(\omega_2)} A_k^{(\omega_1)} \mathcal{E}_j^{(\omega_2)} \mathcal{E}_k^{(\omega_1)} \qquad (20)$$

The coupling equation for co-directional three wave mixing is obtained from (18) after substituting for non-linear polarisation

$$\frac{dA_\mu^{(\omega_3)}}{dz} = -i\kappa e^{i\Delta\beta z} A_j^{(\omega_2)} A_k^{(\omega_1)} \quad \text{where,} \qquad (21)$$

$$\Delta\beta = \beta^{(\omega_3)} - (\beta^{(\omega_2)} + \beta^{(\omega_1)}) \text{ is the phase-mismatch factor.} \qquad (22)$$

κ is the coupling constant and has value

$$\kappa = \frac{\omega_3}{4}\varepsilon_0 \frac{\chi^{(2)}_{ijk}}{2} \int_{-\infty}^{\infty}\int_{-\infty}^{\infty} \mathcal{E}_j^{(\omega_2)} \mathcal{E}_k^{(\omega_1)} \cdot\mathcal{E}_{\mu t}^{(\omega_3)*}\, dxdy \qquad (23)$$

In the subsequent analysis the special case of SHG is considered from a νth mode at fundamental frequency, $\omega/2$, by coupling to μ^{th} mode at frequency, ω. In this case $\omega_3 = \omega$; $\omega_1 = \omega_2 = \omega/2$; $\Delta\beta$ and κ become

$$\Delta\beta = \beta^{(\omega)} - 2\beta^{(\omega/2)} \tag{24}$$

$$\kappa = \frac{\omega}{4} \varepsilon_0 \hat{d}_{ijk} I_0 \tag{25}$$

where,
d_{ijk} is the pertinent SH tensor coefficient and $\hat{d}_{ijk} = \frac{1}{2} \chi^{(2)}_{ijk}$.
I_0 is the overlap integral and is given by

$$I_0 = \int_{-\infty}^{\infty}\int_{-\infty}^{\infty} \mathcal{E}_\nu^{(\omega/2)} \mathcal{E}_\nu^{(\omega/2)} . \mathcal{E}_\mu^{(\omega)*} \, dxdy \tag{26}$$

Assuming weak coupling, ie the amplitude of the fundamental mode is constant with distance, the SH conversion efficiency, η, is given by integration of (21)

$$\eta = \left| \frac{A_\mu^{(\omega)}(L)}{A_\nu^{(\omega/2)}(0)} \right|^2 = \kappa^2 L^2 P^{(\omega/2)} \,\mathrm{Sinc}^2 \left[\frac{\Delta\beta L}{2}\right] \tag{27}$$

where, $P^{(\omega/2)}$ is the fundamental power in watts.

This expression is similar to the formula for SHG in bulk crystals with the exception of the difference in the detail of the coupling constant. For efficient frequency doubling,it is necessary to phase-match the fundamental and SH waves ie $\Delta\beta = 0$. The SH conversion efficiency for strong coupling is derived by also considering the coupled wave equation for the fundamental ie

$$\frac{dA_\nu^{(\omega/2)}}{dz} = -i\kappa^* A_\nu^{(\omega/2)} A_\mu^{(\omega)} \tag{28}$$

where, κ^* is the coupling constant for generating mode at the fundamental frequency by modes at SH and fundamental frequency. It can be derived in similar manner to κ and $|\kappa| = |\kappa^*|$.

Invoking the condition for conservation of power ie

$$\frac{d}{dz}\left[\left|A_\nu^{(\omega/2)}\right|^2 + \left|A_\mu^{(\omega)}\right|^2\right] = 0$$ and integrating (21) gives

$$\eta = \tanh^2 \left(\kappa \sqrt{P_{(o)}^{(\omega/2)}} L\right) \tag{29}$$

The expressions for efficiencies of the other three wave mixing processes can be derived in a similar manner to the above formalism for SHG.

4.1 Overlap Integral

The overlap integral determines the modes among which coupling can take place and the strength of coupling. The overlap integral (26), I_0, in cylindrical co-ordinate system can be written as

$$I_0 = \int_0^a \int_0^{2\pi} \left[\mathcal{E}_\nu^{(\omega/2)}\right]^2 \cdot \mathcal{E}_\nu^{(\omega)*} \, r dr \, d\phi \tag{30}$$

After substituting field expressions (7) it can be written as

$$I_0 = C_\nu^2 \, C_\mu \, I_r \, I_\phi \tag{31}$$

where,
i. I_r gives the magnitude of coupling strength and is given by

$$I_r = a^2 \int_0^1 J_\nu^2 \, (U^{(\omega/2)}\alpha) \, J_\mu \, (U^{(\omega)}\alpha)\alpha \, d\alpha \tag{32}$$

it has maximum value for identical mode fields. This integral has to be numerically solved.
ii. I_ϕ determines the azimuthal symmetry of the modes which can couple and is given by

$$I_\phi = \int_0^{2\pi} \begin{Bmatrix} \text{Cos}^2 \, \nu\phi \\ \text{Sin}^2 \, \nu\phi \end{Bmatrix} \begin{Bmatrix} \text{Cos} \, \mu\phi \\ \text{Sin} \, \mu\phi \end{Bmatrix} d\phi \tag{33}$$

for $I_\phi = 0$ no coupling can take place.
iii. C_ν and C_μ are constants and given by (13).

5. Phase-Matching

In the above analysis for three wave mixing it was shown that for an efficient build up of SH signal it is necessary to phase-match the fundamental and SH modes. The phase-matching condition requires equal phase velocities, and hence equal effective mode indices for both modes.

This is the same condition as in bulk materials, where it is rather difficult to satisfy because of chromatic dispersion. In bulk anisotropic materials with birefringence greater than dispersion, it is possible to phase-match using the 'Angle Phase-Matching' technique[27]. In this method fundamental and SH waves are arranged to be of orthogonal polarisations, and the direction of propagation is so chosen that both the waves 'see' the same refractive index.

In a multimode waveguide, as discussed in section 2, 'modal dispersion' can result in optical signals at different frequencies having the same effective index when they propagate in different mode orders. This property of optical waveguides can be used for phase-matching in non-linear processes. In a practical waveguiding structure it is also necessary to consider chromatic dispersion of the core and cladding materials. A consequence of modal dispersion is that it is not possible to phase-match signals at different frequencies for the same mode. Phase-matching between the lowest order mode at the fundamental frequency and a higher order SH mode is qualitatively shown in fig 4. In this case, if the SH mode order is high then the overlap integral has a small value, thereby resulting in a small or no advantage of using waveguide.

In a similar manner to 'Angle Phase-Matching' in bulk materials, for waveguides fabricated using some anisotropic materials it is possible to phase-match

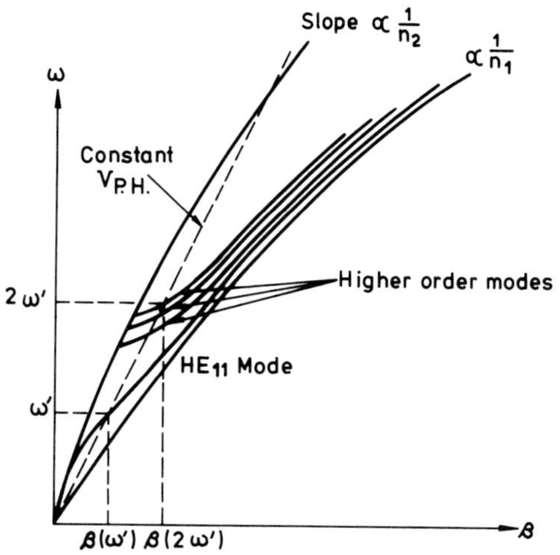

Fig 4 Dispersion diagram showing phase-matching implementation for guided wave SHG

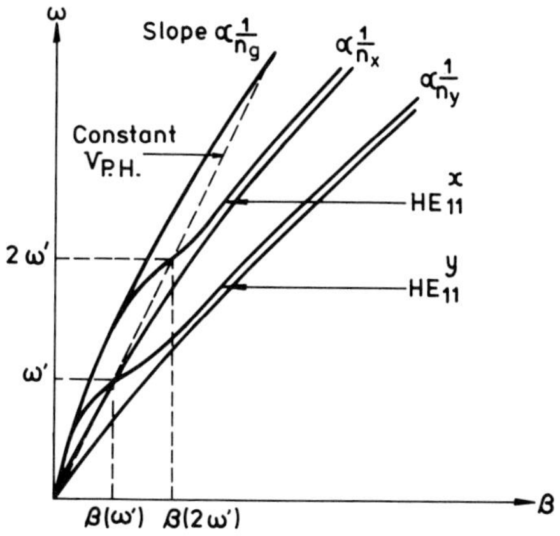

Fig 5 Dispersion diagram showing implementation of phase-matching for the fundamental and SH HE_{11} modes in an anisotropic crystal core or an elliptical core.

fundamental and SH frequencies by arranging them to propagate as lowest order orthogonal modes. A qualitative implementation of this scheme for a crystal-cored fibre is shown in fig 5. This form of phase-matching has been successfully used to give efficient SHG in Ti:$LiNbO_3$ waveguide[28]. The requirement for non-degeneracy of the lowest order modes in a waveguide with an isotropic core can be achieved by fabricating elliptical cored fibres, and thereby exploiting shape-induced birefringence.

An important consideration with regard to phase-matching is that it is necessary to maintain it over the entire waveguide length. As a limit, it is required that $\Delta\beta \ll L^{-1}$. This condition implies very tight tolerances for waveguide thickness and index homogeneity along the entire waveguide length, and has limited interaction lengths for SHG in Ti:$LiNbO_3$ waveguides to about 3cm.

A simpler phase-matching configuration is to couple the SH to the radiation field, thereby allowing the SH to propagate in the cladding. In this case it is only necessary to arrange that the effective mode index, $n_e^{(\omega)}$, is less than $n_2^{(\omega)}$. The chromatic dispersion, δn, required in the cladding glass to implement this form of phase-matching, has been shown[20] to be given by

$$\delta n \geqslant n_e^{(\omega/2)} - n_2^{(\omega/2)} \qquad \text{where,} \tag{34}$$

$$\delta n = n_2^{(\omega)} - n_2^{(\omega/2)} \tag{35}$$

The SH radiation exits from the fibre core at an angle, $\alpha = \mathrm{Cos}^{-1}\,[n_e^{(\omega)}/n_2^{(\omega)}]$. For $n_e^{(\omega)}$ is slightly less than $n_2^{(\omega)}$ it is of the order of few degrees. This form of phase-matching has been demonstrated in benzil crystal-cored fibre by generating SH of 1.064 μm wavelength[20].

The theoretical analysis for this form of coupling shows that the SH conversion efficiency is proportional to the interaction length rather than the square of interaction length. Also, in this case the overlap integral is small.

6. Conclusions

In this chapter fabrication, propagation and three wave mixing in optical fibres with organic crystalline cores has been described. The use of highly non-linear organic materials in this guided wave configuration has the advantage of relative simplicity of fabrication, protection from chemical attack and mechanical damage, uniform guiding region dimensions for phase-matching and compatibility with optical fibres and semiconductor diode lasers. The growth behaviour of organic material in glass capillaries is dependent on the material properties. However, the studies on growth kinetics, void/dislocation formation and on the growth of a number of organic materials in glass capillaries has provided sufficient understanding to make the task of growing 'new' materials less difficult.

The wave propagation behaviour for weakly guiding isotropic cored fibres has been presented. In the special case of uniaxial crystal-cored fibres with crystal axis along the fibre axis, the analytical solutions show that the effect of 10% birefringence is to change the effective mode index by only 1.72%. This gives confidence for the use of the weakly-guiding approximation in anisotropic fibres. For anisotropic cores with arbitrary orientation it is necessary to obtain numerical solutions to study phase-matching.

The theory of three wave mixing developed in this chapter can be used to design efficient devices. As an example, a 1cm-long crystal-cored fibre with crystal-core tensor coefficient, $d = 3.7 \times 10^{-12}$ m/V, gives a conversion efficiency of 12.4%, for the 1W of fundamental power and 5 μm core diameter, in the case of coupling from the fundamental HE_{11} mode into the SH HE_{12} mode.

7. Acknowledgements

I wish to acknowledge S.M. Al-Shukri, F.H. Babai, J.R. Cozens and R.B. Dyott for many useful discussions during the course of my stay at Imperial College, London. I would also like to thank R Kashyap and K.I.White of British Telecom Research Laboratories for discussions and comments on the manuscript. Acknowledgement is made to the Director of Research of British Telecom for permission to publish this paper.

8. References

1. S. Singh in Handbook of Lasers with Selected Data on Optical Technology, Ed. R.J. Pressely (The Chemical Rubber Co., Cleaveland, 1971).
2. D.J. Williams: Optical Properties of Organic and Polymeric Materials (American Chemical Society, Washington DC, 1983).
3. A. Carenco, J. Jerphagnon and A. Perigaud: J. Chem. Phys., 66, 3806 (1977).
4. K. Jain, J.I. Crowley, G.H. Hewig, Y.Y. Cheng and R.J. Twieg, Optics and Laser Tech., 297 (1981).
5. B.F. Levine, C.G. Bethea, C.D. Thurmond, R.T. Lynch and J.L. Bernstein: J. Appl. Phys., 50, 2523 (1979).
6. J.L. Oudar and R. Hierle: J. Appl. Phys., 48, 2699 (1977).
7. G.R. Meredith: "Design and characterization of molecular and polymeric non-linear materials: success and pitfalls", in ref 2, p32.
8. G.H. Hewig and K. Jain: Optics Commn., 47, 347 (1983).
9. S. Tomaru, M. Kawachi and M. Kobyashi: Optics Commn., 50, 154 (1984).
10. P. Vidakovic, J. Badan, R. Hierle and J. Zyss: "Highly efficient organic structures for wave-guided nonlinear optics", PD-C5, OSA 1984 Proc. XIII Quant. Electron. Conf. (post deadline papers).
11. J.L. Stevenson and R.B. Dyott: Electron. Letts., 10, 449 (1974).
12. J.L. Stevenson: J. Crystal Growth, 37, 116 (1977).
13. F.H. Babai, R.B. Dyott and E.A.D. White: J. Mats Sci., 12, 869 (1977).
14. F.H. Babai and E.A.D. White: J. Crystal Growth, 49, 245 (1980).

15. D.W.C. Ballentyne and S.M. Al-Shukri: J. Crystal Growth, 48, 491 (1980).
16. Catalogue of Optical Glasses, Jenaer Glaswerk, Schott and Gen., Mainz, W. Germany.
17. Hoya Optical Glass Catalogue, Hoya Corporation, Akishima-shi, Tokyo, Japan.
18. D.W.C. Ballentyne and S.M. Al-Shukri: J. Crystal Growth, 68, 651 (1984).
19. B.K. Nayar: "Optical second harmonic generation in crystal-cored fibers", in OSA Digest 6th Topical Mtg. on Integrated and Guided Wave Optics, ThA2, 1982.
20. B.K. Nayar: "Nonlinear optical interactions in organic crystal-cored fibres", in ref 2.
21. D. Marcuse: Theory of Dielectric Optical Waveguides (Academic, New York 1974).
22. J.E. Midwinter: Optical Fibers for Transmission (Wiley, New York).
23. J.R. Cozens: Electronics Letts., 12, 413 (1976).
24. D. Gloge: Appl. Optics, 10, 2252 (1971).
25. A.W. Snyder: IEEE Trans. Microwave Theory and Tech., MTT-17, 1310 (1969).
26. A. Yariv: Quantum Electronics, (Wiley, New York 1975), chap. 19.
27. P.D. Maker, R.W. Terhune, M. Nisenoff and C.M. Savage: Phys. Rev. Letts., 18, 21 (1962).
28. N. Uesugi and T. Kimura: Appl. Phys. Letts., 29, 572 (1976).

Nonlinear Optics at Surfaces and in Composite Materials

D. Ricard

Laboratoire d'Optique Quantique, Ecole Polytechnique,
F-91128 Palaiseau, Cedex, France

1 Introduction

It is well known that surfaces and interfaces play a very important role, for example in chemistry, biology ... In physics, they put a limit on the dimensions of crystals,for example, therefore breaking the translational invariance usually assumed, leading to the existence of surface states. One could cite many more examples. They are used in industrial processes such as mineral extraction or heterogeneous catalysis. They have already been studied for a long time by chemists using sophisticated tools such as electron diffraction [1]. However it is only recently that they entered the nonlinear optical domain.

We will concentrate on two areas and leave aside other active fields such as surface polaritons. First, the presence of interfaces may lead to dramatic effects such as the resonances that are encountered in Surface Enhanced Raman Scattering or in linear properties (the dielectric anomaly) of metal colloids. In this paper, we will arbitrarily consider spontaneous Raman scattering as a nonlinear process on the basis that it is a two photon transition. Secondly, the nonlinear properties of a surface or of an adsorbed molecular monolayer are interesting as such. For example, second harmonic generation has recently proven, as we shall see, to be a powerful tool to study surfaces.

This chapter is organized as follows. In section 2, we will deal with Surface Enhanced Raman Scattering. Section 3 will be devoted to Second Harmonic Generation by surfaces and molecular monolayers and its applications. Finally, in section 4, we will discuss the nonlinear optical properties of composite materials, restricting ourselves to the recent studies of optical phase conjugation in metal colloids and in semiconductor-doped glasses. Although the properties of composite materials are often strongly related to the presence of interfaces between the two-component media, specific models have been developed to explain their properties and the field of nonlinear optics in composite materials will undoubtedly grow in the next years.

2 Surface Enhanced Raman Scattering

2.1 Description of SERS and possible explanations

The Raman spectrum of a molecular monolayer is too weak to be observed under normal conditions. In order to observe it, a $\sim$ 3 orders of magnitude enhancement is needed. In the early 1970's, people were thinking of two possibilities to circumvent this difficulty, (1) increasing the surface area and consequently the number density, (2) utilizing resonant Raman scattering thus increasing the scattering cross section. The first observation of the Raman spectrum of a molecular monolayer (pyridine adsorbed on anodized silver) was reported by Fleishmann, Hendra and McQuillan in 1974 [2]. They believed they had increased the number density by roughening the substrate. The fact that a new effect (SERS) had been discovered was realized independently by Jeanmaire and Van Duyne [3] and Albrecht and Creighton [4] in 1976-77. In fact, the Raman cross section is enhanced by about 6 orders of magnitude in this case.

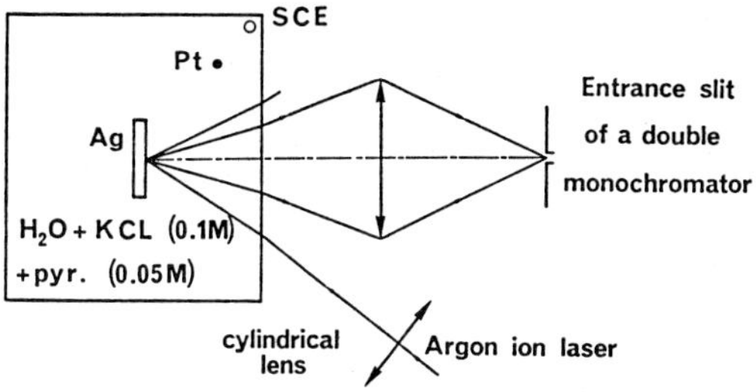

Fig.1. Part of a setup with which SERS can be observed. The silver substrate is placed inside an electrolytic cell containing an aqueous KCl solution (0.1 M) and pyridine (0.05 M). SCE is a saturated calomel electrode.

The enhancement phenomenon is fairly easy to observe with the setup shown in fig.1. A freshly polished silver electrode is immersed in an electrolytic cell together with a platinum counter-electrode and a reference saturated calomel electrode (SCE). The electrolyte is a 0.1 M KCl aqueous solution. One then performs an oxidation - reduction cycle. For example, the silver electrode being first positively polarized, a $\sim$ 500 μA per cm^2 of electrode current is run. Silver is oxidized according to

$$Ag + Cl^- \rightarrow AgCl + e^-$$

After 2 minutes, the applied voltage is inverted : silver chloride is then reduced with complete charge recovery

$$AgCl + e^- \rightarrow Ag + Cl^-$$

This oxidation - reduction cycle with a charge transfer of $\sim$ 60 mC/cm^2 has roughened (anodized) the silver surface. Pyridine is then added to the solution (0.05 mole/liter) and the silver electrode negatively biased in order to have $V_{Ag-SCE} = -0.6V$ where V_{Ag-SCE} is the measured voltage difference between the silver electrode and the SCE one. An Ar^+ laser beam (typically 30 mW at λ = 0.5145 μm) is focused on the silver electrode by a cylindrical lens. The laser spot (a narrow vertical line) is imaged onto the entrance slit of a double monochromator equipped with a conventional detection system for Raman spectroscopy. The lines at 1006 and 1037 cm^{-1} are then readily observable. Several other lines may also be observed. For example, at $V_{Ag-SCE} = -0.8$ V, the line at 1215 cm^{-1} is also very intense. To get an estimate of the enhancement effect, one may shift the silver electrode out of the laser beam. Then, only the pyridine molecules contained in the solution contribute to the Raman signal which is $\sim$ 500 to 1000 times weaker. To sum up, pyridine molecules adsorbed on a rough silver surface have Raman scattering cross sections about 6 orders of magnitude larger than the same molecules dissolved in water. This is, as it was first observed, the new effect now known as SERS.

It was soon realized that many other molecules exhibit the same effect when adsorbed on anodized silver. Other metallic substrates also lead to a similar enhancement, for example, gold and copper in the red part of the spectrum [5]. Most studies have been performed in an electrochemical environment but the effect may be observed under other conditions, for example for pyridine adsorbed on cold and rough silver in a vacuum chamber or for the silver - CN system at room temperature and in ambient air.

When the existence of this new effect was generally accepted, people started thinking about what could be the origin of the enhancement. The surface area increases only by a small factor, on the order of 2 to 3, on roughening. Therefore it is the Raman cross section which is enhanced. In the simple case of an isotropic molecule, the cross section $d\sigma/d\Omega$ is given by :

$$\frac{d\sigma}{d\Omega} = \frac{\hbar}{2\mu\omega_v} \frac{\omega_L \omega_S^3}{c^4} f(\omega_L)^2 f(\omega_S)^2 \left(\frac{d\alpha}{dq}\right)^2 \tag{1}$$

where ω_v is the eigenfrequency of the vibrational mode, μ and q are the corresponding reduced mass and normal coordinate, ω_L and ω_S are the laser and Stokes angular frequencies, α is the polarizability and $f(\omega)$ is the local field factor at frequency ω. f is the ratio of the field at the molecule position E_1 to that of the incident laser beam E_0 :

$$f = E_1/E_0 \tag{2}$$

The local field effect in a dense medium is due to the fact that the incident field polarizes the molecules surrounding the one of interest. The dipoles induced on the neighboring molecules create a field which adds up to E_0 to give E_1. This leads to the familiar $(n^2+2)/3$. In the case of SERS which is related to the surface roughness, the metal bump to which the molecule is adsorbed plays the role of the surrounding molecules, as we shall see.

There are three possible causes of the enhancement factor of SERS :

a) the local field effect
b) the image field effect
c) the molecule-metal interaction.

Causes a) and b) are of electromagnetic origin while cause c) is chemical in nature. The image field effect is the following : the incident field polarizes the molecule creating to zeroth order the dipole $\vec{\mu}_o=\alpha\vec{E}_o$. $\vec{\mu}_o$ polarizes the neighboring metal surface which in turn polarizes the molecule...The resulting effect may be calculated self-consistently. The molecule - metal interaction acts upon the $(d\alpha/dq)^2$ term. By distorting the electron configuration, it may increase α and/or lead to a shift in the position of the energy levels, possibly leading to resonant Raman scattering. In the case of pyridine on silver, a charge transfer state lying 1.9 eV above the ground state has been reported [6].

2.2 Some decisive experiments and the theory of the local field effect

Experiments aimed at studying the possible existence of a resonant Raman effect by varying the frequency of the laser beam were rather inconclusive [7-10]. On the other hand, two experiments clearly pointed out the important role of the local field effect.

The first one studied the enhancement factor as a function of the molecule - metal distance [11]. The system under study was pyridine on silver. Whereas in an electrochemical cell at room temperature, only one molecular monolayer is adsorbed, when working with cold (135 K) substrates, adsorption of a second, third... layer is possible. The second,third,...etc... layers are said to be physisorbed. With such multilayered systems, by measuring the Raman signal as a function of the number of adsorbed layers, one can study the variation of the enhancement factor as a function of the molecule - metal distance. It was found that the enhancement factor decreases slowly and is still quite large at a distance of 50 Å. This indicates that the major factor is of electromagnetic origin. It was also observed that the optimum roughness corresponds to bumps having a diameter of about 500 Å.

The second one addresses the question of the local field effect directly by considering second harmonic generation by a metal sample [12]. SHG is known to be due in this case to the surface (we will come back to this point later) and the local field factor involved in the nonlinear polarization $P^{NLS}(2\omega)$ is $f(2\omega)f(\omega)^2$. The SH signal is porportional to $|P^{NLS}(2\omega)|^2$ and consequently to $f(2\omega)^2f(\omega)^4$. The experiment measures the SH generated by a Q-switched Nd : YAG laser (λ = 1.064 μm) impinging on a silver surface and compares the SH intensity generated by a rough surface to that due to a smooth surface. An enhancement factor of 10^4 was measured for silver and somewhat smaller enhancement factors were also observed for gold and copper surfaces. In the present situation, the molecule-metal interaction and the image field effect are irrelevant and the importance of the local field effect is clearly demonstrated.

The calculation of the local field factor was first carried out in the case of a metallic sphere [13] of dielectric constant $\varepsilon = \varepsilon' + i\varepsilon''$ and radius a embedded in a dielectric medium of dielectric constant ε_0 as illustrated in fig.2.

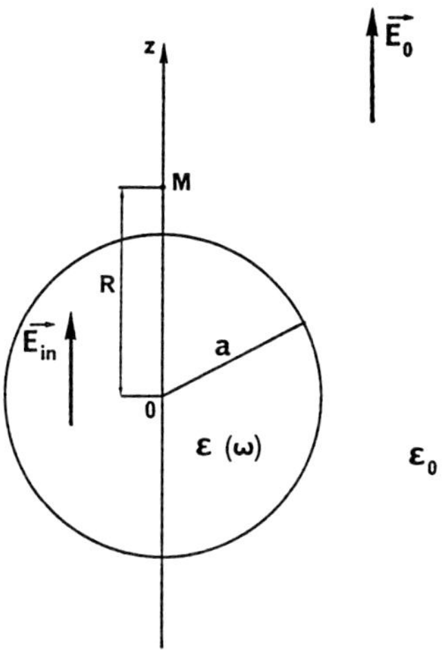

Fig.2. Sphere with dielectric constant $\varepsilon(\omega)$ located in a medium of dielectric constant ε_0 with a uniform applied field $\vec{E}_0$. $\vec{E}_{in}$ is the resulting uniform field inside the sphere.

ε_0 can be assumed frequency independent, whereas ε usually has a strong frequency dependence. The size of the sphere being small (on the order of 500 Å) compared with the laser wavelength ($\lambda \sim 5000$ Å or larger), retardation effects are neglected and the electrostatic approximation is used. The calculation of the field modification, which would otherwise be uniform and equal to $\vec{E}_0$, due to the presence of a dielectric sphere is a classical problem in electrostatics and may be found in textbooks [14]. The field inside the sphere is uniform and given by :

$$\vec{E}_{in} = \frac{3\varepsilon_0}{\varepsilon + 2\varepsilon_0} \vec{E}_0 \tag{3}$$

The field outside the sphere is the sum of $\vec{E}_0$ and of the field due to dipole moment $\vec{p}$:

$$\vec{p} = \frac{\varepsilon - \varepsilon_0}{3} a^3 \vec{E}_{in} \tag{4}$$

assumed located at the origin. For example, at point M on the Oz axis (Oz parallel to $\vec{E}_0$) whose distance from the origin is R>a, the local field is :

$$\vec{E}_l = \vec{E}_0 + 2 \frac{a^3}{R^3} \frac{\varepsilon - \varepsilon_0}{\varepsilon + 2\varepsilon_0} \vec{E}_0 \tag{5}$$

Note the R^{-3} dependence. More importantly, the second term in (5) shows a resonant behavior when :

$$\varepsilon'(\omega) = -2\varepsilon_o \quad (6)$$

This resonance is known as the surface plasma resonance or the localized surface plasmon. For silver and for $\varepsilon_o = 1$, it occurs at $\lambda \sim 355$ nm whereas experiment gives the largest enhancement in the green part of the spectrum.

One may reconcile theory and experiment if one realizes that the surface plasma resonance is shape dependent by considering an ellipsoidal metal bump. The case of an ellipsoid of revolution or spheroid [15] is exactly soluble using spheroidal coordinates. Fig. 3 shows a prolate spheroid of major axis a (along the axis of revolution Oz) and minor axis b, of dielectric constant $\varepsilon(\omega)$ embedded in a medium of dielectric constant ε_o with a uniform applied field $\vec{E}_o$ parallel to Oz. The spheroidal coordinates are defined by [16] :

$$x = f[1 - \xi^2)(\eta^2 - 1)]^{1/2}\cos\phi \quad (7,a)$$

$$y = f[(1 - \xi^2)(\eta^2 - 1)]^{1/2}\sin\phi \quad (7,b)$$

$$z = f\xi\eta \quad (7,c)$$

with : $\xi = \cos v \quad -1 < \xi < 1$

$\eta = \text{ch } u \quad 1 < \eta < \infty$

$f = (a^2 - b^2)^{1/2}$

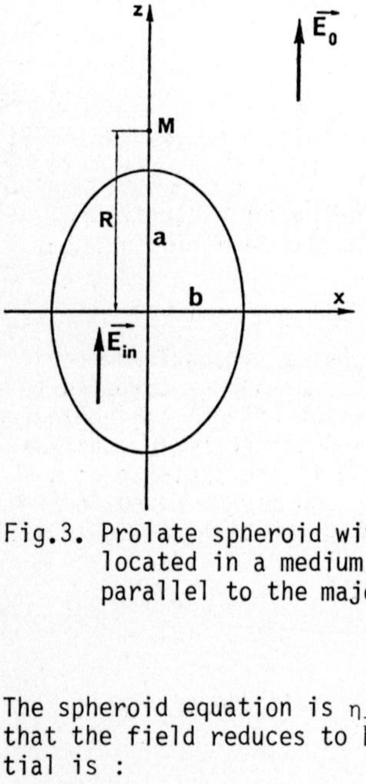

Fig.3. Prolate spheroid with major axis a, minor axis b and dielectric constant ε located in a medium of dielectric constant ε_o with a uniform field $\vec{E}_o$ applied parallel to the major axis.

The spheroid equation is $\eta = \eta_o$ where $\eta_o = a/f$. With the boundary condition implying that the field reduces to $\vec{E}_o$ far from the spheroid, the external electrostatic potential is :

$$V_{out} = \xi\ [-E_o f\eta + B'Q_1(\eta)] \quad (8)$$

where $Q_1(\eta) = \frac{1}{2}\eta \ln\frac{\eta+1}{\eta-1} - 1$ is the second kind Legendre function of order 1. Assuming the field to be uniform inside the spheroid, the internal electrostatic potential is :

$$V_{in} = -E_{in} f\xi\eta \quad (9)$$

The boundary conditions, V and the normal component of $\vec{D} = \varepsilon\vec{E}$ required to be continuous at the interface, allow calculation of E_{in} and B'

$$\vec{E}_{in} = \vec{E}_o \left(1 + \frac{\varepsilon - \varepsilon_o}{\varepsilon_o} A\right)^{-1} \qquad (10)$$

$$B' = \frac{\varepsilon - \varepsilon_o}{\varepsilon_o} (\eta_o^2 - 1) E_o f \eta_o \left(1 + \frac{\varepsilon - \varepsilon_o}{\varepsilon_o} A\right)^{-1} \qquad (11)$$

where the depolarization factor A is given by

$$A = (\eta_o^2 - 1) \left(\frac{1}{2} \eta_o \ln \frac{\eta_o + 1}{\eta_o - 1} - 1\right) \qquad (12)$$

The external field may be calculated from (8). For example, at point M (see fig.3), one has :

$$\vec{E}_1 = \vec{E}_o - \frac{\frac{\varepsilon - \varepsilon_o}{\varepsilon_o} (\eta_o^2 - 1)\eta_o \left(\frac{1}{2}\ln \frac{\eta_1+1}{\eta_1-1} + \frac{1}{2} \frac{\eta_1}{\eta_1+1} - \frac{1}{2} \frac{\eta_1}{\eta_1-1}\right)}{1 + \frac{\varepsilon - \varepsilon_o}{\varepsilon_o} A} \vec{E}_o \qquad (13)$$

with $\eta_1 = R/f$ showing the molecule-metal distance dependence of E_1. The surface plasma resonance now occurs when :

$$\varepsilon'(\omega) = \varepsilon_o \left(1 - \frac{1}{A}\right) \qquad (14)$$

When the aspect ratio a/b is 3, A = 0.1087 and for silver in air or vacuum ($\varepsilon_o = 1$), the resonance occurs in the vicinity of λ = 500 nm. Finally note that at the tip of the spheroid (R = a), the local field is :

$$\vec{E}_1 = \vec{E}_o + \frac{\varepsilon - \varepsilon_o}{\varepsilon_o} (1 - A)\left(1 + \frac{\varepsilon - \varepsilon_o}{\varepsilon_o} A\right)^{-1} \vec{E}_o \qquad (15)$$

Comparing the results for the sphere and the prolate spheroid, one may define a lightning rod factor [17]

$$\gamma = \frac{3}{2} \frac{a^2}{b^2} (1 - A)$$

The lightning rod effect acts in conjunction with the resonance effect to give the local field of (15) according to which the enhancement factor of SERS could reach 10^{11}. In fact, the dipole induced in the spheroid radiates and radiative damping must be taken into account [18], especially for large spheroids.

In ref. [15], the image field effect is also calculated but we will content ourselves with the fact that it is negligible when the molecule-metal distance is larger than $\sim$ 3 Å.

2.3 Our present understanding of SERS

The previous calculations of the local field effect were very nicely confirmed by an experiment performed by Liao et al.[19] using a very neat substrate consisting of silver ellipsoids evaporated on top of SiO_2 posts fabricated by microlithography and arranged in a square array. The adsorbed species was CN and the aspect ratio of the ellipsoids could be varied. By varying the laser frequency ω_L, the resonance was observed at a frequency close to the one expected from the calculation. It was also observed that the resonance shifts to longer wavelengths when the dielectric constant ε_o of the ambient medium is increased as predicted by theory. Finally, the enhancement factor was estimated to be $\sim 10^7$.

In view of the numerous studies of this effect [20], our present understanding of SERS is the following. Three causes may contribute to the enhancement factor : the local field effect, the image field effect and the molecule-metal interaction. The predominant contribution comes from the local field effect. The image field effect is usually thought to be negligible. Finally, the molecule-metal interaction is believed to have a small but finite contribution.

It was sometimes suggested that conventional surface plasmons may contribute significantly to the enhancement effect. But it was shown theoretically [21] and confirmed experimentally [22] that such is not the case.

3 Second Harmonic Generation by Surfaces and Monolayers

When a laser beam of frequency ω is incident on a semi-infinite noncentrosymmetric medium, it gives rise to the simplest of the nonlinear processes : Second Harmonic Generation (SHG) [23]. There is a (usually intense) transmitted harmonic beam (at frequency 2ω) but due to the presence of the front surface, there is also a weak reflected SH beam. Reflection SHG from noncentrosymmetric media was studied since the early days of nonlinear optics [24]. It is particularly relevant for absorbing media. Somewhat later, reflection SHG from centrosymmetric media, namely metals, was reported [25] and subsequently thoroughly studied by Bloembergen et al. [26].

3.1 SHG by a monolayer and by the substrate

A monolayer of noncentrosymmetric molecules also radiates SH waves when irradiated by a laser beam. The calculation of the radiated field proceeds as follows. Assume we have a layer of nonlinear material of thickness a sitting between two media labeled 1 and 2 in fig.4. A plane wave incident from the left creates a nonlinear polarization

$$\vec{P}_{NL} = \vec{p}\, P_o\, e^{i(\vec{k}_s.\vec{r} - 2\omega t)} \tag{16}$$

inside the nonlinear slab where $\vec{p}$ is a unit polarization vector and $\vec{k}_s = 2\vec{k}_\omega$. The solution to Maxwell's equations consists of a driven wave inside the slab :

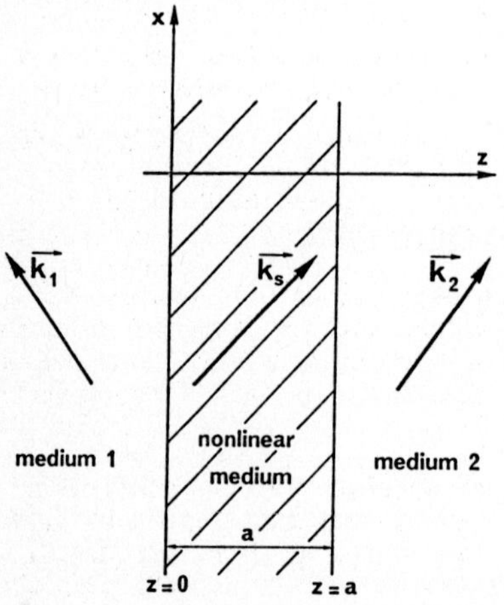

Fig.4. Model used to calculate the waves radiated by a monolayer of dipoles oscillating at frequency 2ω. One starts with a slab sitting between media 1 and 2. The wave radiated in these two media is calculated and the limit $a \to 0$ is taken

$$\vec{E}_f = \frac{1}{k_s^2 - k_i^2} \frac{4\pi(2\omega)^2}{c^2} \left| \vec{p} - \frac{\vec{k}_s(\vec{k}_s.\vec{p})}{k_i^2} \right| P_o \, e^{i\vec{k}_s.\vec{r} - 2i\omega t} \tag{17}$$

a plane wave propagating to the right in medium 2

$$\vec{E} = E_{2x}(1,0, -\frac{k_x}{k_{2z}}) \, e^{i(k_x x + k_{2z} z) - 2i\omega t} \tag{18}$$

a plane wave propagating to the left in medium 1

$$\vec{E} = E_{1x} \, (1,0, \frac{k_x}{k_{1z}}) \, e^{i(k_x x - k_{1z} z) - 2i\omega t} \tag{19}$$

and two plane waves propagating freely inside the slab, one to the right and one to the left. k_x is the x-component of $\vec{k}_s$, k_i is the modulus of $\vec{k}_{2\omega}$ inside the slab. The different wavevectors usually have different z-components but of course have the same x-component. The four unknown amplitudes are determined by the boundary conditions, two for z = 0 and two for z = a. The general solution to the problem may be found in ref. [27]. In the case of a monolayer, the limit $a \to 0$ is taken and only the leading term in a is kept. $P_o a$ is the dipole moment per unit area which we will also denote P_{NS}.

With N_a molecules per cm^2, each having the second-order polarizability $\alpha^{(2)}(\omega,\omega)$, one has

$$\vec{P}_{NS} = \chi_S^{(2)} \, \vec{E}(\omega)\vec{E}(\omega) = N_a \, \alpha^{(2)} \, \vec{E}(\omega)\vec{E}(\omega) \tag{20}$$

$\alpha^{(2)}$ and $\chi_S^{(2)}$ are third rank tensors. If, for example, media 1 and 2 are identical, the intensity of the SH wave radiated to the left is given by

$$I_{2\omega} = \frac{32\pi^3\omega^2}{c^3} \sec^2\theta \, |\vec{e}_{2\omega}.\chi_S^{(2)} : \vec{e}_\omega \, \vec{e}_\omega|^2 \, I_\omega^2 \tag{21}$$

where I_ω is the intensity of the incident wave, $\vec{e}_\omega$ and $\vec{e}_{2\omega}$ are unit polarization vectors for the incident and harmonic waves and θ is the angle of incidence. The question is : is $I_{2\omega}$ large enough to be observable ? The problem is that the molecular monolayer is adsorbed on a substrate. We should then consider reflection SHG from the substrate.

The calculation of the reflected SH wave created in the bulk of the substrate is even simpler than the calculation outlined above. In addition to the driven wave, there are two free waves whose unknown amplitudes are derived by the use of the boundary conditions. A layer having a thickness on the order of the wavelength contributes to the SH generated by the bulk. If the substrate were noncentrosymmetric, this contribution would be overwhelming compared with that of the monolayer. Therefore only centrosymmetric substrates may be used. In a centrosymmetric material, the second-order susceptibility $\chi^{(2)}$ is zero in the electric dipole approximation. The first nonvanishing terms are of electric quadrupole and magnetic dipole character. The exact form of these terms will be given in next subsection. They contribute a nonzero but weak SH signal and the existence of the boundary must be explicitly taken into account. At the boundary, inversion symmetry breaks down and we have a true electric dipole allowed $\chi_S^{(2)}$ leading to a narrow sheet of nonlinear polarization $\vec{P}_{NS}$. The field radiated by this sheet is calculated in exactly the same way as for a monolayer.

3.2 The tensorial aspects of the nonlinear response

Let us consider first the response of the bulk centrosymmetric substrate. Being of electric quadrupole and magnetic dipole character, it may be cast in the general form

$$P_i^{NL} = \sum_{j,k,l} A_{ijkl} E_j \nabla_k E_l \tag{22}$$

where A_{ijkl} has the same symmetry properties as $\chi^{(3)}$. The most anisotropic material we are going to consider being silicon, we limit ourselves to this case of cubic symmetry where only 4 components are independent [28] : xxxx, xyxy, xyyx and xxyy where xxxx stands for A_{xxxx} and so on. Here, x,y and z are the crystallographic axes parallel to the edges of the conventional cube. The x component of $\vec{P}^{NL}$ then reads:

$$P_x^{NL} = (xxxx - xxyy - xyyx - xyxy)\, E_x \frac{\partial}{\partial x} E_x + xxyy\, (\vec{\nabla}.\vec{E})E_x + xyyx\, (\vec{E}.\vec{\nabla})E_x + \frac{1}{2}\, xyxy \frac{\partial}{\partial x}(\vec{E}.\vec{E}) \tag{23}$$

In a more symmetric material, e.g.,isotropic, one has

$$xxxx = xxyy + xyyx + xyxy \tag{24}$$

The nonlinear polarization in such a medium, for example,a metal or a piece of glass, assumes the simple form :

$$\vec{P}^{NL} = a(\vec{\nabla}.\vec{E})\vec{E} + b(\vec{E}.\vec{\nabla})\vec{E} + c\vec{\nabla}(\vec{E}.\vec{E}) \tag{25}$$

With a single incident plane wave, the first two terms do not contribute. Silicon, in addition, has the anisotropic term :

$$P_i^{NL} = d\, E_i \nabla_i E_i \tag{26}$$

the index i being either x, y,or z.

In the case of the surface or of a monolayer, the most anisotropic case of interest to us will be the (111) face of Si which is of 3m symmetry. We will use the reference frame x'y'z' where z' is normal to the surface. $\chi^{(2)}$ then assumes the form [28]

$$\begin{bmatrix} 0 & 0 & 0 & 0 & 0 & x'z'x' & x'x'z' & \overline{y'y'y'} & \overline{y'y'y'} \\ \overline{y'y'y'} & y'y'y' & 0 & x'x'z' & x'z'x' & 0 & 0 & 0 & 0 \\ z'x'x' & z'x'x' & z'z'z' & 0 & 0 & 0 & 0 & 0 & 0 \end{bmatrix}$$

where $\overline{y'y'y'} = - y'y'y'$. Since we deal with $\chi^{(2)}(\omega,\omega)$, $\chi^{(2)}$ is unchanged under permutation of the last two indices. This implies that x'z'x' = x'x'z' and we have only 4 independent components : y'y'y' , x'z'x', z'x'x' and z'z'z'. The components of $\vec{P}_{NS}^{(2)}$ are given by

$$P_{x'}^{(2)} = 2x'z'x'\, E_{z'} E_{x'} - 2y'y'y'\, E_{x'} E_{y'} \tag{27,a}$$

$$P_{y'}^{(2)} = -y'y'y'(E_{x'} E_{x'} - E_{y'} E_{y'}) + 2\, x'z'x'\, E_{z'} E_{y'} \tag{27,b}$$

$$P^{(2)}_{z'} = z'x'x' \ (E_{x'} E_{x'} + E_{y'} E_{y'}) + z'z'z' \ E_{z'} E_{z'} \qquad (27,c)$$

If the surface is more symmetric,as for example in the case of the (100) face of Si or in the case of axial symmetry about z' (two dimentional isotropy), then the y'y'y' component vanishes, $\chi^{(2)}$ then has only 3 independent components which we denote $\chi^{(2)}_{\parallel\perp\parallel} = 2x'z'x'$, $\chi^{(2)}_{\perp\parallel\parallel} = z'x'x'$, and $\chi^{(2)}_{\perp\perp\perp} = z'z'z'$ with obvious meanings considering the simple form (27) then assumes. In the case of the (111) face of Si, one has in addition the anisotropic term :

$$P^{(2)}_{x'} = - 2y'y'y' \ E_{x'} E_{y'} \qquad (28,a)$$

$$P^{(2)}_{y'} = - y'y'y' \ (E_{x'} E_{x'} - E_{y'} E_{y'}) \qquad (28,b)$$

$$P^{(2)}_{z'} = 0 \qquad (28,c)$$

y' is the projection on the surface of the [100] axis.

The monolayers which have been studied up to now all possess two-dimensional isotropy and the second harmonic wave they generate is p polarized ($\vec{E}$ in the plane of incidence) when the incident laser beam is either purely s or p polarized. The same is true for the bulk contribution of an isotropic substrate. Monolayers deposited by the Langmuir-Blodgett technique are sometimes two-dimensional solids and their $\chi^{(2)}$ would be strongly anisotropic.

3.3. Observation of SHG by an adsorbed monolayer

Going back to the question of the detectability of the SH wave generated by a molecular monolayer adsorbed on a substrate, we must consider the appropriate formula. For example, for a monolayer adsorbed on silver, the relevant modification of (21) leads to the following numerical estimate : assuming $\alpha^{(2)} = 10^{-29}$esu, $N_a=4\times10^{14}cm^{-2}$, $I_\omega=1$ MW/cm^2 at $\lambda=1.064\mu m$, a beam cross-section area of 0.2 cm^2,a pulse duration of 10 ns and a p polarized laser beam incident at 45°, one expects a SH signal of about 1.5×10^3 photons per pulse. Under the same conditions, a silver surface would yield $\sim$ 20 SH photons per pulse [26]. This means that a monolayer having a $\chi^{(2)}_s = N_a\alpha^{(2)}$ of a few 10^{-15} esu would give a SH signal large enough to be detected even in the presence of the silver contribution. The numerical estimate just mentioned above applies to the case of a smooth silver surface. In the case of a rough silver surface, both contributions, that of the monolayer and that of the substrate would be enhanced by the same factor.

If the nonlinear susceptibility of the monolayer is smaller, the use of a dielectric substrate giving rise to a much weaker SH signal is strongly recommended. Historically, the first observation of SHG by a monolayer was reported in 1973, the system under study being a Na-covered Ge surface [29]. But the true significance of this observation was not fully understood by the authors. The first real observation of SHG by a monolayer [30] was performed with a silver electrode as the substrate in the same device as shown in fig.1. The second harmonic signal was observed as a function of time during an oxidation-reduction cycle starting with a rough surface : it increases suddenly at the beginning of the cycle and decreases also abruptly at the end of the cycle. The rise and fall correspond to the formation or reduction of 1 - 2 monolayers of AgCl. The SH signal also increases dramatically upon addition of pyridine to the KCl electrolytic solution and when the silver electrode is negatively biased.Fig.5 which shows the voltage dependence of the signal in the absence and in the presence of pyridine clearly demonstrates that the signal is due to the pyridine monolayer.

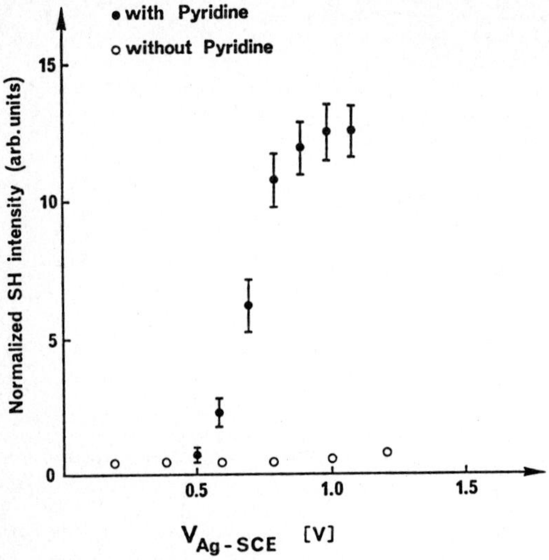

Fig.5. Second harmonic signal versus V_{Ag-SCE} with and without pyridine in the electrolytic solution (after Chen et al.[30]).

Apart from AgCl and pyridine, SHG due to a monolayer adsorbed on rough silver was also observed for several other molecules including pyrazine, aniline, phenylenediamine and CN. It was also observed for the pyridine-copper system. Since then, a large number of systems have been studied successfully. Some involve rather large molecules such as various rhodamines on glass or p-nitrobenzoic acid on sapphire or on fused silica. But others involve simple molecules or even atoms such as O_2, CO and Na on rhodium [31] . The face of a single crystal was used as the substrate in this last instance. SHG from a clean silicon substrate has also been studied and the contribution of the surface layer has been identified [32]. We will come back to several of these examples in the next subsection.

It should be stressed that it is because second-order processes are electric dipole forbidden in centrosymmetric media that they are so sensitive to surfaces and interfaces.

3.4. Applications

We will see, by examining several examples, that SHG is a very useful tool in the study of surfaces and monolayers.

a) Adsorption isotherms

In the case of a solid-liquid interface, an adsorption isotherm is the curve giving the number of adsorbed molecules per unit area N_a as a function of the concentration c of the species under study in the solution. Second harmonic generation has been used in conjunction with SERS to study the system pyridine-water-silver [33]. The Raman signal is a priori simply proportional to N_a. In the case of SHG, the adsorbed molecules and the substrate both contribute to the nonlinear polarization giving rise to a signal.

$$P(2\omega) = |A + B\ N_a|^2 \qquad (29)$$

It turns out that A and B are in phase and N_a is obtained from the SH signal in the following way :

$$N_a = (\sqrt{P(2\omega)} - A)/\ B$$

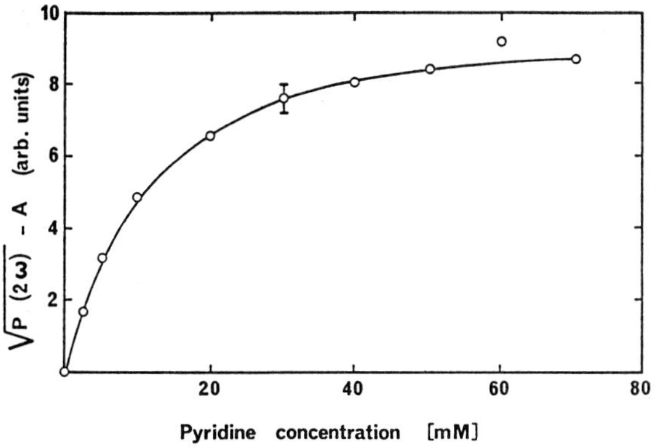

Fig.6. Second harmonic signal (after processing to subtract the substrate contribution) versus pyridine concentration. The solid line is a Langmuir isotherm (after Chen et al. [33]).

The adsorption isotherm obtained in this way is shown in fig.6. The data points are very well fitted by a curve of equation

$$N_a = \frac{c}{K+c} N_{as} \qquad (30)$$

known as a Langmuir isotherm. N_{as} is the saturated value of N_a and K is related to the adsorption free enthalpy ΔG. When c and K are in moles/liter, $K=55\exp(-\Delta G/RT)$ where R is the ideal gas constant. From the data shown in fig.6, one calculates the value $\Delta G = 5.1$ kcal/mole. The Langmuir isotherm is frequently encountered in solid-liquid systems.

It turns out that the adsorption isotherm obtained from the Raman measurements is slightly different, but it is also a Langmuir isotherm with $\Delta G=5.7$ kcal/mole. The difference between the results obtained by the two techniques is certainly due to the fact that SHG is much more orientation dependent than SERS. This may also be related to the different voltage dependence of the SH and Raman signals.

b) The molecule - metal interaction

The molecule - metal interaction sometimes leads to a dramatic modification of the second-order polarizability $\alpha^{(2)}$ of the adsorbed molecule. From the signal obtained with pyridine on silver, a numerical value of $\sim 2\times10^{-29}$ esu was deduced for pyridine. Such a value is rather large, a typical value for substituted benzene molecules being on the order of 10^{-30} esu. It would be more appropriate to assign the aforementioned numerical value to "adsorbed pyridine" or to the pyridine-silver system.

The consequences of the chemical interaction are even more clearly demonstrated in the case of the pyrazine-silver [34] and the sodium-rhodium [31] systems. The pyrazine molecule as well as the sodium atom are centrosymmetric systems and consequently have a vanishing $\alpha^{(2)}$ in the electric dipole approximation. However, when adsorbed on silver and rhodium respectively, they give rise to a fairly large SH signal. For example, the signal observed in the pyrazine on silver case leads to $\alpha^{(2)} \sim 10^{-29}$ esu, to be assigned to the pyrazine-silver system (i.e., half the corresponding value for pyridine-silver).

This clearly shows the importance of the molecule-metal interaction which breaks the inversion symmetry of the adsorbed species. In the case of pyrazine, this is in agreement with Raman (SERS) observations. It was observed that for pyrazine, IR active modes which are Raman inactive for the free molecule become Raman active upon adsorption on silver [35]. The enhancement of $\alpha^{(2)}$ for the adsorbed species is now thought to arise from an important charge transfer between the adsorbate and the substrate, especially in the case of a metal. This leads to an increase or a depletion of the charge density of the substrate surface layer.

c) Nonlinear spectroscopy of adsorbed monolayers

The SH signal due to a p-nitrobenzoic acid monolayer on sapphire is much weaker than that due to the pyridine-silver system when the laser wavelength is $\lambda=1.064\mu m$. It is much larger when the laser wavelength is $\lambda=532$ nm. Large signals were also obtained with rhodamine 6G adsorbed on glass starting with either $\lambda=1.064\mu m$ or $\lambda=683$ nm. This is due to a resonance effect. The second-order polarizability is given quantum mechanically by :

$$\alpha^{(2)}_{\mu\alpha\beta}(\omega,\omega) = \frac{1}{\hbar^2} \sum_{i,j} \left\{ \frac{<g|P_\mu|j><j|P_\alpha|i><i|P_\beta|g>}{(\omega_{jg}-2\omega)(\omega_{ig}-\omega)} + 5 \text{ similar terms} \right\} \qquad (31)$$

where P_μ, P_α, P_β are components of the electric dipole moment operator, $|g>$ is the ground state, $|i>$ and $|j>$ are excited states and $\hbar\omega_{ig} = E_i - E_g$. $\alpha^{(2)}$ therefore shows a resonant behavior whenever there exists an excited state $|e>$ such that

$$\omega_{eg} \simeq \omega \qquad \text{or} \qquad \omega_{eg} \simeq 2\omega$$

provided that the relevant matrix elements do not vanish.

The aforementioned enhancements are due to a two-photon resonance since p-nitrobenzoic acid absorbs near 266 nm while the first and second absorption peaks of rhodamine 6G fall near 532 nm and 341 nm, their exact position depending on the solvent.

One may take advantage of these resonances to study the nonlinear spectroscopy of adsorbed monolayers. The $S_2 \leftarrow S_0$ transition of rhodamine 6G and rhodamine 110 was studied in this way using a dye laser as the tunable fundamental source [36]. Close to a two-photon resonance the dominant resonant part of $\alpha^{(2)}$ reads

$$\alpha^{(2)}_R = \frac{1}{\hbar^2} \sum_i \frac{<g|P|e><e|P|i><i|P|g>}{(\omega_{eg}-2\omega+i\Gamma)(\omega_{ig}-\omega)} \qquad (32)$$

where (homogeneous) broadening has been taken into account. Fig.7 shows the first electronic nonlinear spectrum of a monolayer ever obtained. The dye laser was tuned from 600 to 730 nm and the figure shows clearly the $S_2 \leftarrow S_0$ transition and probably the wing of the $S_3 \leftarrow S_0$ one. From (32), it is clear that the transition has to be both one-photon and two-photon allowed.

A theoretically straightforward but experimentally difficult extension of this technique to sum-frequency generation would allow vibrational spectroscopy of adsorbed monolayers. By shining an intense fixed-frequency visible beam together with an IR tunable laser on the sample, the sum-frequency signal plotted versus IR wavelength or wavenumber would yield the vibrational spectrum of the monolayer. This would be a powerful technique of high sensitivity (submonolayer) and high resolution (only limited by the laser linewidth). This will probably be the most important application of the SHG technique to the study of surfaces.

d) Studying the orientation of adsorbed molecules

In the case of parallel molecules, the tensor $\chi^{(2)}$ expressed in the xyz reference frame of fig.4 would be related to the tensor $\alpha^{(2)}$ via a rotation. Knowledge of $\chi^{(2)}$ and $\alpha^{(2)}$ would then lead to the knowledge of the molecular orientation. Even when the orientation is not uniform, very useful information may be obtained. For all monolayers studied up to now, the SH signal did not vary when the sample was rotated about its normal : these monolayers show two-dimensional isotropy. Let us consider the case of p-nitrobenzoic acid [37] which is a rod-like molecule as shown

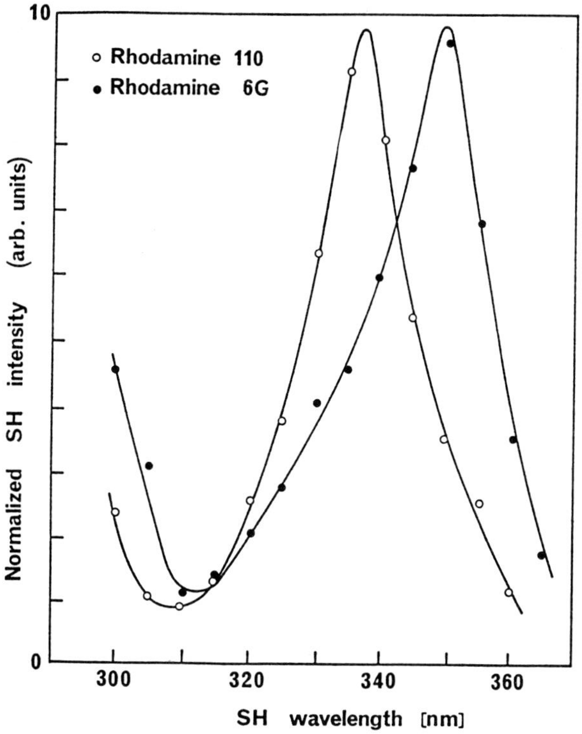

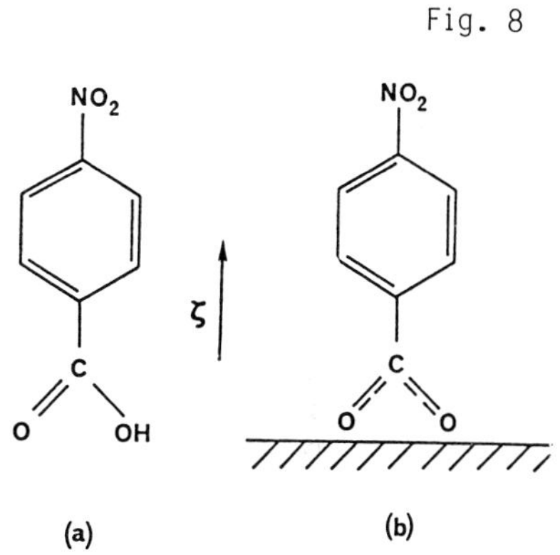

Fig.7. Nonlinear spectra ($S_2 \leftarrow S_0$ transition) of rhodamine molecules adsorbed at submonolayer coverage on a fused silica substrate (after Heinz et al. [36])

Fig.8. Structure of p-nitrobenzoic acid (a) as a free molecule and (b) as adsorbed on a fused silica surface. The ζ axis is shown (after Heinz et al. [37])

in fig.8. The molecular axis is labeled ζ. Its orientation is defined by the two angles θ and ψ. The axial symmetry implies that ψ is randomly distributed.

The only important component of the $\alpha^{(2)}$ tensor is $\alpha^{(2)}_{\zeta\zeta\zeta}$. The independent components of $\chi^{(2)}$ are easily calculated

$$\chi^{(2)}_{\parallel\perp\parallel} = \chi^{(2)}_{\perp\parallel\parallel} = \frac{1}{2} N_a \langle\cos\theta \sin^2\theta\rangle \alpha^{(2)}_{\zeta\zeta\zeta} \quad (33)$$

$$\chi^{(2)}_{\perp\perp\perp} = N_a \langle\cos^3\theta\rangle \alpha^{(2)}_{\zeta\zeta\zeta} \quad (34)$$

where θ is the angle between the molecular axis ζ and the surface normal z. To determine the average molecular orientation, one need only measure the ratio of the two independent components of $\chi^{(2)}$. This ratio is independent of N_a and $\alpha^{(2)}_{\zeta\zeta\zeta}$. One can, for example, measure the SH signal for two different settings of the laser polarization, p and s for example, or one can use a more elegant and more precise nulling technique.

In the case of p-nitrobenzoic acid, assuming a narrow θ distribution, the average θ was found to be 38° at the fused silica-ethanol interface and 70° at the fused silica-air interface. The molecules are then farther from the normal direction in air than in ethanol. This may be interpreted in terms of the dielectric screening of the interaction between the static dipole moments of the molecules or in terms of the solvation energy of p-nitrobenzoic acid in ethanol.

e) Two-dimensional phase transitions

The determination of molecular orientation may prove to be very useful in studying two-dimensional phase transitions. In an analogous way to the pressure - volume diagrams in 3 dimensions, a surface pressure (π) versus area per molecule ($A=1/N_a$) diagram may show 2D - phase transitions of a monolayer. Usually, for surfactant monolayer systems, the π-A curve in the pressure range 0.1 to 30 m N/m exhibits a discontinuity attributed to an orientational phase transition between an expanded and a condensed liquid phase.

For the SDNS surfactant on water, it was found [38] that the orientation changes continuously as a function of surface pressure π. No evidence of any phase transition was thus observed in agreement with the π-A curve. Despite this negative result, the SHG technique will certainly prove to be a useful tool in this field.

f) Structure of a surface

The polarization dependence of the SH signal may also be used to study the structural symmetry of a clean surface, i.e.,without any adsorbate. This was done for silicon [32] for which we already discussed the tensorial form of the bulk and surface layer nonlinear response. In particular, the study of the anisotropic components is of interest. These anisotropic components are the only ones to contribute to an s-polarized SH wave when the incident beam is either purely s- or p-polarized. For example, for a p-polarized beam incident on the (111) face, the s component of the SH field is

$$E_s^{(111)} \propto (e + \lambda d) \cos 3\theta \tag{35}$$

where e=y'y'y', λ is a constant determined by the linear properties and θ is now the angle through which the projection of the [100] crystal axis on the surface is rotated away from the normal to the plane of incidence. e gives the surface contribution and d the bulk one. For a p-polarized beam incident on the (100) face, the s component of the SH field is

$$E_s^{(100)} \propto d \sin 4\,\theta \tag{36}$$

where now θ measures the angle through which the [100] axis has been rotated away from the normal to the plane of incidence (the y axis). The surface layer does not contribute in this case.

When the sample is rotated about its normal, the SH signal, proportional to $|E|^2$, shows six identical lobes when θ is varied from 0 to 2π in the case of the (111) surface and eight identical lobes in the case of the (100) surface. The dependence of the SH intensity on θ shows the symmetry of the surface. The two measurements also allow comparison of the bulk and surface contributions : in this case, they are of comparable magnitude.

If the polarizer placed in front of the detector is no longer set for the s component, one has in addition a contribution from the isotropic part of the nonlinear response and a wealth of data may be accumulated by varying the polarization of the incident or harmonic beams.

When a single crystal is cleaved, the surface atoms usually do not assume the same position as in the bulk ; they relax to new positions. This phenomenon is known as surface reconstruction. In the case of the Si (111) surface, two such reconstructions are known, the 2x1 one which is metastable at room temperature and the 7x7 one. If they have different symmetries, they may be differentiated by polarization-dependent SHG measurements. This was found to be the case [39] with the 2x1 reconstruction showing only one mirror plane while the familiar 7x7 reconstruction has the full 3m symmetry. The transformation from the 2x1 reconstruction to the 7x7 one during a heat treatment could also be followed.

g) The adsorption sites

In some cases, a given molecule may adsorb at different sites on a surface. This is known to be true for CO on the Rh (111) surface. For a fractional coverage $\theta < 1/3$, CO bonds only to the top sites; for $1/3 < \theta < 3/4$, CO bonds to both the top sites and the bridge sites. If the nonlinear polarizability $\alpha^{(2)}$ is different for the two kinds of sites, the adsorption sites may be studied by the SHG technique. This is indeed what was observed for CO adsorption on the Rh(111) surface [31] by monitoring the SH signal as a function of time, i.e.,of surface coverage. A change in the slope of the curve was observed near $\theta = 1/3$. This experiment was performed in an UHV chamber equipped with conventional tools such as LEED and Auger electron spectroscopy. SHG is sensitive to adsorption of 5 % of a monolayer of O_2 or CO in this case.

In summary, second harmonic generation is a very powerful tool for surface studies. Numerous applications have already been demonstrated, sometimes in conjunction with other more conventional techniques. Compared to LEED, FIM, AES, EELS,... it does not require ultrahigh vacuum and it may consequently be used to study interfaces in their real-life "dirty" environment : the solid-liquid interface is such an example.

4 Optical phase conjugation in composite media

A composite medium is made of (at least) two different component media A and B, but it is not an alloy. They are not mixed at the atomic level. On the contrary , large pieces of either material A or B exist in the composite. One may find particles of A with sizes typically in the range 100 to 1000 Å embedded in material B or vice-versa. Composite media have been studied extensively from the point of view of the mechanical or electrical properties. For example, composites made of a dielectric and a metal behave as metals (good conductors) when the metal volume fraction exceeds a certain value known as the percolation threshold. A lot of work has dealt with that problem .

The linear optical properties of composite media have also been studied for a long time. The beautiful colors of metal colloids which consist of metal particles floating in a dielectric medium have been used in the art of stained glass for churchies or cathedrals. The first model aiming at describing the linear optical properties of such media was developed at the turn of the century [40]. Until very recently however, composite materials were not studied as nonlinear optical materials and the field of nonlinear optics in composite materials is still in its infancy.

We will restrict our discussion to the studies of optical phase conjugation in metal colloids and semiconductor-doped glasses. Because of the random orientation of the constituting particles, composite materials should not show sizable second-order effects and their main nonlinear response is third order. Optical phase conjugation in the degenerate four wave mixing configuration is an example of a third-order phenomenon [41]. Basically, three beams at the same frequency ω are incident on the nonlinear medium : two counterpropagating pump beams and a probe beam usually propagating at a small angle to the forward pump beam. The third-order response $\chi^{(3)}(\omega,-\omega,\omega)$ leads to the creation of a conjugate wave at frequency ω propagating in the direction opposite to that of the probe beam and the simultaneous amplification of the probe wave. This process is automatically phase-matched and reflectivities larger than 1 may be obtained.

4.1 Metal colloids

The linear optical properties of metal colloids which consist of small spherical metal particles embedded in a dielectric medium are quite different from that of the component materials. The embedding medium may be glass, water or any other dielectric. The work to be described was performed in aqueous suspensions or hydrosols [42]. In this case, a gold colloid is ruby red with an absorption band peaking at $\sim$ 520 nm and a silver colloid is yellow, the absorption band peaking at 400 nm. Their color is due to the surface plasma resonance which we already encountered in section 2.2.

Let us denote the metal dielectric constant $\varepsilon_m = \varepsilon'_m + i\varepsilon''_m$ and that of the dielectric ε_d. The latter is almost constant whereas ε_m is strongly frequency dependent. The linear properties of this composite material may be described in terms of an effective dielectric constant $\tilde{\varepsilon}$. Several models have been suggested [43] to calculate $\tilde{\varepsilon}$, but in the case of a small metal volume fraction p, they are all equivalent and we will use the Maxwell-Garnett result [40] :

$$\frac{\tilde{\varepsilon} - \varepsilon_d}{\tilde{\varepsilon} + 2\varepsilon_d} = p \, \frac{\varepsilon_m - \varepsilon_d}{\varepsilon_m + 2\varepsilon_d} \tag{37}$$

This result may be obtained in several ways, for example, using a local field argument [44]. It turns out to be very often a good description of the observed properties. For $p << 1$, we will content ourselves with the approximate form

$$\tilde{\varepsilon} = \varepsilon_d + 3p\,\varepsilon_d \, \frac{\varepsilon_m - \varepsilon_d}{\varepsilon_m + 2\varepsilon_d} \tag{38}$$

correct to first order in p. The absorption coefficient of the colloid, related to the imaginary part of $\tilde{\varepsilon}$, is easily deduced from (38) :

$$\alpha = 9\,p\,\frac{\omega}{c}\,\varepsilon_d^{3/2}\,\frac{\varepsilon''_m}{(\varepsilon'_m + 2\varepsilon_d)^2 + \varepsilon''^2_m} \tag{39}$$

In the case of a hydrosol, the incident beam may either be absorbed or scattered (Mie scattering). It turns out that when the diameter d of the spheres is small compared to the wavelength, which is the case since $d \sim 100$ Å, the scattering losses are negligible. The absorption (or extinction) coefficient given by (39) could also be obtained as the leading term of the Lorentz-Mie theory [45]. From the form of (39), the absorption coefficient shows a resonance when

$$\varepsilon'_m = -2\varepsilon_d \tag{40}$$

which is the same as(6) with our new notations. In fact, in the vicinity of this resonance, ε''_m is roughly constant and ε'_m varies in a first approximation linearly with ω so that the absorption band is roughly Lorentzian. The absorption spectra of the gold and silver colloids are shown in fig.9. The structure observed on the high frequency side is due to interband contributions to $\varepsilon_m(\omega)$ but the main peak is due to the surface-mediated resonance and is a translation to the visible of the resonance that normally occurs at zero frequency (free electrons).

The spectrum of the gold colloid can be well accounted for in terms of (39) and the measured [46] dielectric constant of pure gold. For the silver colloid, a correction had to be introduced as the spectrum calculated from (39) was sharper than the measured one. This is due to the fact that ε''_m is larger (here by a factor 5.5) for a small silver particle than in bulk silver. The limited size effect may be taken into account quantum mechanically [47] (quantum size effect) or classically (limited mean free path), leading to the same conclusion [48]. The classical argument, for example, leads to a collision time τ given by

$$\frac{1}{\tau} = \frac{1}{\tau_b} + 2\,\frac{v_F}{d} \tag{41}$$

where τ_b is the bulk value for a large specimen and v_F is the Fermi velocity. τ enters the formula giving the free electron contribution to ε_m as

$$\varepsilon_m(\omega) = 1 - \frac{\omega_p^2}{\omega(\omega + i/\tau)}$$

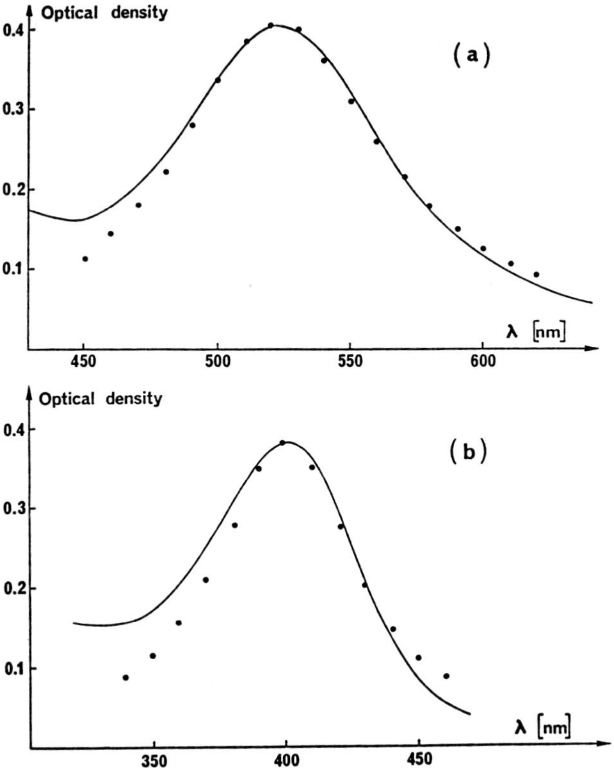

Fig.9. Absorption spectra of (a) the gold colloid and (b) the silver colloid. The solid line corresponds to the measurement and the circles to the calculation. In the calculation, only the free electron contribution is taken into account.

where ω_p is the plasma frequency. The best fit to the silver colloid absorption spectrum is obtained for a diameter d $\sim$ 40 Å. Transmission electron micrographs show gold and silver particles of 100 and 80 Å diameter respectively.

We now consider optical phase conjugation which is related to a change in the dielectric constant $\tilde{\varepsilon}$. This change $\delta\tilde{\varepsilon}$ is due to a change $\delta\varepsilon_m$ of ε_m. Differentiating (38), we obtain

$$\delta\tilde{\varepsilon} = f_2 \, p \, \delta\varepsilon_m \tag{42}$$

with

$$f_2 = \left(\frac{3 \, \varepsilon_d}{\varepsilon_m + 2 \, \varepsilon_d}\right)^2 \tag{43}$$

This formalism is completely general, even when $\delta\varepsilon_m$ is due to a slow, not truly $\chi^{(3)}$, process. When $\delta\varepsilon_m$ is due to a $\chi^{(3)}$ process, it is given by :

$$\delta\varepsilon_m = 12\pi \, \chi_m^{(3)} \, E_{in}^2 \tag{44}$$

E_{in} being the field inside the metal sphere. Applying (3), we have

$$E_{in} = \frac{3 \, \varepsilon_d}{\varepsilon_m + 2\varepsilon_d} \, E_o = f_1 \, E_o \tag{45}$$

where E_o is the applied field and f_1 is the local field factor of interest here since we deal with the metal nonlinearity. The cubic nonlinear polarization is thus found to be :

$$P^{(3)} = 3p\, f_1^2\, f_2\, \chi_m^{(3)}\, E_o^3 \tag{46}$$

Since $f_2 = f_1^2$, we recover in this way the usual [49] local field correction $(f_1(\omega))^4$ since all fields oscillate at ω. To be more specific, the nonlinear polarization giving rise to the conjugate wave is

$$P^{(3)}(\omega) = 3\, p[f_1(\omega)]^4\, \chi_m^{(3)}(\omega,-\omega,\omega)\, E_f\, E_p^*\, E_b \tag{47}$$

where E_f, E_p and E_b are the amplitudes of the forward pump, probe and backward pump fields respectively.

Optical phase conjugation being a coherent process, the radiated power is proportional to $|P^{(3)}(\omega)|^2$. The intensity of the phase conjugated wave is thus enhanced by the factor $|f_1(\omega)|^8$. It is clear from (45) that f_1 shows a resonant behavior when $\varepsilon'_m + 2\varepsilon_d = 0$, i.e., once again at the surface plasma resonance frequency. In fact, simple numerical estimates show that close to resonance, $|f_1(\omega)| > 1$ and the nonlinear response of the small metallic particles is enhanced, whereas away from resonance, especially far from resonance, $|f_1(\omega)| < 1$ and the nonlinear response of the metal particles is then attenuated. When we compare a resonant situation to a nonresonant one, we may define a relative enhancement factor F :

$$F = |f_1^R|^8 \,/\, |f_1^{NR}|^8 \tag{48}$$

R and NR standing for resonant and nonresonant respectively.

Close to resonance, the incident and conjugate beams will suffer absorption losses. In such a case, when the reflectivity of the phase conjugate mirror is small, this reflectivity is smaller by the factor [50]

$$4\, e^{-\alpha L} \,/\, (\alpha L + 2)^2 \tag{49}$$

where α is the absorption coefficient given by (39) and L is the interaction length.

The experiment was performed on gold and silver colloids [44] starting with a single laser pulse delivered by a mode-locked Nd:YAG laser of pulse duration 28 ps using the conventional setup shown in fig.10. Variable delay lines allow the experimentalist to delay any one of the three incoming pulses with respect to the other two and to time resolve the nonlinear response. The volume fraction p was 5×10^{-6} for the gold colloid and 1.5×10^{-6} for the silver one. These hydrosols were placed in a 2mm thick glass cell and all measurements were performed relative to a CS_2 reference cell. The gold colloids were studied at resonance at λ=532 nm and off resonance at λ=1.064μm. For the silver colloid, the wavelengths λ=396 nm (resonant) and λ=532 nm (nonresonant) were used.

Close to resonance, the nonlinear response is dominated by the metal spheres and the reflectivity is large (comparable to that of CS_2). On the other hand, away from resonance, the metal spheres contribution is negligible compared to that of water. Taking into account the absorption losses according to (49) and also the dispersion of the CS_2 nonlinear susceptibility, the relative enhancement factor F is found to be larger than 3×10^3 for the gold colloid and larger than 5×10^3 in the case of silver. These lower bounds are in agreement with theory which predicts a relative enhancement factor of 1.3×10^{10} for gold and 3.6×10^6 for silver. The conjugate signal intensity was measured as a function of the delay of any one of the pump pulses as shown in fig.11. The nonlinear response is found to be predominantly fast.

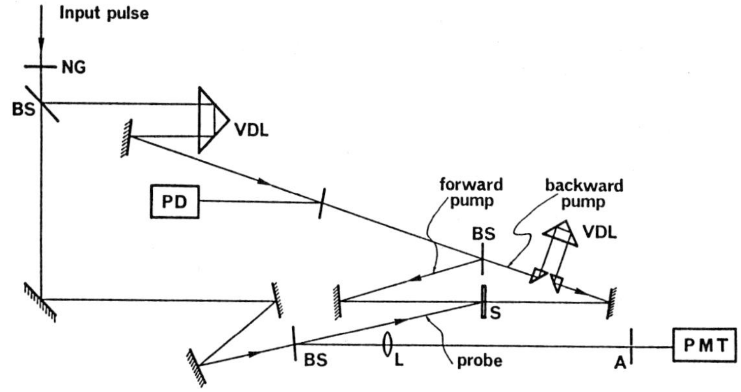

Fig.10. Typical experimental setup usd in optical phase conjugation studies. NG : neutral density filters, BS : beam splitters, VDL : variable delay lines, S : sample, PD : reference photodiode, PMT : photomultiplier tube.

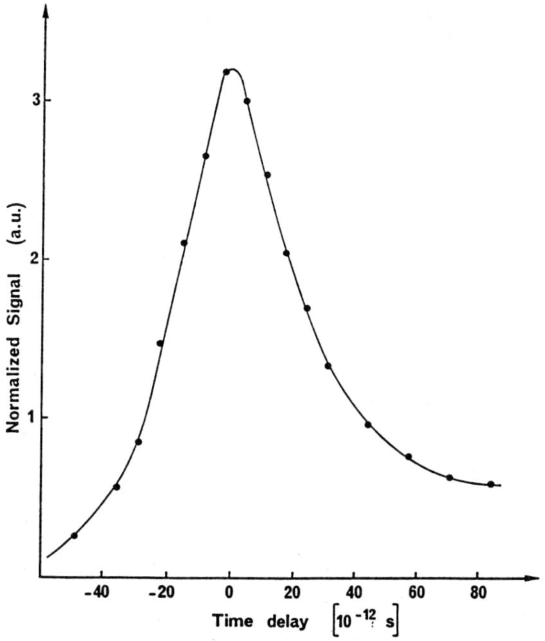

Fig.11. Normalized conjugated signal from the gold colloid as a function of the forward or backward pump pulse delay.

Using (47) and known values of $\chi^{(3)}$ for CS_2 [51] , the nonlinear susceptibility $\chi_m^{(3)}(\omega,-\omega,\omega)$ of the metal spheres was deduced from the measurements and the following results were obtained at resonance :

$$\chi_{Au}^{(3)}(\omega,-\omega,\omega) = 1.5 \times 10^{-8} \text{ esu}$$

$$\chi_{Ag}^{(3)}(\omega,-\omega,\omega) = 2.4 \times 10^{-9} \text{ esu}$$

for gold and silver respectively. These values are about 2 orders of magnitude larger than $\chi^{(3)}$ values obtained by third harmonic reflection [52]. Theory says that the nonlinearity of free electrons is small but, in the case of small particles, because of the quantum size effect, it may become sizable.

The surface-mediated enhancement which was observed in this work has important implications. In the case of optical phase conjugation, the enhancement is a maximum since all four waves have the same frequency ω. On the other hand, in the case of third harmonic generation, for example, the signal at 3ω would be "enhanced" by the factor $|f_1(3\omega)|^2 \times |f_1(\omega)|^6$ and even with a resonance at 3ω the fact that $|f_1(\omega)| < 1$ would certainly lead to an attenuation instead of an enhancement effect. Such difficulties should be kept in mind.

4.2. Semiconductor-doped glasses

Some color filter glasses such as the sharp cutoff long-wavelength bandpass filters are composite materials. They are made of glass containing semiconductor microcrystals. When the cutoff wavelength is in the yellow to red part of the spectrum, the semiconductor is $CdS_x Se_{1-x}$ [53] . By varying the composition parameter x, the band gap of the mixed semiconductor and consequently the corresponding cutoff frequency may be tuned continuously from the CdS value, 2.55 eV, to the CdSe one, 1.82 eV (at low temperature). They are available from different manufacturers. The first nonlinear optical study of a composite material was on one of these glasses : the Corning 3.68 glass [54]. Absorption saturation had also been observed in this type of glass [55].

Jain and Lind [54] studied optical phase conjugation in the crystal CdS and in the Corning 3.68 glass using a setup similar to the one shown in fig.10 but using nanosecond pulses from Q-switched lasers. For the Corning 3.68 glass, we have $x \sim 0.9$, and the cutoff wavelength is close to that of the frequency-doubled Nd : YAG laser. At room temperature, these glasses show a fairly broad exponential tail in the absorption spectrum. This lack of structure makes plausible a very weak frequency dependence of the dielectric constant of the microcrystals and we will not worry about the local field factor f_1, simply assuming it to be about constant. We will concentrate on the nonlinearity of the semiconductor microcrystals [56] .

In isotropic media, the nonlinear polarization responsible for optical phase conjugation may be written :

$$\vec{P}^{NLS} = A(\theta)\,(\vec{E}_f\cdot\vec{E}_p{}^*)\,\vec{E}_b + A(\pi-\theta)\,(\vec{E}_b\cdot\vec{E}_p{}^*)\,\vec{E}_f + B(\vec{E}_f\cdot\vec{E}_b)\,\vec{E}_p{}^* \qquad (50)$$

where A and B are related to the nonzero components of $\chi^{(3)}$ and θ is the angle between the forward pump and the probe wavevectors. The first two terms correspond to holographic gratings : the forward pump and the probe beams create a large-spaced index grating which diffracts the backward pump (first term), the backward pump and the probe beams create a small-spaced index grating which diffracts the forward pump (second term). The third term corresponds to a nonlinear index which has no spatial modulation and which oscillates at frequency 2ω.

Starting with parallel polarizations for the three input beams, one may, by crossing the polarization of one of them, assess the relative importance of the 3 terms in (50). An important finding in Ref. [54] is that in pure CdS, the only important contribution comes from the first term whereas in glass, the third term does not contribute but the first two give similar contributions to the conjugate signal. This is due to the fact that free carriers (which are responsible for the index gratings) diffuse rapidly in a semiconductor. Diffusion causes the small-spaced grating to disappear rapidly in the pure CdS sample whereas in the composite, diffusion is hampered by the small size of the microcrystals (less than 1000 Å) and the small-spaced grating contributes about as much as the large-spaced one.

The response time of the nonlinearity was also studied and found to be fast, namely subnanosecond. The luminescence decay time was also found to be very short,

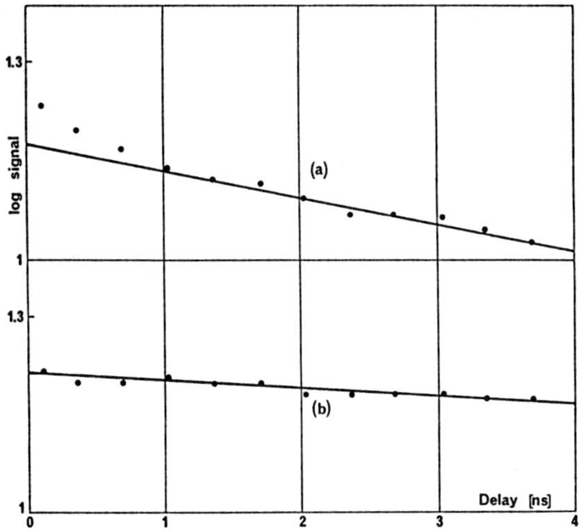

Fig.12. Semi-log plot of the conjugate signal as a function of the backward pump pulse delay for (a) the Corning 3.68 glass and (b) the Schott OG 530 glass

also subnanosecond. Similar findings have also been obtained recently for various similar glasses [57] . The reflectivity of the phase conjugate mirror was also studied as a function of the pump beam intensity or fluence. The two pump beams being derived from the same incident beam, the reflectivity normally shows a quadratic dependence as a function of the fluence. In the case of CdS, this quadratic dependence was indeed observed up to a certain level where the reflectivity saturates. In the case of the Corning 3.68 glass, the quadratic dependence was not observed, the range under study - from ∿ 1 to ∿ 100 mJ/cm^2 - corresponding almost completely to the saturated regime. The saturated reflectivity is about 10 %.

More recently [58], optical phase conjugation was studied in the 3.68 glass and in the corresponding Schott glass OG 530 using picosecond pulses from a frequency-doubled mode-locked Nd : YAG laser. The time behavior of the nonlinear response was studied by delaying the backward pump pulse . The response was found to be slow as shown in fig.12. In the case of the Schott OG 530 glass, an exponential decay of the signal is observed with a time constant of ∿40 ns. In the case of the Corning 3.68 filter, the response is mainly slow with a time constant of ∿ 11 ns and also shows a small faster component (∿ 0.5 ns). Due to experimental uncertainties, the slow time constants could even be longer. The time behavior of the luminescence from these glasses was also studied. For the 3.68 filter, there is a fast component but also a slow one with a time constant of ∿ 40 ns. The OG 530 filter only shows a slow nonexponential luminescence with a typical time constant of ∿ 400 ns.

In the same study, the reflectivity was measured as a function of the pump fluence for small fluences, down to 15 μJ/cm^2. The unsaturated quadratic regime was clearly observed as well as the onset of saturation which occurs at a lower fluence for the Corning 3.68 than for the Schott OG 530 as can be seen in fig.13. By measuring the transmission of the filter at λ=532 nm as a function of the incident fluence, absorption saturation was also observed for these two glasses. Absorption and reflectivity seem to start saturating at the same fluence. This fact together with the similar slow luminescence and slow nonlinear response led to a tentative interpretation in terms of optical phase conjugation by a saturable absorber. This process had already been considered theoretically [59,60].

Several groups are presently pursuing the study of optical nonlinear properties of semiconductor-doped glasses. Among others, the following topics are currently under investigation : the frequency dependence of the effective Kerr susceptibi-

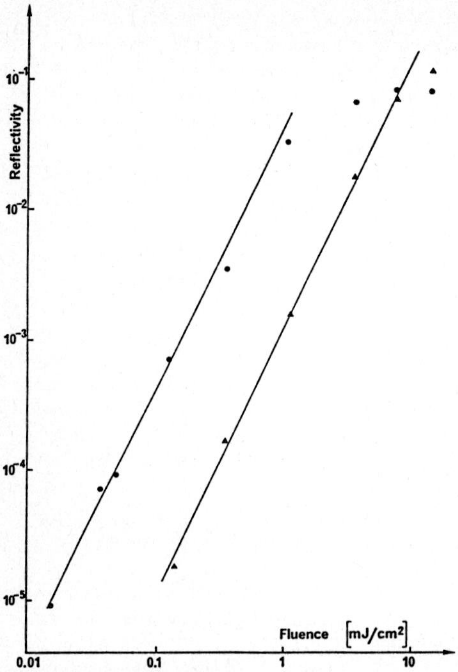

Fig.13. Log-log plot of the reflectivity of the phase conjugate mirror versus the pump fluence for the Corning 3.68 glass (circles) and the Schott OG 530 glass (triangles). Straight lines of slope 2 showing the unsaturated behavior are also indicated

lity of these media, their possible use as bistable devices or guided nonlinear optical devices. The results reported up to now were all obtained at room temperature. It would be of interest to extend these studies to low-temperature conditions.

In conclusion to this section, it is clear from the first results of nonlinear studies of composite materials that they show interesting properties but that several points remain to be elucidated before one has a clear and complete understanding of these materials. For these reasons, the field of optical nonlinearities of composite materials will undoubtedly grow in the next few years. As an example, a nonlinear spectroscopic study of colloidal gold has recently been reported [61].

Acknowledgements The author gratefully acknowledges fruitful collaboration with C.K. Chen, T.F. Heinz, Y.R. Shen, P. Roussignol, J. Lukasik, C. Flytzanis and F. Hache.

References

1 See for example G.A. Somorjai : Chemistry in Two Dimensions : Surfaces (Cornell University Press, Ithaca, 1981)
2 M. Fleischmann, P.J. Hendra and A.J. McQuillan : Chem.Phys.Lett. 26, 163 (1974)
3 D.L. Jeanmaire and R.P. Van Duyne : J. Electroanal.Chem. 84, 1 (1977)
4 M.G. Albrecht and J.A. Creighton : J.Am.Chem.Soc. 99, 5215 (1977)
5 U. Wenning, B. Pettinger and H. Wetzel : Chem.Phys.Lett. 70, 49 (1980)
B. Pettinger, U. Wenning and H. Wetzel : Surf.Sci. 101, 409 (1980)
C.S. Allen, G.C. Schatz and R.P. Van Duyne : Chem.Phys.Lett., 75, 201 (1980)

6 J.E. Demuth and P.N. Sanda : Phys.Rev.Lett. 47, 57 (1981)
7 R.P. Van Duyne, in Chemical and Biochemical Applications of Lasers, vol IV, C..B Moore ed. (Academic, New York, 1979)
8 M. Moskovits : J.Chem.Phys. 69, 4159 (1978)
9 B. Pettinger, U. Wenning and D.M. Kolb : Ber.Bunsenges Physik.Chem. 82, 1326 (1978)
10 J. Creighton, M. Albrecht, R. Hester and J. Matthew : Chem.Phys.Lett. 55, 55 (1978)
11 J.E. Rowe, C.V. Shank, D.A. Zwemer and C.A. Murray : Phys.Rev.Lett. 44, 1770 (1980)
12 C.K. Chen, A.R.B. de Castro and Y.R. Shen, Phys.Rev.Lett. 46, 145 (1981)
13 S.L. Mc Call, P.M. Platzman and P.A. Wolff : Phys.Lett. 77A, 381 (1980)
14 J.D. Jackson : Classical Electrodynamics (Wiley, New York, 1975)
C.J.F. Böttcher : Theory of Electric Polarization (Elsevier, Amsterdam, 1973)
15 J. Gersten and A. Nitzan: J.Chem.Phys. 73, 3023 (1980)
16 G. Arfken : Mathematical Methods for Physicists (Academic, New York 1970)
17 P.F. Liao and A. Wokaun : J.Chem.Phys. 76, 751 (1982)
18 A. Wokaun, J.P. Gordon and P.F. Liao : Phys.Rev.Lett. 48, 957 (1982)
19 P.F. Liao, J.G. Bergman, D.S. Chemla, A. Wokaun, J. Melngailis, A.M. Hawryluk and N.P. Economou : Chem.Phys.Lett. 82, 355 (1981)
20 R.K. Chang and T.E. Furtak : Surface Enhanced Raman Scattering (Plenum, New York, 1982)
21 W.H. Weber and G.W. Ford : Opt.Lett. 6, 122 (1981)
22 H.W.K. Tom, C.K. Chen, A.R.B. de Castro and Y.R. Shen : Solid State Comm. 41, 259 (1982)
23 Y.R. Shen : The Principles of Nonlinear Optics (Wiley, New York, 1984)
24 J. Ducuing and N. Bloembergen : Phys.Rev.Lett. 10, 474 (1963)
25 F. Brown, R.E. Parks and A.M. Sleeper : Phys.Rev.Lett. 14, 1029 (1965)
26 N. Bloembergen, R.K. Chang, S.S. Jha and C.H. Lee : Phys. Rev. 174, 813 (1968)
27 N. Bloembergen and P.S. Pershan : Phys.Rev. 128, 606 (1962)
28 P.N. Butcher : Nonlinear Optical Phenomena (Ohio State University, Columbus, 1965)
29 J.M.Chen, J.R. Bower, C.S. Wang and C.H. Lee : Opt.Comm. 9, 132 (1973)
30 C.K. Chen, T.F. Heinz, D. Ricard and Y.R. Shen : Phys.Rev.Lett. 46, 1010 (1981)
31 H.W.K. Tom, C.M. Mate, X.K. Zhu, J.E. Crowell, T.F. Heinz, G.A. Somorjai and Y.R. Shen : Phys.Rev.Lett. 52, 348 (1984)
32 H.W.K. Tom, T.F. Heinz and Y.R. Shen : Phys.Rev.Lett. 51, 1983 (1983)
33 C.K. Chen, T.F. Heinz, D. Ricard and Y.R. Shen : Chem.Phys.Lett. 83, 455 (1981)
34 T.F. Heinz, C.K. Chen, D. Ricard and Y.R. Shen : Chem.Phys.Lett. 83, 180 (1981)
35 G.R. Erdheim, R.L. Birke and J.R. Lombardi : Chem.Phys.Lett. 69, 495 (1980)
36 T.F. Heinz, C.K. Chen, D. Ricard and Y.R. Shen : Phys.Rev.Lett. 48, 478 (1982)
37 T.F. Heinz, H.W.K. Tom and Y.R. Shen : Phys.Rev.A. 28, 1883 (1983)
38 Th. Rasing, Y.R. Shen, M.W. Kim, P. Valint and J. Bock : Phys.Rev.A. 31, 537 (1985)
39 T.F. Heinz, M.M.T. Loy and W.A. Thompson : Phys.Rev.Lett.54, 63 (1985)
40 J.C. Maxwell-Garnett : Philos-Trans.R.Soc.London 203, 385 (1904) and 205, 237 (1906)
41 R.A. Fisher : Optical Phase Conjugation (Academic, New York, 1983)
42 D. Ricard, P. Roussignol and C. Flytzanis : to be published
43 J.A.A.J. Perenboom, P. Wyder and F. Meier: Physics Reports 78, 173 (1981)
44 R. Landauer in Electrical Transport and Optical Properties of Inhomogeneous Media J.C. Barland and D.B. Tanner eds. (American Institute of Physics, New York, n°40)
45 G. Mie : Ann.Phys. (Leipzig) 25, 377 (1908)
46 P.B. Johnson and R.N. Christy : Phys.Rev. B 6, 4370 (1972)
47 R. Ruppin and H. Yatom : Phys.Stat.Sol. (b) 74, 647 (1976)
48 R.H. Doremus : J.Chem.Phys. 42, 414 (1965)
49 J.A. Armstrong, N. Bloembergen, J. Ducuing and P.S. Pershan : Phys.Rev. 127, 1918 (1962)
50 D.M. Pepper and A. Yariv : in Optical Phase Conjugation, R.A. Fisher ed. (Academic, New York, 1983) p. 41

51 R.W. Hellwarth : Prog. Quant. Electr. 5, 1 (1977)
52 W.K. Burns and N. Bloembergen : Phys. Rev. B 4, 3437 (1971)
53 L. Levi : Applied Optics (Wiley, New York, 1980) vol 2, p 31 and references therein
54 R.K. Jain and R.C. Lind : J.Opt.Soc.Am. 73, 647 (1983)
55 G. Bret and F. Gires : Comptes Rendus Acad. Sci. 258, 3469 (1964)
56 see R.K. Jain and M.B. Klein in Optical Phase Conjugation, R.A. Fisher ed. (Academic, New York, 1983) p 307
57 S.S. Yao, C. Karaguleff, A. Gabel, R. Fortenberry, C.T. Seaton and G.I. Stegeman : Appl. Phys.Lett. 46, 801 (1985)
58 P. Roussignol, D. Ricard, K.C. Rustagi and C. Flytzanis : to be published in Optics Communications
59 R.L. Abrams and R.C. Lind : Opt. Lett. 2, 94 (1978) and erratum 3, 205 (1978)
60 Y. Silberberg and I. Bar-Joseph : IEEE J.Quantum Electron. QE-17, 1967 (1981)
61 E.J. Heilweil and R.M. Hochstrasser : J.Chem.Phys. 82, 4762 (1985)

Part IV

Optical Bistability and Instabilities in Nonlinear Optical Devices

Semiconductor Optical Bistability: Towards the Optical Computer

B.S. Wherrett

Department of Physics, Heriot-Watt University,
Edinburgh EH14 4AS, Scotland, U.K.

1. INTRODUCTION

Optical computing was first discussed seriously in the 1960's. In the context of the nonlinear optical devices available at that time, however, it was concluded that digital optical computational techniques could not hope to compete with electronics. Since that time there have of course been great advances in electronic machines. Nevertheless the possibility of developing viable optical computing systems returned to prominence in the early 1980's.

One key to the renewed interest in optical systems has been the recognition that there are materials and devices for which one can mimic (and expand upon) the responses of electronic transistors, using optical beams at low power levels and employing small-sized samples. In particular, optical bistability and associated transphasor action have been shown to occur in semiconductor devices with milliwatt and microwatt holding and switching beams respectively. In addition, it is now recognised that the major restriction to electronic computer power now lies in the lack of ability to interconnect in parallel to the vast numbers of electronic switches that can be fabricated on semiconductor chips. The ability of optical beams to interconnect between any two planes in free space is seen as of major significance in this respect.

In this article the fundamental properties of semiconductor optically bistable elements are discussed, the experimental observations of bistable switching of cw laser beams in both the infrared and visible spectral regions are summarised and device applications, including some speculation for optical computer architectures, are presented.

There are many parameters that one might wish to optimise in a given bistable device - switching irradiance, power, energy, speed, packing density, power dissipation, cascadability, etc. In chapters 2, 3 we shall concentrate on the achievement of minimum switching irradiance and minimum power levels - by optimising the semiconductor material, the device configuration and the operational laser frequency. The physical origins of the optical responses are introduced as they become vital to such optimization.

Response characteristics required for digital computation are schemed in figure 1. An inverter (NOR or NAND gate), nonhysteretic response is used for logic operation. A hysteresis loop is used for memory; for the same input level two stable outputs are possible. The actual output at a given time depends on the history of the device; hence the term bi-stability. In

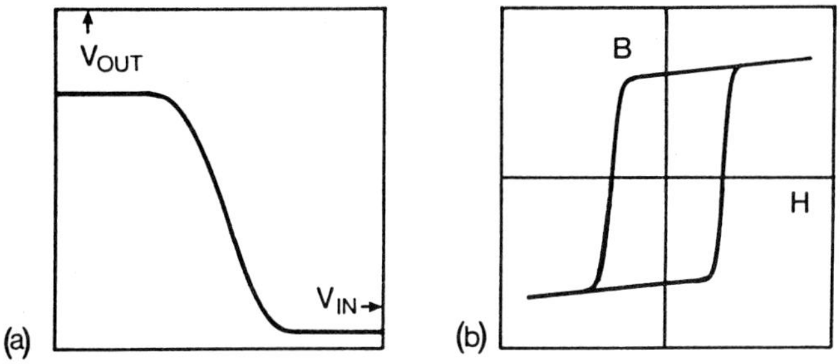

Fig. 1. Input/output inverter and switch characteristics

electronics the inverter and memory parameters are voltage levels or perhaps magnetic field and induction.

It was in the early 1960's [1] that the first suggestions of optical responses of a bistable nature were put forward; the first analysis of optical bistability was not made until 1969 [2]. Observations of intrinsic optical bistability, involving Fabry-Perot feedback (optically) rather than hybrid feedback via electronics, began to be reported in the late 1970's, firstly in gases [3] then solids [4] and liquids [5]. Table 1 shows a large number of those materials in which intrinsic bistability has now been observed. Significantly, and particularly in small semiconductor devices, most observations have required laser irradiance levels in excess of 1 kW cm^{-2}, by implication demanding pulsed laser sources.

The first observation of intrinsic optical bistability under low power, cw conditions in a small sample was for the narrow gap semiconductor indium antimonide, (InSb); 5 μm CO laser irradiation was used [6], figure 2. The ability to hold cw (indefinitely) on one branch within the bistable loop and then switch to the other branch on demand using a signal beam gives great flexibility to the device potential of a bistable system. In this article the two cw systems that have been demonstrated to operate most successfully to date (InSb at 5 μm and 77 K and ZnSe interference filters at .5 μm and room temperature) are used as example cases for bistability discussion.

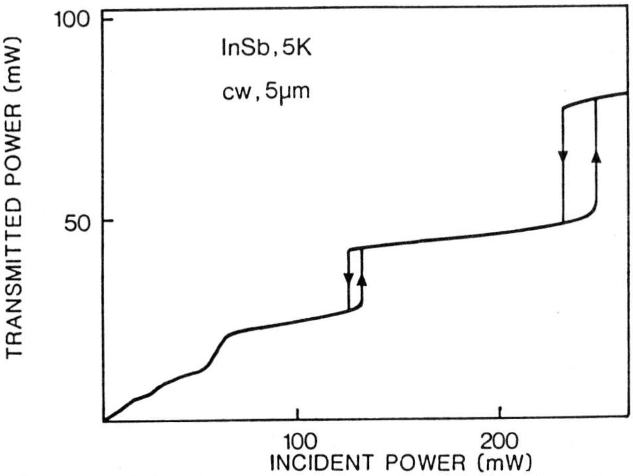

Fig. 2. First cw observation of semiconductor optical bistability

Table 1. Semiconductors in which Intrinsic Optical Bistability has been observed

Semiconductor	Operating Temperature	Operating Wavelength λ	Irradiance	Reference
InSb	5 K	5.4 μm	.3 kW/cm^2	a
InSb	77-120	5.5	.030	b
InSb	77	5.5	.06	c
InSb	300	10.6	200	d
GaAs	5-120	0.82	100	e
GaAs	300	0.857	100	f
GaAs/GaAlAs	300	0.83	100	f
GaAs/GaAlAs	300	0.857	.9	g
CdS	2	0.487	.3	h
CdS	2	0.489	600	i
CdS	2-50	0.487	1	j
$Cd_{0.23}Hg_{0.77}Te$	77	10.6	4	k
$Cd_{0.23}Hg_{0.77}Te$	300	10.6	100	l
InAs	77	3.1	.075	m
$CuCl_2$	17	0.3900	10 K	n,o
Te	300	10.6/5.3	10 K	p
Si	300	1.06	1	q
$GeSe_2$	300	0.6328	.1	r
ZnSe	300	0.5145	.25	s
ZnSe	300	0.5145	1	t
ZnS	300	0.5145	10	u

References for Table 1

a D.A.B. Miller, S.D. Smith and A.M. Johnston
Appl. Phys. Lett., 35, 658 (1979).

b F.A.P. Tooley
Ph.D. Thesis, Heriot-Watt University (unpublished) 1984.

c B.S. Wherrett, F.A.P. Tooley and S.D. Smith
Opt. Commun., 52, 301 (1984).

d A.K. Kar, J.G.H. Mathew, S.D. Smith, B. Davis and W. Prettl
Appl. Phys. Lett., 42, 4 (1983).

e H.M. Gibbs, S.L. McCall, T.N.C. Venkatesan, A.C. Gossard, A. Passner and W. Wiegmann
Appl. Phys. Lett., 35, 6 (1979).

f H.M. Gibbs, J.L. Jewell, N. Peyghambarian, M.C. Rushford, M. Tai, S.S. Tarng, D.A. Weinberger, A.C. Gossard, W. Wiegmann and T.N.C. Venkatesan
Proc. Discussion Meeting "Optical Bistabiity, Dynamical Nonlinearity and Photonic Logic", London, 1984, Ed. B.S. Wherrett and S.D. Smith, p. 55, The Roy. Soc. (1985).

g D.A.B. Miller, A.C. Gossard and W. Wiegmann
Opt. Lett., 9, 162 (1984).

h M. Dagenais and H.G. Winful
Appl. Phys. Lett., 44, 6 (1984).

i K. Bohnert, H. Kalt and C. Klingshirn
Appl. Phys. Lett., 43, 12 (1983).

j M. Dagenais and W.F. Sharfin
Appl. Phys. Lett., 45, 3 (1984).

k A. Miller, G. Parry and R. Daley
IEEE JQE, QE-20, 7 1984).

l J.G.H. Mathew, D. Craig and A. Miller
To be published Appl. Phys. Lett. (1984).

m C.D. Poole and E. Garmire
Appl. Phys. Lett., 44, 4 (1984).

n N. Peyghambarian, H.M. Gibbs, M.C. Rushford and D.A. Weinberger
Phys. Rev. Lett., 51, 18 (1983).

o R. Levy, J.Y. Bigot, B. Hönerlage, F. Tomasini and J.B. Grun
Solid State Commun., 48, 8 (1983).

p G. Staupendahl and K. Schindler
Opt. Quant. Electron., 14, 157 (1982).

q H.J. Eichler
Opt. Commun., 45, 1 (1983).

r J. Hajto and I. Janossy
Phil. Mag. B, 47, 4 (1983).

s S.D. Smith, J.G.H. Mathew, M.R. Taghizadeh, A.C. Walker, B.S. Wherrett and A. Hendry
Opt. Commun., 51, 357 (1984).

t A.K. Kar and B.S. Wherrett
To be published JOSA (1985).

u F.V. Karpushko and G.V. Sinitsyn
J. Appl. Spect. (U.S.S.R.), 29 (1978).

2. MACROSCOPIC THEORY OF NONLINEAR REFRACTION AND FABRY-PEROT OPTICAL BISTABILITY

2.1 Nonlinear refraction through self-defocussing

During 1976, following a lengthy study of the so-called spin-flip Raman process in InSb, it was realised that the Raman efficiency was dramatically influenced by nonlinear optical effects which were strongly enhanced at frequencies just below the band edge of this narow gap semiconductor. In particular, an incident cw beam of Gaussian spatial profile, whilst being transmitted as a Gaussian at very low powers began to break up spatially at moderate levels of around 100 mW, figure 3a. The interpretation and analysis of this observation led us to recognise the giant nonlinear refraction of InSb. The local semiconductor refractive index (n) at position r is taken to contain a nonlinear contribution, proportional to the local laser irradiance level (I):

$$n(\underset{\sim}{r}) = n_0 + n_2\, I(\underset{\sim}{r}) \tag{1}$$

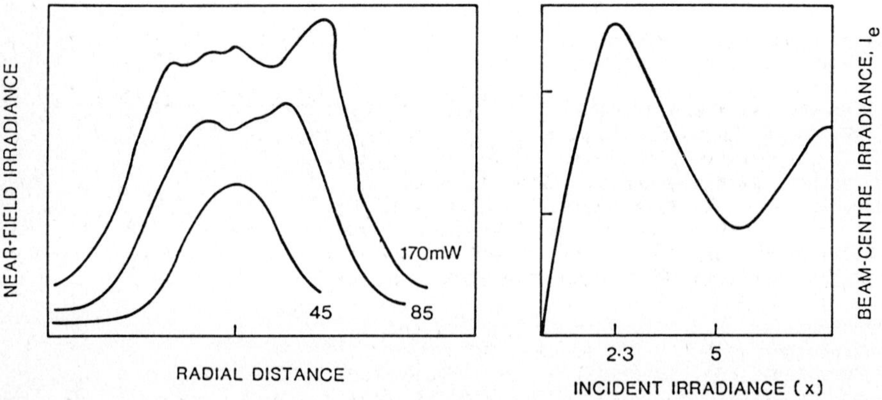

Fig. 3. a) Input-power dependence of near-field InSb transmission profiles; b) Beam-centre irradiance of defocussed Gaussian beam

In Chapter 3 the physical origin of the nonlinear coefficient n_2 is described; for the present we note that it is negative in the InSb case and that we are interested in n_2I values of order 10^{-3} n_0. I(r) has a Gaussian cross-section; thus in principle an index that diminishes towards the beam centre is established. This corresponds to a concave lens effect, and induces a beam defocussing. In practice, providing thin samples are used the defocussing is not strong within the medium but a phase-variation is established across the beam at the emergent face of the sample [7]. The emergent field is

$$E_e = E_0 \exp(-\alpha D/2) \exp(-r^2/r_0{}^2) \exp(ik_eD) \tag{2}$$

$$k_e = (2\pi n_0/\lambda_v)+(2\pi n_2/\lambda_v)\ (I_0a/\alpha D) \exp(-2r^2/r_0{}^2)$$

$$= k_0 + x \exp(-2r^2/r_0{}^2) \tag{3}$$

Here E_0 is the incident field in the medium, α the absorption coefficient, D the thickness, we also use $a = \exp(-\alpha D)$ throughout. λ_v is the radiation vacuum wavelength, r_0 the Gaussian irradiance $1/e^2$ beam radius, k_0 the linear wavevector in the medium, and x is the change of wavevector at the beam centre due to the nonlinearity. Equation (3) expresses the spatially-varying phase-change, $k_e(r)D$. It is simple to follow the emergent field as it propagates beyond the sample by expanding

E_e as a series of Gaussians of differing spot-sizes, because the propagation of Gaussians is well understood [8].

$$E_e = E_0 \exp\left(\frac{-\alpha D}{2} + ik_0 D\right) \sum_m \frac{(ix)^m}{m!} \exp\left\{\frac{-r^2(2m+1)}{r_0^{\,2}}\right\} \qquad (4)$$

Numerical summation of the propagating Gaussians produces precisely the beam break-up in the near field that is obtained experimentally, for a negative n_2. More usefully, by observing the input-irradiance dependence of the on-axis transmission at some distance z from the emergent face, one can obtain a direct measure of the index n_2. The on-axis propagation of a single Gaussian, from [8], is

$$E_e(z) = E_e(o)\ (1 + z^2/z_0^{\,2})^{\frac{1}{2}} \exp(i \tan^{-1} \frac{z}{z_0}) \quad , \qquad (5)$$

with $z_0 = r_0^{\,2}\, k_0/2$. Thus in the far-field ($z \gg z_0$), from (4),

$$I_e(z) = I_0 a \frac{z_0^{\,2}}{z^2} \sum_m \left| \frac{(ix)^m}{m!\ (2m+1)} \right|^2 \quad ,$$

Remembering that I_0 is proportional to x, the form of $I_e(z)$ is shown in figure 3b; the measured irradiance level at the first peak (I_p) gives n_2:

$$n_2 = .37\ \alpha\lambda_v \ \{n_0\ I_p\ (1-a)\} \quad . \qquad (6)$$

Using this technique, n_2 values of order 10^{-6} per W cm^{-2} were deduced. Significantly, these were six orders of magnitude greater than previously discussed similar nonlinear coefficients in InSb (even greater variations are discussed in Chapter 3). Further, the result means that refractive index-changes of order 10^{-3} can be achieved at cw irradiance levels of just 1 kW cm^{-2}. Using a Fabry-Perot sample with feedback, rather than the anti-reflection coated sample in which the defocussing work was carried out, is therefore expected to lead to highly dramatic transmission and/or reflection responses, as will now be discussed.

2.2 Introductory Fabry-Perot theory

The simplest Fabry-Perot cavity is a plane-parallel sided, linear medium, in air, with no absorption; figure 4. The electric fields (E_1' to E_2''') are evaluated in terms of the incident field E_1 by solving Maxwells wave equation for the linear propagation

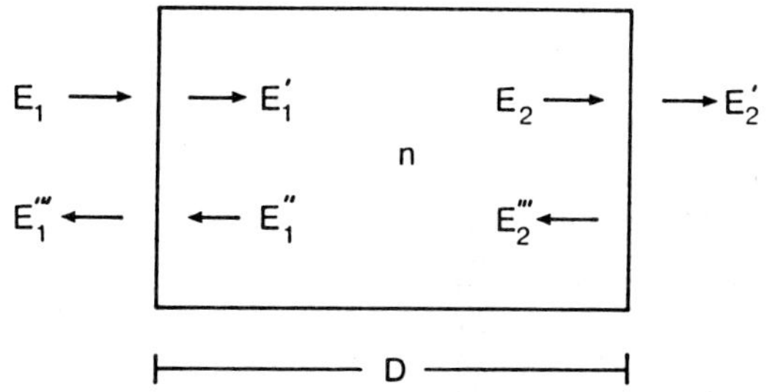

Fig. 4. Boundary fields in a Fabry-Perot cavity

inside the medium, with continuity of the tangential components of the E and H fields at the two semiconductor-air interfaces. Thus

$$E_2 = E_1 \exp(ik_0D) \quad , \quad E_1'' = E_2''' \exp(ik_0D) \quad . \tag{7}$$

$$E_1 + E_1''' = E_1' + E_1'' \quad , \quad E_2 + E_2''' = E_2' \quad . \tag{8}$$

$$E_1 - E_1''' = nE_1' - nE_1'' \quad , \quad nE_2 - nE_2''' = E_2' \quad . \tag{9}$$

With the field definition $E(t) = \{E(\omega) \exp(-i\omega t) + c.c\}$ the irradiance in a medium of index n is

$$I = nc \;\; E(\omega) \;^2/2\pi \quad , \tag{10}$$

and the solutions for the transmitted and internal irradiances are:

$$I_t = \frac{I_0}{1 + F \sin^2 k_0D} \quad , \quad I = \left(\frac{1 + R}{1 - R}\right) I_t \quad , \tag{11}$$

with surface reflectivity $R = \{(n-1)(n+1)\}^2$ and $F = 4R/(1-R)^2$. Thus for example in InSb, with an index n = 4; the natural surface reflectivity is 36% and the internal irradiance 'on-resonance ' is twice the incident level. Figure 5 shows, for InSb, the classic Fabry-Perot transmission as a function of the single-pass cavity phase, $\phi = k_0D$. Resonance occurs, as usual in bounded systems, when the cavity length is a half integer number of optical wavelengths, $D = (M/2)\ \lambda_v/n$.

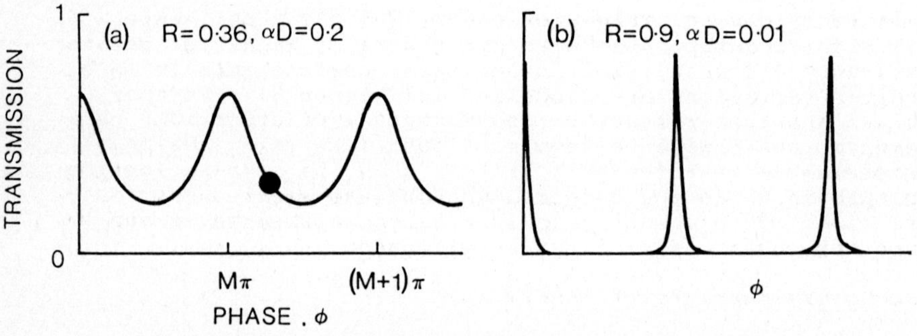

Fig. 5. Fabry-Perot transmission for a) InSb, b) an R = .9 cavity

For more complicated cavities, described by front and back-face reflectivities R_f, R_b, and including absorption, one obtains:

$$I_t = \frac{(1-R_f)(1-R_b)a}{(1-R_\alpha)^2} \frac{I_0}{1+ F_\alpha \sin^2\phi} \quad , \quad I = \frac{(1+R_b a)(1-a)}{(1-R_b)\ \alpha D a} I_t \quad , \tag{12}$$

with $R_\alpha = (R_bR_f)^{\frac{1}{2}}a$, $F_\alpha = 4R_\alpha/(1-R_\alpha)^2$, $a = \exp(-\alpha D)$, and where ϕ includes phase changes at the surfaces (ϕ_f, ϕ_b),

$$2\phi = 2k_0D + \phi_f + \phi_b \tag{13}$$

A significant point to note is that the internal irradiance may be far higher than the incident for suitable cavity parameters. Thus with $\alpha D = .01$, $R_b \simeq 1$, $R_f = R_b a$ the internal level is 50 times the incident, and any nonlinearities should be considerably enhanced by the cavity feedback.

2.3 The nonlinear Fabry-Perot cavity

The Fabry-Perot response is far more sensitive to changes in n as manifested in the phase ϕ, than as manifested in the reflectivities R_f, R_b. Thus in the nonlinear case, consider a cavity detuned at zero irradiance from the M-th order resonance (where $\phi_M = M\pi$) by a phase δ. For finite irradiance,

$$\phi(I) = M\pi + \delta + 2\pi D\, n_2 I/\lambda_v \quad . \qquad (14)$$

We will treat the case for negative nonlinear index n_2, and positive δ (depicted by the circle on figure 5). As I is increased, by increasing the incident irradiance, then $\phi(I)$ will diminish. But as this takes the cavity closer to its resonance, it is possible that more of the incident light will get into the cavity. Thus I increases further than would be expected from purely linear transmission. This effect can avalanche the cavity onto resonance - a dramatic combination of the nonlinearity and feedback.

To see this graphically, consider the transmission ratio I_t/I_0 in terms of I, as given by (12,14):

$$\frac{I_t}{I_0} = \frac{A_1}{1+F_\alpha \sin^2(2\pi D|n_2| I/\lambda_v - |\delta|)} \quad ; \quad \frac{I_t}{I_0} = \frac{A_2}{I_0} I \quad . \qquad (15)$$

Consistent solutions are intersections of the Airy curve, given as a function of I now by the former of these expressions, and the set of straight lines of gradients proportional to I_0^{-1}, given by the latter expression. The possibility of either one or three solutions is shown clearly in figure 6a and example solutions are plotted in figure 6b.

Whenever three solutions are obtained the central one proves to be unstable. There may thus be ranges of I_0 values over which two solutions - bistable operation - is possible. The system switches sharply from the lower to upper transmission branches (or vice versa) as I_0 is increased (or decreased), with hysteresis as depicted in figure 7.

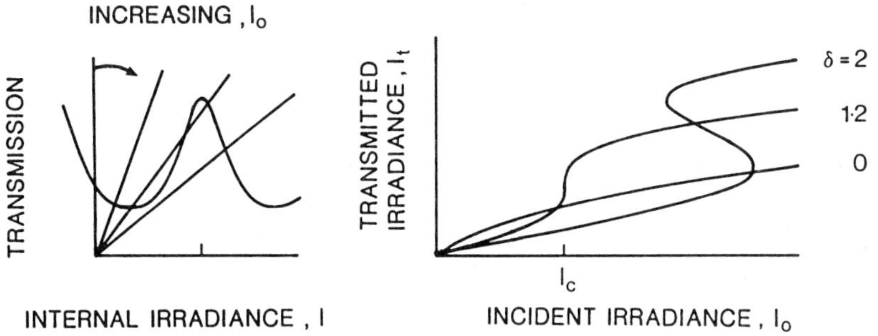

Fig. 6. a) Graphical solution for transmission of a nonlinear Fabry-Perot. b) Dependence of transmitted irradiance on initial detuning.

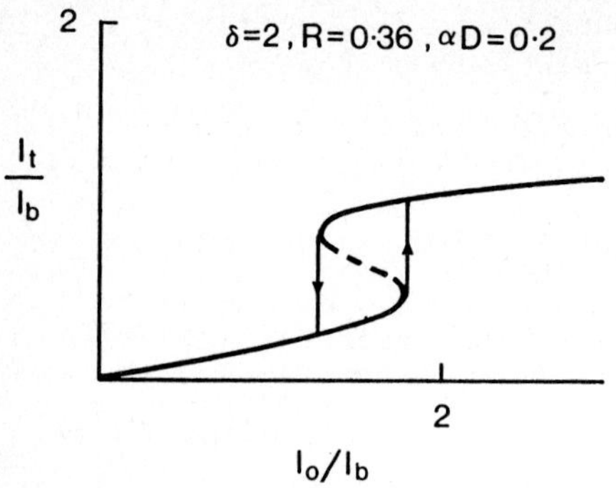

Fig. 7. First-order optical bistability region for InSb

A particularly useful technique for obtaining plots such as these, avoiding the necessity of finding the intersects on figure 6a, is the dummy-variable method. For any given internal irradiance there is just one solution for the transmitted irradiance, and just one initial irradiance that could have produced the internal value. Define a characteristic irradiance I_b and a scaled internal irradiance x by:

$$I_b = \alpha\lambda_v(2\pi|n_2|)^{-1} \quad , \quad x = I\alpha D/I_b \quad . \tag{16}$$

The incident, transmitted and reflected irradiances are all single-valued functions of x, [9]:

$$I_0 = I_b\{(1-R_\alpha)^2/(1-R_f)(1+R_b a)(1-a)\}x\{1+F_\alpha \sin^2(x-\delta)\},$$

$$I_t = I_b\ \{(1-R_b)a/(1+R_b a)(1-a)\}\ x$$

$$I_r = I_0 - I_t - x\ I_b \quad . \tag{17}$$

Hence, using x as a dummy-variable one can plot I_t or I_r versus I_0 for given cavity parameters, these irradiances being scaled characteristically to I_b, figure 8.

For low irradiance bistability one wishes to minimise the critical irradiance I_c for which the Fabry-Perot response shows a step-like form. For irradiances greater than I_c then bistability can be achieved providing that the initial detuning is

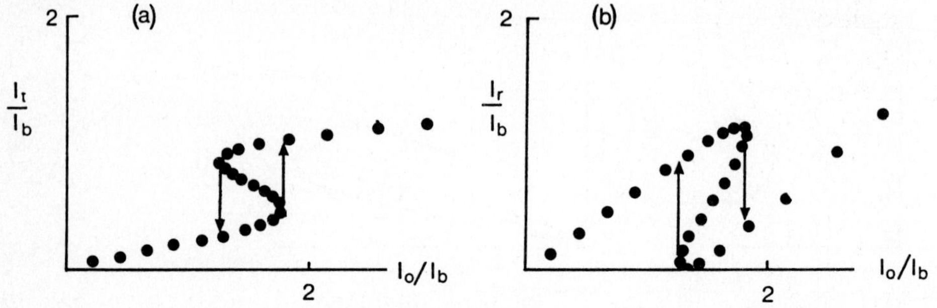

Fig. 8. Transmission a) and reflection b) nonlinear response using the dummy-variable method. Dot separations correspond to equal x differences

greater than the corresponding critical detuning δ_c. For lower irradiances bistability is never observed.

The critical condition occurs for example when the straight line in figure 7a, intercepts the curve tangentially at its point of inflexion. Rewriting (15) as

$$y_1 = A_1/\{1+F_\alpha \sin^2(x-|\delta|)\} \quad ; \quad y_2 = A_2\, I_b\, x\, (I_0\ \alpha D)^{-1} \tag{18}$$

then at the critical point: $y_1=y_2$, $dy_1/dx = dy_2/dx$, $dy_1{}^2/dx^2 = 0$ [10]. The solution is key to the optimization procedures considered in this article:

$$I_c = I_b\ f(R_b, R_f, \alpha D) \quad . \tag{19}$$

The cavity factor f should be thought of as a function of R_f, R_b, D (the latter being scaled to the absorption length α^{-1}). I_b depends on the material and on the radiation wavelength. Provided that n_2 is not influenced by the cavity configuration then I_b is independent of the cavity parameters.

$$f = \frac{\alpha D(1-R_\alpha)^2}{(1-R_f)(1+R_b a)(1-a)}\ \frac{\surd 2}{16}\ \frac{\{3(F_\alpha+2) - \surd\}^2}{\{(F_\alpha+2)\surd-(F_\alpha+2)^2-2F_\alpha{}^2\}^{\frac{1}{2}}} \quad , \tag{20}$$

$$\surd = \{(3F_\alpha+2)^2 - 8F_\alpha\}^{\frac{1}{2}}$$

The associated critical detuning depends only on the cavity finesse ($F_i = \pi F_\alpha{}^{\frac{1}{2}}/2$) through F_α:

$$\delta_c = \frac{\surd 2}{4}\ \frac{\{3(F_\alpha+2) - \surd\}}{\{(F_\alpha+2)\surd-(F_\alpha+2)^2-2F_\alpha{}^2\}} + \sin^{-1}\left(\left\{\frac{3F_\alpha+2-\surd}{4F_\alpha}\right\}^{\frac{1}{2}}\right). \tag{21}$$

Figure 9 shows how f varies with the cavity reflectivities, for example D values of $1/\alpha$ and $.01/\alpha$. In order to achieve low irradiance bistability one requires a small sample thickness combined with particular reflectivities. As D is decreased, the tolerance on the range of R_f, R_b that gives low f becomes finer and finer, the limit is therefore set by fabrication ability.

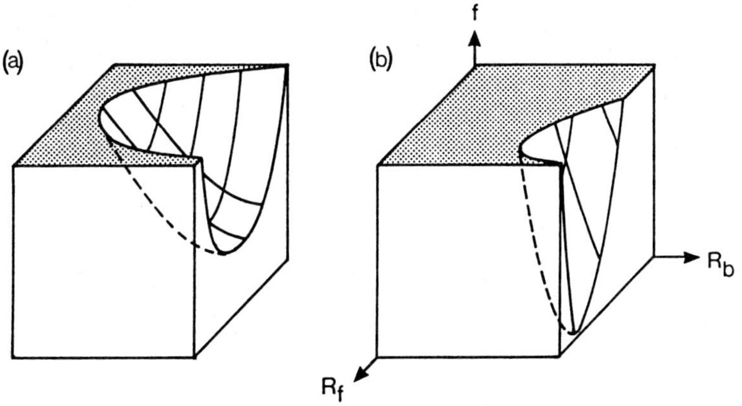

Fig. 9 Reflectivity-dependence of the cavity factor f, for $\alpha D = 1$ (a), and 0.01 (b)

It may at first sight be surprising that small D can lead to low incident irradiance requirements, given that the cavity must be brought through resonance from the initial detuning by a change of phase ($2\pi|n_2|I\,D/\lambda_v$) that is proportional to D. The point is that in a high finesse cavity the Fabry-Perot resonance peaks are considerably sharper (figure 5) and the critical detuning is smaller ($\delta_c \simeq \pi\sqrt{3}/2F_i$ for the optimum conditions at the minima of figure (9) and at high finesse). Hence the phase-change to be brought about by the internal irradiance is independent of D to a first approximation. A small cavity has a lower I_c because the required, D-independent internal level can be achieved using a lower incident irradiance, as discussed in the section on linear Fabry-Perot characteristics.

The required radiation-induced phase-change tells us the refractive index variation, $\Delta n \simeq n_2 I_c$, that must be obtained. Under optimized conditions for optical bistability [11]:

$$\Delta n_{OB} \simeq \alpha\lambda_v/5 \tag{22}$$

Typically, for semiconductors, one therefore requires $\Delta n_{OB} \gtrsim 10^{-3}$.

2.4 Cavity optimization for InSb and ZnSe-filter examples

Within the limits of the plane wave approximation, the optimization conditions described above apply to those materials such as InSb in which the nonlinear refraction is of electronic origin, that is, where the generation of carriers dominates (Chapter 3). In contrast, in zinc selenide interference filter studies the nonlinearity mechanism in sample heating [12] and the geometric configuration influences the effective n_2 coefficient through the style of thermal diffusion.

A typical filter consists of a semiconductor spacer region of a micron or less thickness surrounded by alternating layers of low and high refractive index materials each of just $\lambda_v/4$ optical thickness. These $\lambda_v/4$-stacks form high reflectivity coatings and Fabry-Perot theory can be taken over, with slight modifications to account for the stack phase changes on reflection, in order to describe the nonlinear responses. The semiconductor layers are deposited on transparent substrates for mechanical stability. For laser spot sizes large compared to the total thickness of all the dielectric layers, any heat generated as a consequence of radiation absorption diffuses primarily into the substrate. To a good approximation

$$\Delta n = (\partial n/\partial T)\Delta T \quad , \quad \Delta T \simeq \alpha I\, r_0 D/\kappa_s \quad , \tag{23}$$

where ΔT is the spacer temperature-rise, κ_s the substrate thermal conductivity. We are therefore dealing with an effective thermal nonlinear refractive index,

$$n_2^T \simeq (\alpha\, r_0 D/\kappa_s)\ \partial n/\partial T \quad , \tag{24}$$

that depends on the cavity thickness D.
The optical bistability critical condition still applies, but now the cavity factor must be modified to account for the above D-dependence.

$$I_c^T \simeq \frac{\lambda_v\ \alpha\ \kappa_s}{r_0\ \partial n/\partial T}\ \frac{f(R_f,R_b,\alpha D)}{\alpha D} \quad . \tag{25}$$

The contrast between cavity optimization in the two cases is manifested in the plots of figure 10. In both cases, for a given pair of reflectivities and given absorption coefficient there is an optimum D for minimising the critical irradiance. In practice this optimization proves to be the most significant in producing a bistable element for low-power operation.

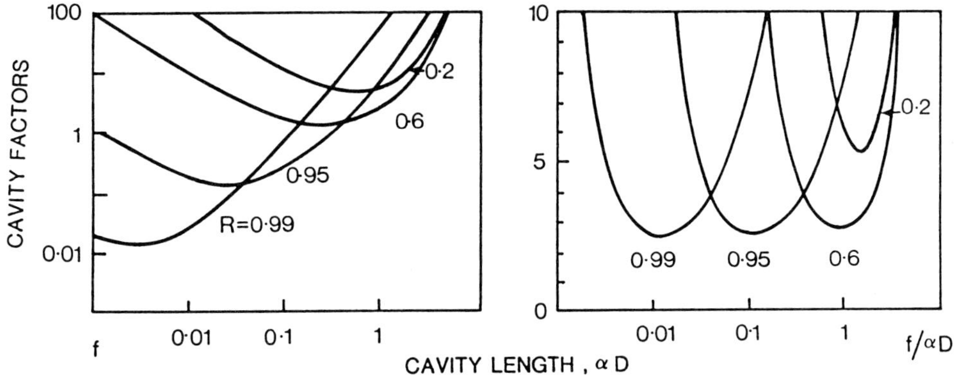

Fig. 10. Cavity factors for a) electronic and b) thermal nonlinearity.

At high finesse the two optimum conditions are:

$$D = (2-R_f-R_b)/4 \quad , \quad D^T = (2-R_f-R_b) \quad . \tag{26}$$

The main difference in the two cases is, however, the variation of the optimum cavity factor as one moves toward higher finesse. In the thermal case there is no obvious advantage in using many layers to build up high reflectivity stacks; $f/\alpha D$ has a minimum value of the order of 2 that is achievable with just one low-high layer pair in each stack. Only when one reduces the laser spot-size to the order of D so that transverse heat conductivity becomes significant ($\Delta T \propto \{\alpha\, r_0{}^2/\kappa\}\partial n/\partial T$) does one regain the advantages of high reflectivity. The reason is that whilst for small spots the spacer temperature-rise depends on the absorbed energy per unit transverse distance, at large spots the total absorbed energy is relevant.

Under optimized cavity conditions the critical switching power for bistability is, in the electronic case,

$$P_{cm} = \frac{\surd 3}{2}\ \frac{\lambda_v \alpha}{|n_2|}\ \alpha D\, r_0{}^2 \quad . \tag{27}$$

Using the InSb case as an example: $\lambda_v \simeq 5\ \mu m$, $\alpha \simeq 10\ cm^{-1}$, $n_2 \simeq -10^{-3}$ per W cm^{-2}, $D \simeq 100\ \mu m$, if $r_0 \simeq 10\ \mu m$ then $I_c \simeq 2\ W\ cm^{-2}$, $P_{cm} \simeq 10\ \mu W$.

For the large-spot thermal nonlinearity

$$P^T_{cm} = \frac{\lambda_v\ \alpha}{\partial n/\partial T}\ \alpha r_0 \quad . \tag{28}$$

The parameters for a ZnSe-spacer filter on a float-glass substrate are $\lambda_v \simeq 0.5\ \mu m$, $\kappa_s \simeq .01\ W\ cm^{-1}\ K^{-1}$, $\partial n/\partial T \simeq 2 \times 10^{-4}\ K^{-1}$, $\alpha \simeq 10^3\ cm^{-1}$, $r_0 \simeq 50\ \mu m$ giving $P^T_{cm} \simeq 10$ mW. The effective n_2 for such a configuration is 5×10^{-4} per W cm^{-2}. It is a fairly general conclusion that either due to fabrication difficulties in the electronic case, or due to substrate heat sinking in the thermal case, the practical power level for achievement of optical bistability of typical 50 μm laser spots will not reduce much below 10 mW. One must tend to the diffraction-limited spot-sizes and use optimized cavities, to produce submilliwatt switching - this has not yet been achieved.

Whilst cavity optimization for given absorption has now been described,we now wish to determine the optimum α value itself. This will depend on the semiconductor material and the operational radiation frequency. We must understand the origins of both the absorption and of the nonlinear refractive coefficient to achieve material optimization, this is the subject of Chapter 3. It is also important to bear in mind that if refractive index-changes of order 10^{-3} are to be achieved under cw conditions and preferably at irradiance levels of less than 1 kW cm^{-2} (10 mW on a 100 μm spot),then the magnitude of n_2 should exceed 10^{-6} per W cm^{-2}.

3. MICROSCOPIC THEORY FOR OPTICAL BISTABILITY

3.1 Nonlinear susceptibilities

In nonlinear optics it has been conventional to describe nonlinearities by appropriate susceptibilities, through the polarization in the medium that is established by the incident field/s. Thus at the frequency ω we are interested in a polarization $P(t) = \{P(\omega)\exp(-i\omega t) + c.c.\}$ where,

$$P(\omega) = \chi^{(1)} E(\omega) + \chi^{(3)} |E(\omega)|^2 E(\omega) + \ldots \qquad (29)$$

This contains those components of all powers of E(t) that have $\exp(-i\omega t)$ time-dependence. $\chi^{(n)}$ is known as an n-th order nonlinear susceptibility, it depends on material and frequency, but not on the field strength. One can define a generalized dielectric constant, which will be used to compare $\chi^{(3)}$ with the index coefficient n_2:

$$\varepsilon(\omega,E) = 1 + 4\pi P(\omega)/E(\omega) \qquad . \qquad (30)$$

By expressing ε in terms of irradiance-dependent refraction and extinction coefficients,

$$\varepsilon^{\frac{1}{2}} = (n_0+n_2 I+\ldots) + i(\alpha_0+\alpha_2 I+\ldots)c/2\omega \qquad , \qquad (31)$$

one obtains relations between the susceptibilities and these coefficients. In particular

$$n_2 = \frac{4\pi^2}{n_0{}^2 c} \ \mathrm{Re}\ \chi^{(3)} \qquad (32)$$

For bistability the required n_2 values of 10^{-6} per W cm^{-2} imply $\chi^{(3)}$ values of order 10^{-3} esu (where 1 esu $\equiv 1\ cm^3\ erg^{-1} \equiv 1.4 \times 10^{-8}\ m^2\ V^{-2}$). It is intriguing to compare this demand with some of the many measurements of $\chi^{(3)}$ that were reported prior to the 1979 observation of cw semiconductor bistability.

In a review of those nonlinear optical phenomena that had been considered by 1963, just three years after the first laser demonstration, Franken and Ward reported that "intensity-dependent" refraction should occur, but that it had not then been observed [13]. In the previous year, however, an associated third-order nonlinearity due to third-harmonic generation in calcite was reported at a value of 10^{-15} esu [14]. The first refractive susceptibilities, measured by birefringence studies in liquids, were of order 10^{-16} esu (H_2O) to 10^{-12} esu (CS_2) [15,18]. The CS_2 value is today used as a standard for self-focussing (positive) nonlinear refraction. By 1966, third-order frequency-mixing experiments in crystalline solids produced from 5×10^{-15} esu in large bandgap materials (12 eV in LiF) to 8×10^{-10} esu in InSb (.18 eV at room temperature) [17,18]. From 1965-75 considerable interest in self-focussing, primarily in liquids and based on the molecular-orientational Kerr effect, involved typically 10^{-13}-10^{-11} esu $\chi^{(3)}$ values [19]. Only one prediction, in 1966 by Javan and Kelley [20], suggested that the saturation of anomalous dispersion should produce much larger susceptibilities; refractive index-changes of 10^{-6} to 10^{-3} were predicted.

A revival of interest in semiconductor nonlinear refraction occurred in the 1970's as a consequence of transient holography [21] and dynamic phase grating studies [22], leading to phase conjugation [23]; also as a consequence of semiconductor saturation investigations [24,25]. The 1978 beam-profiling experiments discussed in section 1 implied 10^{-2} esu values in InSb [7], phase conjugation and degenerate four-wave mixing work in Si and HgCdTe produced 8×10^{-8} and 5×10^{-2} esu respectively by 1980 [26,27] and 1979 optical bistability observations implied $\chi^{(3)}$ values of 10^{-5} esu in GaAs [28] and 1 esu in InSb [29]. Since that time large numbers of experiments, in which the implied $\chi^{(3)}$ is 10^{-6} esu or above, have been reported.

There is presently a demand for both highly nonlinear and highly linear materials. In the context of glasses for laser fusion studies, and for optical communications fibre materials, a recent paper notes a $\chi^{(3)}$ of value 8×10^{-16} esu in an SiO_2 glass sample. It is an interesting question as to how one fabricates different materials with nonlinearities that can differ by fifteen orders of magnitude. It is also challenging from a fundamental physics point of view to try to understand why such a range of magnitudes exists for the same basic phenomenon, and to describe how, when one gradually changes the material and/or radiation conditions, the fifteen orders are bridged. Before calculations for the giant bistability nonlinearities are presented (section 3.3), we shall discuss this vast range of the magnitudes of $\chi^{(3)}$ (or n_2).

3.2 First estimates of $\chi^{(3)}$ and n_2

It is useful to remind ourselves of the contribution to the first-order susceptibility, $\Delta\chi^{(1)}$, and the linear refraction, Δn, associated with particular (two-level) absorption transitions in any medium.

$$\Delta n = 2\pi \mathrm{Re}\, \Delta\chi^{(1)}/n_0 \quad , \quad \Delta\chi^{(1)} = \Delta P(\omega)/E(\omega) \ . \tag{33}$$

Here $\Delta P(\omega)$ is the linear polarization per unit volume associated with the transitions, which are supposed to give a small contribution to the total refractive index n_0. Consider N

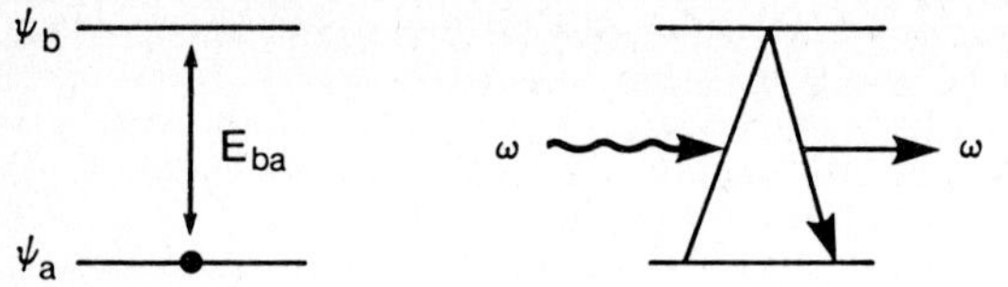

Fig. 11. Two-level atom a) and refractive two-stage transitions b)

two-level systems per unit volume, e.g. electronic levels in atoms in which electrons occupy the ground states initially, figure 11a.

The radiation interaction Hamiltonian in the electric-dipole approximation (H') is to be included in the time-dependent Schrödinger equation:

$$H' = -er\,E(\omega)\,e^{-i\omega t} + c.c. \quad ; \quad (H_0+H')\phi = i\hbar\,\partial\phi/\partial t\,. \tag{34}$$

The expectation value of er, with respect to the perturbed state, gives us the polarization ΔP.

$$\phi(t) \simeq \phi_a \exp(-iE_a t/\hbar) + c_b(t)\,\phi_b \exp(-iE_b t/\hbar),$$

$$c_b(t) \simeq er_{ba}\,E(\omega)\,\frac{\exp\{i(E_{ba}-\hbar\omega)\,t/\hbar\}}{(E_{ba}-\hbar\omega)}$$

Here the so-called rotating wave approximation has been used to obtain just the dominant term in $c_b(t)$, and damping terms have been removed. The contribution to $\chi^{(1)}$ is thus,

$$\text{Re}\,\Delta\chi^{(1)} = N\,\frac{|er_{ba}|^2}{(E_{ba}-\hbar\omega)}\,. \tag{35}$$

The form of this expression may be understood by considering the 'event' to which it corresponds, namely: the temporary removal of a photon of energy $\hbar\omega$ from the incident field with the simultaneous electron excitation from level-a to level-b, followed by the re-emission of a photon accompanied by de-excitation back to the ground state (figure 11b). The dipole moment $|er_{ba}|$ represents the strength of each transition; the denominator is the energy mismatch in the intermediate state of the system, in which a photon is removed and an electron is excited. The time-interval for which this energy mismatch can be sustained is $\Delta t \simeq \hbar/(E_{ba}-\hbar\omega)$, from the uncertainty principle. It is not surprising that the effect of the 'event' on the radiation, as expressed by $\chi^{(1)}$, is proportional to this 'virtual-state' lifetime. The factor N is equally to be expected.

Turning to the nonlinear coefficient: $\chi^{(3)} \simeq \Delta P(\omega)/E^3(\omega)$ in essence, and one must use third-order perturbation theory to obtain the polarisation component of third-order in E. From a quasi-dimensional analysis therefore, the susceptibility contribution from N transition systems must be of the form,

$$\chi^{(3)} \propto N\,|er_{ba}|^4\,\hbar^A\,E_{ba}^{\ B}\,F(\hbar\omega/E_{ba})\,. \tag{36}$$

It is straightforward to show that A = 0, B = -3. The dimensionless function F will contain various energy denominators associated with virtual state lifetimes. If we set F = 1 then a characteristic base-value for $\chi^{(3)}$ is obtained. In the atomic case with $N \simeq 10^{19}$ cm^{-3}, $r_{ba} \simeq 10^{-8}$ cm, $E_{ba} \simeq 10$ eV one obtains a base value of 10^{-18} esu!

A similar analysis may be applied to the semiconductor case. Firstly however, the coordinate matrix elements should not be used for semiconductors, where the electronic Bloch states are extended, periodic functions. Instead, one uses momentum matrix elements. We will concentrate on interband transition contributions to $\chi^{(3)}$.

$$|er_{cv}| \to |ep_{cv}/m\omega_{cv}| \simeq eP/E_{cv} \quad , \qquad (37)$$

where P is known as the Kane momentum parameter, it has dimensions of $[ML^3T^{-2}]$ and a value close to 10^{-19} esu in most materials. Secondly, the atomic density must be replaced by a sum over k-states per unit volume, and the two-level energy replaced within the sum by k-dependent energies. The latter energies are determined by the reduced effective mass for the interband transitions, which in turn is dominated by the band-interaction between the conduction and valence bands, $H'_{band} = \hbar k.p/m$. The only parameters that appear in the summation are therefore matrix elements of p/m, and the fundamental gap E_g, so that finally:

$$\chi^{(3)} \propto e^4 \hbar^A E_g^B P^C F(\hbar\omega/E_g) \quad , \qquad (38)$$

where F differs from that in (36), and one finds A = 0, B = -4, C = 1. One has in effect used (37) plus the substitution,

$$N \to E_g^3/P^3 \quad , \qquad (39)$$

this being the number of k-states per unit volume in an energy-range of order E_g.

For the small-gap material InSb the base value for $\chi^{(3)}$ obtained through the above analysis is 10^{-9} esu. Compared to the atomic case four orders of magnitude improvement is obtained because the transition dipole moment is thirty times larger, and a further five orders because the InSb gap is .02 times that of our atomic model. In larger-gap materials (ZnSe for example) the base electronic $\chi^{(3)}$ is 10^{-13} esu - the bandgap and transition dipole moment factors are reduced compared to InSb, the density-of-states factor actually increases.

To estimate the factor F is not simple. Third-order nonlinearities are described by four rather than two electronic transitions, and by reference to figure 12 there are four-stage

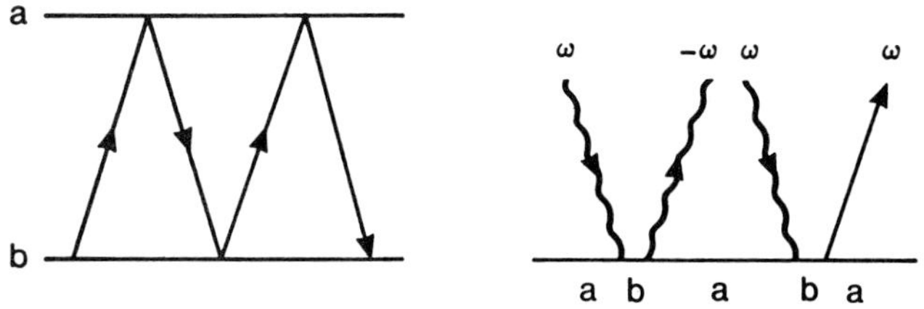

Fig. 12. Four-stage transition scheme contribution to nonlinear refraction

events for which energy is conserved exactly at intermediate states; apparently Δt is infinite for such states and F diverges. A detailed study of the virtual transition schemes produces a partial cancellation of the apparently divergent terms. For $\hbar\omega$ close to a discrete two-level transition energy one obtains [30],

$$F(\hbar\omega/E_{ba}) \simeq E_{ba}{}^3/(E_{ba}-\hbar\omega)^3 \quad . \tag{40}$$

Averaging over the range of states within E_g of resonance in the semiconductor case reduces this resonance enhancement,

$$F(\hbar\omega/E_g) \simeq E_g{}^{3/2}/(E_g-\hbar\omega)^{3/2} \quad . \tag{41}$$

Typically one might expect a factor of 10^3 enhancement as one approaches the band edge, limited by various non-radiative processes.

This leads us naturally to the role played by damping and by the recombination back to the ground (atomic) or valence electron states. As uncertainty lifetimes become large, then the state lifetimes themselves restrict the duration over which an excited electron can influence the susceptibilities. in the next section we shall see that such processes lead in effect to an additional factor in the susceptibility expressions - given a damping or dephasing time of the excitation, T_2 (associated for example in the atomic case with atom-atom collisions, whether or not they lead to de-excitation) and a recombination lifetime T_1, then

$$\chi^{(3)} \rightarrow (T_1/T_2)\,\chi^{(3)}\ \text{(virtual)} \quad . \tag{42}$$

In semiconductors the lifetime ratio can be as large as 10^5.

3.3 Nonlinear refraction under saturation conditions

Figure 13a shows the Lorentzian, homogeneously-broadened absorption of a two-level system and the associated, so-called

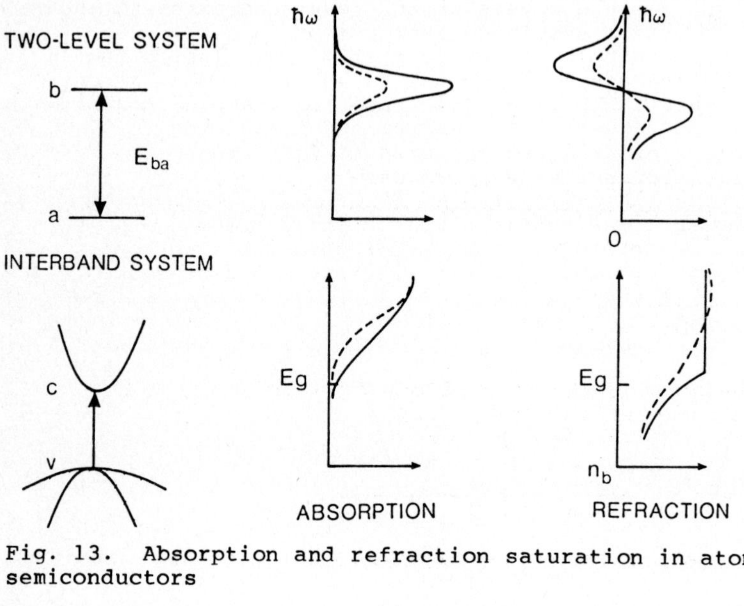

Fig. 13. Absorption and refraction saturation in atoms and semiconductors

anomalous, dispersion. At high excitation the absorption begins to saturate as the ground state becomes depopulated and the excited state populated. The dashed lines depict this saturation of α and simultaneously of the refraction. Note that at radiation frequencies below the transition frequency, the refraction reduces from some hitherto positive value, that is, there is a negative, irradiance-dependent contribution to n. Think of the semiconductor interband absorption as a summation over many two-level transitions: the same argument applies; saturation reduces the effective absorption coefficient, and the interband refraction term diminishes at frequencies close to $E_g/\hbar$, as schemed.

We need only to determine the number of excited electrons in the two cases in order to estimate an effective $\chi^{(3)}$ or n_2. The net number of electrons excited per unit volume per second is

$$\frac{dN}{dt} = \frac{\alpha I}{\hbar\omega} - \frac{\Delta N}{T_1} \quad , \tag{43}$$

where ΔN is the excess excitation level and a simple recombination rate process has been assumed. In equilibrium

$$\Delta N = \alpha\, I\, T_1/\hbar\omega \quad . \tag{44}$$

Equation (35) suffices to tell us the effect of the excitation. A reduction by ΔN of the empty states (b) must give a negative contribution to Re $\chi^{(1)}$, because transitions into these states are removed.

$$\mathrm{Re}\, \Delta\chi^{(1)}{}_b = -\Delta N |e\, r_{ba}|^2/(E_{ba}-\hbar\omega) \tag{45}$$

Similarly, ΔN lower states (a) no longer contain electrons, so that this number of states becomes available for radiative emission processes associated with b-to-a de-excitation, the susceptibility contribution is of the same magnitude and sign as $\Delta\chi^{(1)}{}_b$. Thus, regardless of the recombination process, one obtains a refractive index-change,

$$\Delta n = \frac{2\pi}{n_0}\Delta\chi^{(1)} = \frac{2\pi}{n_0}(-2\Delta N)\frac{|er_{ba}|^2}{E_{ba}}\frac{E_{ba}}{E_{ba}-\hbar\omega} \quad . \tag{46}$$

For T_1-type recombination; the change in $\chi^{(1)}$ or n is proportional to I and can therefore be considered as an effective $\chi^{(3)}$ or n_2 phenomenon.

$$n_2 = \frac{\Delta n}{I} = \frac{-4\pi}{n_0}\frac{|er_{ba}|^2}{E_{ba}^2}\alpha T_1\frac{E_{ba}^2}{\hbar\omega(E_{ba}-\hbar\omega)} \quad . \tag{47}$$

To see how this relates to the expressions of section 3.2 note that the linear two-level absorption is,

$$\alpha = \frac{4\pi}{n_0} N \frac{\hbar\omega}{c}\frac{|er_{ba}|^2\, T_2^{-1}}{(E_{ba}-\hbar\omega)^2+\hbar^2 T_2^{-2}} \quad . \tag{48}$$

Thus for $E_{ba}-\hbar\omega \gg \hbar T_2^{-1}$ the nonlinear index has essential form

$$n_2 \propto -\frac{N|er_{ba}|^4}{c\, n_0{}^2\, E_{ba}^3} \frac{T_1}{T_2} \left[\frac{E_{ba}}{E_{ba}-\hbar\omega}\right]^3 \tag{49}$$

which is T_1/T_2 times the results anticipated from (32,36,40).

Applying the transformation (37), to give the index contribution due to the removal of heavy-hole to conduction transitions caused by the occupancy of ΔN conduction electrons in states at the bottom of the conduction band:

$$\Delta n = \frac{-2\pi}{n_0} \Delta N \frac{|eP|^2}{3E_g^3} \frac{E_g}{E_g-\hbar\omega} \tag{50}$$

The factor 1/3 is included to account for averaging over directions in k-space. Similar Δn contributions arise from the removal of the heavy-hole electrons, and for light-hole to conduction transitions etc. The n_2 coefficient can therefore now be expressed in terms of a sign, a material scaling factor and a resonance enhancement.

$$n_2 \simeq -\frac{2\pi\, e^2}{3} \frac{\alpha_i\, T_1\, P^2}{n_0\, E_g{}^4} \frac{E_g{}^2}{\hbar\omega(E_g-\hbar\omega)} \quad . \tag{51}$$

If the band-tail interband absorption coefficient, α_i, at frequencies just below $E_g/\hbar$, is treated as a result of T_2-broadening of the individual two-level interband systems, then the absorption coefficient is proportional to $1/\{PT_2\surd(E_g-\hbar\omega)\}$ and

$$n_2 \propto \frac{-e^4}{c} \frac{P}{n_0\, E_g{}^4} \frac{T_1}{T_2} \left[\frac{E_g}{E_g-\hbar\omega}\right]^{3/2} \quad . \tag{52}$$

This basic expression gives us some feeling for the form of the low-temperature nonlinear refraction. Note that the P parameter is almost constant over all semiconductors, and the linear refractive index varies by less than a factor of two over most, whilst bandgaps range over more than an order of magnitude. For a large nonlinearity, the small-gap materials are clearly favoured due to the $E_g{}^{-4}$ factor.

Equations (50,52) apply for low carrier excitation and at low temperatures (such that only those states at the band extrema are influenced). A very useful and simple expression may be obtained for the case in which relatively large numbers of carriers are excited, at 77 K or room temperature. In principle (50) should be replaced by:

$$\Delta n = \frac{-2\pi}{n_0} \frac{|eP|^2}{3} \Sigma' \, \Sigma \frac{1}{E_g{}^2(k)\{E_g(k)-\hbar\omega\}} \quad ; \tag{53}$$

Σ' sums over different bands, Σ sums over band states that are occupied by photogenerated carriers. For example

$$\sum_{ck} \rightarrow \frac{1}{\pi^2} \int_0^\infty f_c(k)\, k^2 dk/(E_g-\hbar\omega + \hbar^2 k^2/2m_{ch}) \quad , \tag{54}$$

$$\Delta N_c \to \frac{1}{\pi^2} \int_0^\infty f_c(k)\ k^2/dk \quad . \tag{55}$$

These expressions will give the Δn contribution due to ΔN_c conduction carriers, influencing heavy-hole to conduction transitions (at reduced mass m_{ch}). $f_c(k)$ is the occupancy factor of the conduction states. In the Boltzmann limit,

$$f_c(k) = \exp\{(E_f - E_g - \hbar^2 k^2/2m_c)/k_b T\} \quad . \tag{56}$$

The Fermi energy E_f is determined via the ΔN_c equation, it drops out of (54,55) to give

$$\Delta n_c \simeq - \frac{2\pi}{n_0} \frac{|eP|^2}{3\ E_g^{\,2}} \frac{\Delta N_c}{kT} \frac{m_{ch}}{m_c} \frac{4}{\sqrt{\pi}}\ J\left[\frac{E_g - \hbar\omega}{k_b T}\ \frac{m_{ch}}{m_c}\right] \quad . \tag{57}$$

The function J is shown in figure 14a

$$J(d) = \int_0^\infty y^2\ e^{-y^2}\ dy/(y^2+d) \quad . \tag{58}$$

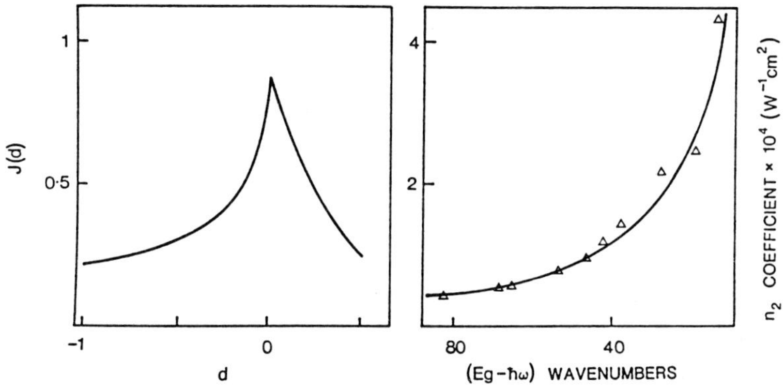

Fig. 14. a) The function J(d), equation 58. b) Experimental and theoretical bandgap enhancement of n_2.

The form of the refractive index coefficient on this basis is

$$n_2 \propto -e^2\ \frac{\alpha_i\ T_1}{n_0} \frac{P^2}{E_g^{\,3}} \frac{1}{k_b T}\ f\left[\frac{\hbar\omega}{E_g}\ ,\ \frac{\hbar\omega}{k_b T}\right] \quad . \tag{59}$$

Again, narrow-gap materials will give favourably large coefficients.

Note that expressions (46,50,57) allow one to extract the refractive cross-section per generated carrier, $\sigma = \partial n/\partial N$. This cross-section is particularly useful in those cases where recombination cannot be described simply by a T_1-process. However, revised cavity optimization is needed under such circumstances.

Figure 14b shows the observed n_2 bandgap enhancement obtained from beam-profiling experiments. The fit to the frequency dependence of the experimental absorption, times that of J is good. Equally, modelling the band tail to a T_2 broadening gives a reasonable fit for $T_2 \approx 10$ ps. The n_2 coefficient fits demand $T_1 \approx 1$ μs, manifesting the large enhancement that the T_1/T_2 factor gives.

Note that in this section the nonlinear coefficient is described by the change in the linear index brought about by the relatively long-term saturation of transitions. We have in effect undertaken a Kramers-Krönig analysis of the linear aborption change [31]:

$$\Delta n(\omega) = \frac{c}{\pi} \int_0^\infty \frac{\Delta\alpha(\omega')}{\omega'^2-\omega^2} d\omega' \quad . \tag{60}$$

An additional but non-resonant contribution to $\Delta\chi^{(3)}$ comes from free-carrier transitions. Following classical Drude theory, applied to ΔN carriers of effective mass m_c in a medium of background refractive index n_0,

$$\varepsilon = n_0^2 \, (1-\omega_p^2/\omega^2) \quad ; \quad \omega_p^2 = 4\pi\Delta Ne^2/m_c n_0^2 \quad . \tag{61}$$

The induced-carrier contribution to the total index may be written in terms similar to (57), by noting that $m_c \propto \hbar^2 E_g/P^2$ in semiconductors,

$$\Delta n_{fc} \approx - \frac{2\pi}{n_0} \Delta N \frac{|eP|^2}{E_g^3} \left[\frac{E_g}{\hbar\omega}\right]^2 \quad . \tag{62}$$

The characteristic free-carrier refractive cross-section $\Delta n_{fc}/\Delta N$ is of order -2×10^{-19} cm^3 in InSb at 5 μm. By comparison the interband cross-section is of order -8×10^{-18} cm^3, a factor of 40 larger. One implication of the above calculations is that to achieve the Δn value of 10^{-3} required for optical bistability, somewhat in excess of 10^{14} carriers/cm^3 must be generated.

In general, it has been shown that the passage of observed $\chi^{(3)}$ values from 10^{-15} esu to 1 esu is to be associated with (i) the reduction of transition energies (typically up to five orders) (ii) increased transition strengths or, in large-gap crystals, a higher density of active states (four-orders) (iii) the approach to resonance (three) and (iv) the long-term excitation on-resonance, subject to recombination (five orders).

3.4 Material and frequency optimization - the electronic case

Equations (57,19,16) allow us to express the critical bistability irradiance as,

$$I_c \approx - \underbrace{\frac{\hbar c}{e^2} k_b T}_{\text{temperature}} \quad \underbrace{\frac{n_0 E_g^2}{P^2 \, T_1}}_{\text{material}} \quad \underbrace{J^{-1} \left[\frac{k_b T}{E_g - \hbar\omega}\right]}_{\text{frequency}} \quad \underbrace{f(R_f, R_b, \alpha D)}_{\text{cavity}} \tag{63}$$

The various factors are emphasised. Note that the proportionality of I_c to α/n_2 and that of n_2 to α serves to remove the absorption-coefficient dependence. (An exception to this result occurs if there is a strong parasitic absorption (α_p) associated with processes that do not generate carriers and therefore do not contribute strongly to Δn. In this eventuality $\alpha = \alpha_i + \alpha_p$ in the analysis of chapter 2; whereas $\alpha = \alpha_i$ in that of section 3.3).

Ignoring parasitic absorption, note firstly that the frequency-dependence of I_c is contained entirely in J^{-1} and in f. The ideal optimization procedure is therefore to select $\hbar\omega = E_g$, as this is where J^{-1} is a minimum. In turn, this will determine the value of α, and one then optimizes the cavity in accord with the ideas of chapter 2.

As far as the irradiance level is concerned, the material factor is dominated by the $E_g{}^2$ dependence. The successful cw operation of large-spot optical bistability only in the narrow-gap material InSb is a direct manifestation of this result. However, minimum power levels ($P_{cm} = \pi\, r_0{}^2\, I_c$) are achieved in the diffraction limit where $r_0 \sim \hbar c/E_g$. E_g drops out of the expression for P_{cm}, the material is represented only through the recombination time T_1. It is hoped that T_1 can be tailored to give various device characteristics. In the next section we shall see that switching speeds are of the order of T_1 at low powers, and that switching energies are of order $P_{cm}T_1$. There is hence a trade-off in power-speed, but the switching energy is dependent only on frequency and cavity optimization.

3.5 The thermal nonlinearity

Given the dominant role of the spacer optical path length in Fabry-Perot bistability, one should in principle consider its relative change as

$$\frac{\Delta(nD)}{nD} = \frac{1}{D}\frac{\partial D}{\partial T}\,\Delta T + \frac{1}{n}\frac{\partial n}{\partial T}\,\Delta T + \frac{1}{n}\frac{\partial n}{\partial N}\,\Delta N \quad . \tag{64}$$

The first term is due to the thermal expansion of the cavity and is small in most systems considered to date [30]. The relative magnitude of the thermal and electronic index terms is equal to $n_2{}^T/n_2$.

$$n_2{}^T/n_2 = (\alpha A\ \partial n/\partial T)/(\kappa n_2) \quad . \tag{65}$$

The factor A is an area that depends on the geometric configuration in the thermal case. There is a critical value of A below which the electronic term dominates and vice-versa. $A = r_0 D$ for the filter example and $\kappa = \kappa_s$; $A = r_0{}^2$ and $\kappa = \kappa_{medium}$ for a thick sample with a small beam diameter. The critical area at which $n_2{}^T = n_2$ is of order 0.1 cm^2 for InSb, 10^{-7} cm^2 for ZnSe filters. This demonstrates emphatically that electronic effects dominate in InSb, thermal effects in the filters, given typically values of A of order 10^{-4} cm^2 and 10^{-5} cm^2 respectively.

The physical origin of $n_2{}^T$ is required if we are to optimize in the thermal case for materials and wavelengths. For near band-gap operation, the thermal movement of the edge dominates; $\partial n/\partial T \simeq \partial n/\partial E_g\ \partial E_g/\partial T$. Because the latter coefficient lies in a range -3×10^{-4} to -7×10^{-4} eV K^{-1} for the majority of materials [32] we are concerned with the form, primarily, of

$\partial n/\partial E_g$ for which a Kramers-Krönig analysis of the band edge should give a good approximation.

$$\frac{\partial n}{\partial E_g} = \frac{1}{\Delta E_g}\frac{c}{\pi}\int_0^\infty \frac{\alpha(E_g+\Delta E_g) - \alpha(E_g)}{\omega'^2 - \omega^2}\, d\omega' \quad . \tag{66}$$

In the scaling form used previously

$$\frac{\partial n}{\partial E_g} \propto \frac{-e^2}{n_0\ P\ E_g} F(\hbar\omega/E_g) \quad . \tag{67}$$

Using (25) for I_c^T, the frequency-dependence lies in $\alpha/(\partial n/\partial T)$; this factor is found to be only weakly enhanced near the band-edge [11]. Note that $\partial n/\partial T$ is positive for $h\omega \quad E_g$. This contrast with the electronic case is physically straightforward. Carrier generation blocks interband transitions and raises the effective band edge, conversely heating reduces the band edge.

Combining the cavity and microscopic considerations one obtains:

$$I_c^T \propto + \frac{E_g\ \kappa_s}{r_0\ |\partial E_g/\partial T|} F\left[\frac{\hbar\omega}{E_g}\right] \frac{f(R_f, R_b, \alpha D)}{\alpha D} \quad . \tag{68}$$

Larger gap materials are now favoured, but not dramatically so; the substrate thermal conductivity plays a similar tailoring role as did the recombination time in the electronic case.

In summary, so far we have obtained a variety of expressions for critical irradiances and powers, containing material and cavity factors. The optimization procedure is to use a knowledge of the microscopic parametric dependences to select materials and frequencies. Either theoretical or, where available, empirical dependences can be used. This determines the operational absorption coefficient. The cavity factor is then optimized in view of this operational coefficient. The microscopic mechanisms described have been exclusively of interband nature in conventional materials. For other, more specific bistability cases, different mechanisms are appropriate - excitonic and biexcitonic, multi-quantum well two-dimensional band-structure effects and excitons, as discussed by other authors in this proceedings, for example. Many of the general principles described here may however be adapted to cover these examples. In the following chapter, we shall show how the predictions fit experimental observations in the InSb and ZnSe-filter systems.

4. EXPERIMENTAL RESULTS

In the previous two chapters the steady-state, plane-wave, n_2 limit has been used in order to indicate trends from material-to-material, frequency-to-frequency and configuration-to-configuration. In practice the situation is far more complicated. (i) One is dealing with switching, which is inherently a time-dependent effect. It may in principle be influenced by cavity round-trip times, cavity lifetimes, laser pulse durations, carrier generation or sample heating times, carrier recombination or sample cooling times. (ii) For Gaussian beams nonlinear refractive lensing, diffraction, and electron or thermal diffusion should be considered. (iii) The nonlinearity, be it electronic or thermal, is not always going to be directly proportional to the irradiance level; also the absorption is not precisely linear as assumed above.

Various such thermal, spatial and high-order nonlinear aspects will be brought out in the following discussion of 'cw-switching' experiments in InSb and ZnSe-filters. The limiting solutions are shown to give a good indication of experimental trends, and are particularly useful in view of the variation of the additional functions between different experimental configurations, and also in the absence of analytic solutions for more precise models.

4.1 Free-carrier effects in InSb

Beyond the first-order bistability, there are possible higher orders, occurring as each successive Fabry-Perot peak is swept through the operational frequency. The cavity phase, ϕ, changes by π between each peak, so that in the n_2-limit one expects equal irradiance separations between each order. Referring to figure 2 this does not appear to be the case. Whilst spatial effects may prove to influence this phenomenon, it has been shown that deviations from n_2 and α_0 behaviour can satisfactorily explain similar observations at 77 K [33]. Figure 16a shows three orders of refractive bistabiilty and in excess of 10^{16} carriers cm^{-3} are considered to be generated at the higher irradiance levels. An empirical fit to lifetime measurements in InSb at 77 K indicates a relaxation of form:

$$\frac{1}{N}\frac{dN}{dt} \simeq (1.5 \times 10^6 + 1.5 \times 10^{-10} N + cN^2)\ s \quad . \tag{69}$$

The constant term is associated with trap-recombination mechanisms, the second with radiative recombination, the third with Auger recombination. Thus, for carrier concentrations N in excess of 10^{16} cm^{-3} the radiative process is significant. The coefficient c is small enough for Auger processes to be ignored in the present analysis. Further, the intraband absorption of the induced free-carriers, which is parasitic in respect of bistability at 5 μm because it reduces the Fabry-Perot finesse but does not contribute significantly to the nonlinear refraction, begins to exceed the linear interband absorption. At 1844 cm^{-1} for example,

$$\alpha_{total} \simeq (10 + 2.5.10^{-15} N)\ cm^{-1} \quad . \tag{70}$$

The combined effect predicted for the bistability is shown in figure 15, in good agreement with the experiment.

4.2 External addressing

Table 2 shows the range of optical sources (signals) that have been used to switch a cw CO laser beam. The role of the signal beam is in each case to generate free carriers and thereby change the refractive index, in principle at all frequencies. The index change will however be particularly pronounced at frequencies just

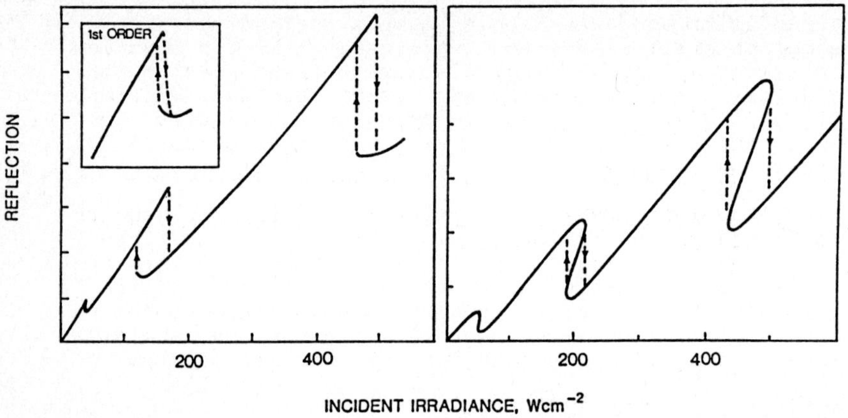

Fig. 15. Three orders of optical bistability: in reflection a) experiment, b) theory

below the band gap, as discussed in chapter 3. Hence a 5.5 μm beam at a (bias) power level such as to give transmission on the lower branch of a bistable loop may be switched to the upper branch by the temporary generation of sufficient carrier numbers. The increased internal 5.5 μm beam power should then be sufficient to maintain the additional carrier generation required for steady-state operation in the upper branch once the signal source is removed.

The most obvious experiment is to separate off part of the CO laser beam itself and reintroduce it as a switching signal having set the main beam at the correct bias point. It has proven possible to bias to within only a few microWatts of switch-up and then to cause switching with a signal of this low-power level [34]. The resultant change in the transmitted level can be thought of as an amplification of the signal. Operating with no hysteresis loop but a step-like characteristic (figure 16a),

Table 2. Modulation and switching of InSb optically bistable devices

Source	Wavelength (μm)	Operation Mode	Switching energy or Power
CO Laser	~ 5.5	Switching	~ 6 μW
Nd:YAG Laser	1.06 and 0.53	Switching	5 nJ
He:Ne Laser	0.6328	Modulation	10 mW
Photographic Flash	Colour temperature 5500 K	Switch 'on'	1 nJ
Xenon Lamp	5500 K	Modulation	~ W
InGaAsP Diode Laser	1.3	Modulation	~ 0.5 mW
PbSSe Diode Laser	5.5	Modulation	~ 50 μW

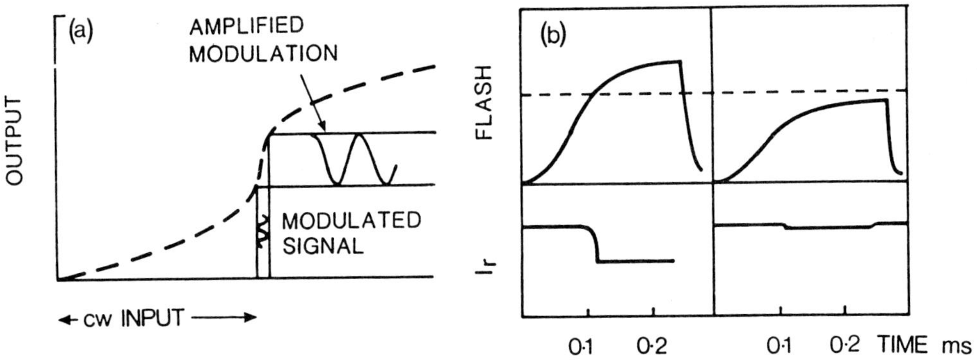

Fig. 16. a) Schematic of transphasor action. b) Reflection switching using an incoherent flash, showing switching at the same irradiance for different flash energies

amplification in excess of 200, with kHz modulation has been achieved. By analogy with voltage amplification using a transistor, this mode of operation has been called 'transphasor' action.

The use of a photographic camera flash as a switching signal has demonstrated both the lack of sensitivity to the wavelength (given $\hbar\omega_s \gtrsim E_g$) of the signal radiation, and the absence of a requirement for a coherent signal [35]. The flash duration is long compared to the carrier recombination time, and is then expected to generate a quasi-equilibrium population as the flash irradiance increases (figue 16b). For a 100 μm thick InSb sample switching indeed occurred at a specific irradiance level. The absorbed flash-energy during the 0.1 ms before switching was 1 nj; this number is however of little relevance in the device context, firstly because even at low transmission a steady absorption loss of order 1 nj per second occurs, furthermore even during the switching time (see section 4.4) the bias absorption is likely to exceed that of the signal.

4.3 Thermal effects

The above photographic flash experiments lead naturally to a discussion of various thermal effects, because similar investigations in a slightly less well-heat sunk 260 μm-thick sample produced dramatically different responses [35]. The main point is that optical bistability is a very sensitive effect. Thus, small bandgap changes (due to heating) will alter significantly the Fabry-Perot nonlinear response (figure 17). Switching was obtained for specific flash fluences rather than irradiances, for the 260 μm sample, indicating an energy-dependent (heating) effect.

Even more dramatic is the effect of the band-edge movement when heat-sinking is almost completely removed (figure 18a). Thus, a small absorption raises the temperature, the band-edge moves and the absorption coefficient at the operational frequency increases. This effect may avalanche to give thermal runaway to a high-temperature, low-transmission condition. The runaway may be gradual with the incident irradiance level, or it may switch suddenly with associated hysteresis as the irradiance is reduced again [36]. No cavity feedback is needed for the effect, which is known are purely-absorptive switching or "bistability by

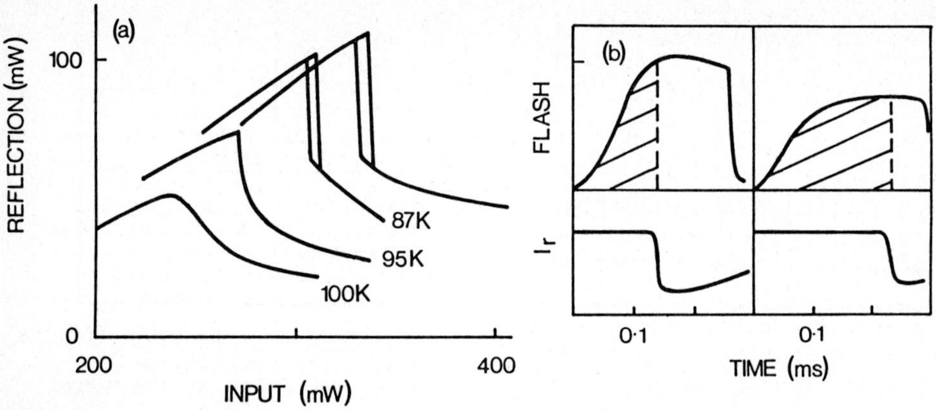

Fig. 17. a) Temperature-dependence of InSb reflection bistability. b) Reflection switching at fixed energy for different irradiances

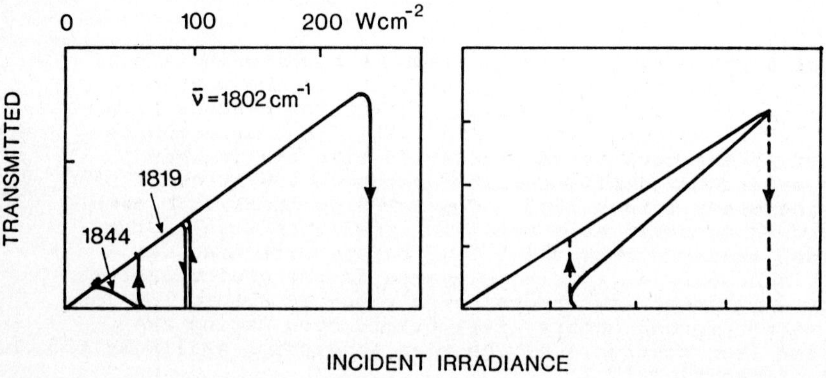

Fig. 18. Absorptive optical bistability a) experiment, b) theory

increasing absorption". This phenomenon has been observed for a number of materials [37].

Analysis of absorptive switching is achieved by appealing to three coupled relations;

$$\alpha = \alpha(\omega,T) \quad ,$$

$$\Delta T \propto I_0(1-a) \quad ,$$

$$I_t \propto I_0 a \quad . \qquad (71)$$

For a given temperature-dependent absorption, the temperature-rise may be used as a dummy-variable in similar manner to the use of I in the refractive problem. Critical-switching conditions occur by analogy. Thus, for example, if the absorption is of form $\alpha(\Delta T) = \alpha(0)\exp(\Delta T/T_0)$ then bistability occurs for $\alpha(0)D < .18$, $I_0 > 2.7\,A\,T_0$, where A is the coefficient of proportionality in (71b). The initial absorption $\alpha(o)D$ is seen to play a similar role in absorptive switching to the Fabry-Perot cavity detuning (δ) in the refractive case.

An example of both absorptive and refractive switching in the same sample is shown in figure 19a. The result was achieved for a 360 μm thick bulk ZnSe sample operated at room temperature and with 476 nm cw radiation from an argon ion laser [37]. The absorptive switching observed in the 250-350 mW region has precisely the same origin as described above. The transmission steps occurring at roughly 30 mW intervals are refractive, associated with small temperature increases in the low-finesse sample cavity. The cavity thickness is approximately 1500 optical wavelengths, so that a refraction change of order 1 part in 3000 is adequate to produce a π phase-change. The thermal refractive coefficient is $\partial n/\partial T \simeq 3 \times 10^{-4}\ K^{-1}$ so that a 2 K temperature-rise is adequate to produce thermal refractive optical bistability. The electronic nonlinearity is insignificant in this relatively large-gap semiconductor. These observations are quite consistent with the predictions of chapters 2, 3.

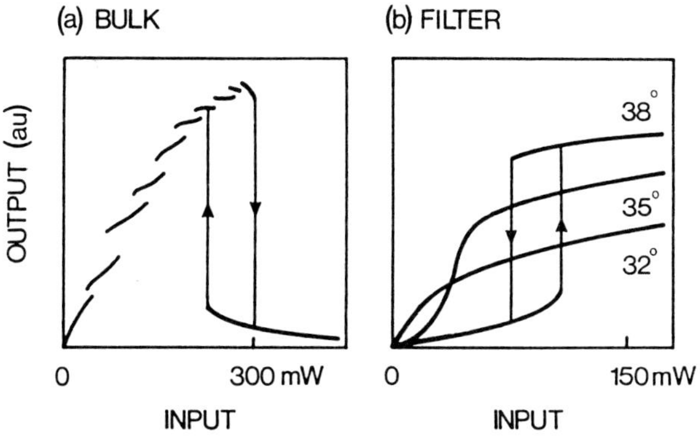

Fig. 19. a) Thermal refractive bistability and absorptive switching in bulk ZnSe. b) Refractive nonlinearity of a ZnSe-filter, for various angles of incidence.

Most ZnSe work has been carried out using dielectric thin film interference filters. A relatively high cavity finesse is achieved using reflective stacks of perhaps 80-90% reflectivity and a spacer of αD-value of order .04. However the filter material, deposited by thermal evaporation methods, is not crystalline, and typically has a loss corresponding to $\alpha \simeq 10^3\ cm^{-1}$ in the (argon) 514 nm region. Whilst the improved finesse compared to bulk ZnSe allows for switching in such small ($\simeq$ 1 μm) spacers at almost identical power levels (20-100 mW) to those in the bulk, the absorbed energy per unit volume is larger, and the temperature-change at switch point is markedly higher (50-100 K). The attraction of filter work lies, however, in the good 2-D uniformity, over square centimetres or more, and the potential for cavity design and thermal engineering. Figure 19b shows filter nonlinear and optically-bistable outputs for different orientations with respect to the incident beam of a narrow-band-pass filter [38]. The orientation effectively determines the initial detuning of the system.

4.4 Switching speed

Of the timescales mentioned at the beginning of this section the cavity round-trip time, in the case of semiconductor samples, is of the order of picoseconds. This is far smaller than the other relevant times, and allows for adiabatic elimination of the optical field time-dependence in any dynamic analysis. The internal irradiance is assumed to adjust instantaneously to any refractive index-variation of the medium, to a good approximation.

$$\frac{I(t)}{I_0} = \frac{(1-R_f)(1+R_b a)(1-a)}{(1-R_\alpha)^2\ \alpha D} \frac{1}{1+F_\alpha \sin^2 \phi(t)} \quad . \tag{72}$$

However, the phase ϕ is determined by the electron concentration (or the sample temperature) so that the response of this material parameter to any input signal changes will dominate the dynamics of switching.

In the electronic case

$$\phi(t) = M\,\pi + \delta + (2\pi D/\lambda_v)(\partial n/\partial N)\ \Delta N(t) \quad , \tag{73}$$

$$\frac{d\Delta N(t)}{dt} = -\frac{\Delta N(t)}{T_1} + \frac{\alpha I(t)}{\hbar\omega} \quad . \tag{74}$$

Thus, for example, consider a switching signal I_s incident after time t_s, in addition to a bias level I_0. Because ΔN is constant under cw conditions, the approach to switching is driven initially by I_s alone. The effective time-scale for the dynamics is the recombination time T_1 under small signal conditions. Approach to switching requires a time proportional to $(I_s-I_{sw})^{-\frac{1}{2}}$ where I_{sw} is the signal irradiance needed to reach the switch point exactly [39,40]. This switching delay is known as critical slowing down; the implication is that energy-loss during switching cannot be minimised by biasing as close as possible to the switch point as noise limitations would allow.

In the thermal case the same phenomenon occurs. Here one needs to solve the thermal equation for the temperature time-dependence.

$$\rho\ c_p\ dT/dt = \kappa\ \nabla^2 T + Q \quad , \tag{75}$$

where Q is the heat dissipated per unit volume per second. For large laser-spot, filter studies, the essential form of this equation, demonstrating the effective cooling (T_1) and heating times is:

$$\frac{d\Delta T(t)}{dt} = -\frac{\Delta T(t)}{(\rho c_p r_0 D/\kappa_s)} + \frac{\alpha I}{\rho c_p} \quad . \tag{76}$$

As an example of the thermal time-scale: (i) Rapid heating to generate a 10 K temperature rise in ZnSe using 10 mW focussed to a 1 μm radius spot would require in principle 10 ns. That is $\Delta t \simeq \rho c_p \Delta T/\alpha I$ with $\rho = 5.3\ \mathrm{gcm^{-3}}$, $c_p = .07\ \mathrm{J\ gm^{-1}K^{-1}}$. (ii) The cooling time, $T_1 \simeq \rho c_p r_0 D/\kappa_s$ is of order 2 μs for $D \sim 5$ μm. $\kappa_s = .01\ \mathrm{W\ k^{-1}\ cm^{-1}}$. (iii) In the electronic case the carrier generation time, $\Delta t \simeq \Delta N\ \hbar\omega/\alpha I$, for generating 10^{15} carriers $\mathrm{cm^{-3}}$ is of order of 30 ns using $\alpha = 10\ \mathrm{cm^{-1}}$, $I = 100\ \mathrm{W\ cm^{-2}}$. (iv) T_1 is typically 1 μs - 100 ns for carrier recombination.

Two types of switching-speed experiment have been carried out; short pulse or stepped signal switching in the bistability mode, and modulated signal switching in the transphasor mode. In the former case for example, 35 ps pulses at 1.06 μm wavelength, from a Nd:YAG laser produced switching in InSb for a 5 μm beam held approximately 0.4 mW from the switch point [41]. The required pulse energy was 5 nj. This corresponds to an average excited free-carrier density in the irradiated portion of the sample of order 10^{15} cm^{-3}, concentrated in the surface region. No attempt was made to determine the time taken for switching. The experiment did demonstrate, however, that sufficient energy could be injected within 35 ps, to cause switching. By using two YAG pulses, the first of one-half the required switching energy, and the second of this energy or above and delayed with respect to the first, a measure of the carrier recombination time was obtained. The measured 100 ns value was consistent with the relatively rapid recombination near the sample surface, where the 1.06 μm radiation is absorbed.

A recent experiment, in which critical slowing down has been dramatically demonstrated, is the use of a step function input power for filter switching [42]. Operating again within a bistable loop, with a 100 μm diameter beam and stepping to a power level marginally above switching, it took some 360 ms before the transmission jumped to a steady high-output level. By comparison use of larger step, to twice the power required for switching, only a 500 μs delay occurred. Use of lower spot-size enabled 30 μs switching to be obtained.

In the transphasor (non-hysteretic) mode of operation, modulated laser diodes have been used to test the response speed of the InSb devices. Both InGaAsP diodes working at 1.3 μm and PbSSe diodes at 5.5 μm have been used [43]. Figure 20 shows typical results. For cw holding in a region of amplification (figure 16a) modulation is accrued on the hitherto cw beam by use of a modulated signal beam; a linear amplification is obtained over a range of signal amplitudes. This amplification factor represents the device gain. As the modulation frequency is increased, the gain eventually falls, figure 20. Different curves on the figure correspond to different initial gain values obtained by changing the initial detuning, and hence the transphasor characteristic.

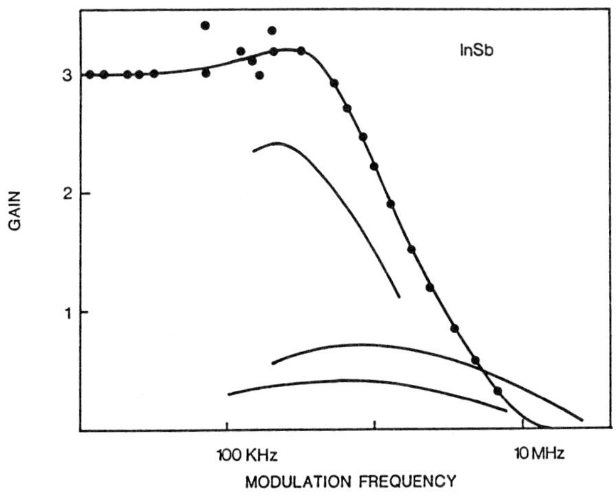

Fig. 20. Modulation-frequency-dependence of transphasor gain

Analysis is achieved using (72-74) in slightly modified form If the cavity feedback on the modulated signal beam is ignored initially (this is valid for 1.3 μm signals) then the dynamic equation for the phase becomes,

$$\frac{d\phi}{dt} = -\frac{\phi}{T_1} + \frac{A_0 I_0}{1+F_\alpha \sin^2\phi} + \phi_m (1+\cos\Omega t) \quad . \tag{77}$$

ϕ_m is proportional to the internal irradiance modulation associated with the signal beam. A_0 contains reflection, absorption, and n_2 factors. One assumes a solution of form,

$$\phi = \phi_{cw} + \phi_1 + \phi_2 \cos(\Omega t+\theta) \quad . \tag{78}$$

where ϕ_{cw} is the internal phase in the absence of the modulation. The solution is

$$\phi_1 = \phi_m \{1 + A_0 I_0 F_\alpha \sin 2\phi_{cw} (1+F_\alpha \sin^2 \phi_{cw})^{-1}\}^{-1} \quad ,$$

$$\phi_2 = \frac{\phi_1}{(1+\Omega^2 T_1{}^2/\phi_1{}^2)^{\frac{1}{2}}} \quad ; \quad \theta = -\tan^{-1}\left[\frac{\Omega T_1}{\phi_1}\right] \quad . \tag{79}$$

The gain of the device is directly proportional to ϕ_2/ϕ_m and depends on the modulation frequency only through the denominator in (79). In particular, this means that the product of the zero-Ω gain and the frequency at which the gain drops to one half of this value - the gain-bandwidth product - is independent of the initial cavity detuning, it is a direct measure of the recombination time. This constant product is found experimentally.

4.4 Spot size dependencies

It was recognised above that in the filter work, lower switching times were obtained at lower spot diameters. In addition, the switching power reduces with spot size. Results are shown for bulk InSb (figure 21a) and for filters (21b). Note that in the

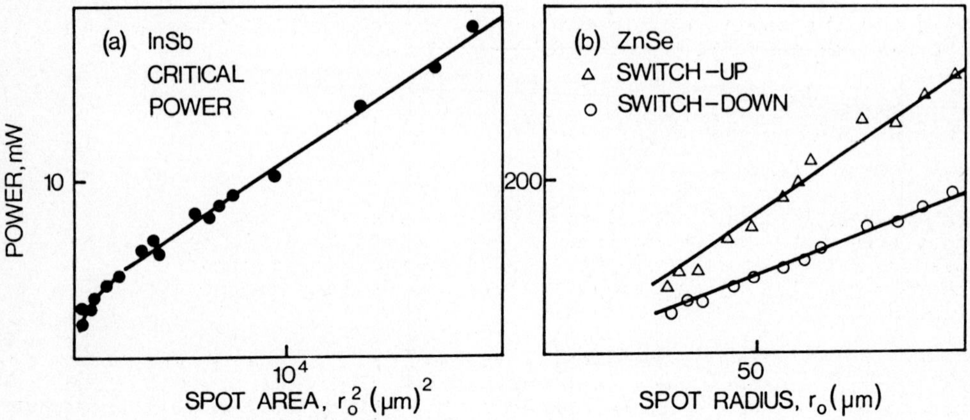

Fig. 21. Spot-size dependence of the critical irradiance I_c in a) InSb, b) a ZnSe-filter

latter case a linear dependence on spot diameter is obtained, in agreement with (28), due to heat loss to the substrate. In the bulk-sample electronic case (27) pertains in the plane-wave limit and the switching power diminishes in proportion to the spot area; this is found experimentally. The relative reduction in power at low-spot radii is associated with electron diffusion [44].

For InSb the proximity within which two beams can be made to switch independently has been studied, as has the limiting process of electron diffusion. A diffusion length of order 60 μm in InSb [45] requires that spot separations in excess of 100 μm are needed to prevent cross-talk. In summary, cw holding beam optical bistability has been observed at 10 mW power levels and 100 W cm^{-2} irradiance levels in both InSb and in ZnSe filters.

In the InSb work a number of switching sources have been employed, with signal power levels down to 3 μW. Photographic flash experiments demonstrate the use of bistable devices for incoherent-to-coherent optical conversion. InSb switching speeds are less than 500 ns and spatial resolution is of order 100 μm. Transmission and reflection bistability and transphasor action has been observed, and a number of logic gates, described in the next section, have been operated. Two-gates on a single sample have been coupled.

InSb has a very large nonlinearity indeed and is relatively fast. The disadvantages of the system are (i) Operation at 5 μm, with relatively difficult experimental alignment problems, low pixel packing density (both due to the large wavelength and to the electron diffusion). (ii) Cryostatic operation with consequent difficulty in coupling devices together and in achieving flexible optical imaging. (iii) InSb itself is not excellent for fabrication purposes, large plane-parallel samples are presently obtained only with great care.

In contrast, the interference filters operate in the visible (green at present), they can operate at room temperature because the band-gap energy is large compared to kT, and with available powerful argon sources. The established filter technology allows for uniform two-dimensional plates. Resolution of order 10-100 μm appears possible, with speeds of order 10 μs or less. Very recently, incoherent-to-coherent conversion has been achieved [46], small arrays of spots have been used with independent switching [47] and the basics of a computational loop circuit, described in the next chapter, have been constructed [48].

5. DEVICE CONSIDERATIONS - TOWARDS AN OPTICAL COMPUTER

5.1 Introduction

In chapter 1 it was stated that an inverter-like response characteristic and an optical hysteretic response are the required basics of an optical computer (see figure 1). The actual characteristics of nonlinear Fabry-Perot devices are schemed in figure 22.

The non-hysteretic reflection characteristic is similar to the inverter and the hysteretic transmission is clearly a memory-like response. In addition, the optically bistable plates are able to operate generally both in transmission and reflection, simultaneously, giving considerable response flexibility. Most

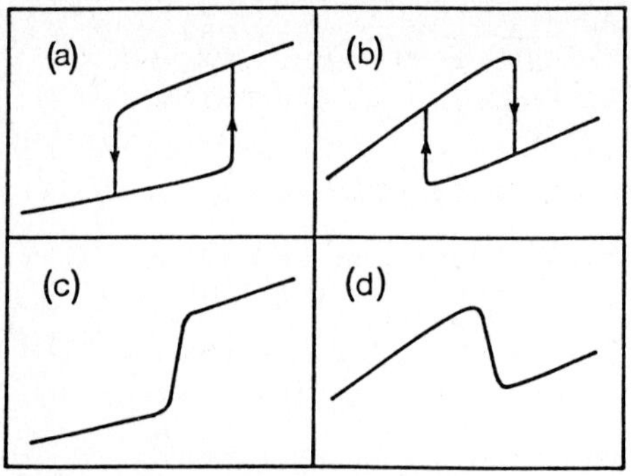

Fig. 22. Nonlinear Fabry-Perot characteristics available for computational elements

importantly it should be possible to operate, independently, large numbers of pixels on a single bistable plate, and to interconnect each of these pixels optically to single or multiple pixels on further plates, using free-space propagation and lens or holographic imaging techniques. This innate parallelism of optics is seen as its fundamental advantage in digital information processing by comparison to electronics. Whilst up to 10^6 electronic gates may presently be fabricated onto a single chip, they operate to a large extent sequentially, and the number of simultaneous interconnects to neighbouring chips is at best just 300. The optical parallelism leads naturally to array processor architectures for digital optical computation, optical image-processing (of either 2-D input images or of sequential optical signals that are converted to array images), spatial light modulation, and displays. Optical fibres are already being introduced into electronic computers, in order to aid the interconnect problems at various levels of the computer architecture (processor-to-processor, board-to-board, chip-to-chip; and for clock broadcasting). Additional device potential for bistable semiconductor devices lies in the areas of directional coupling (transmission or reflection), signal amplification in the transphasor mode (which may also be employed for analogue image-processing) such as is required at fibre communications repeater stations, and power stabilization at incident levels between the first- and second-order switching point, for example.

Figure 23 schemes a real-time spatial-light-modulator configuration. Using the transphasor mode of operation, the input image is in effect amplified and encoded onto the laser beam array. Use of a polarization beam splitter to bring together the laser array and the image avoids insertion loss - note that there is no polarization requirement on the 'switching' signal. The light-modulator may form the front-end of an image processor such as a robot eye would require, the laser-encoded image could then interconnect to Fourier transform/correlation stages etc. Such a system might be used either for the full image processing, or more likely as an intermediary stage prior to electronic processing, using the optical stage to reduce the input information rate in a significant manner to one that can be handled electronically. Alternatively, the modulator could be used in the bistable mode, for digital image-processing - as an input to an optical computer in effect, or simply as a display device. In the latter system the switching signal would be a 'write' light, possibly incoherent

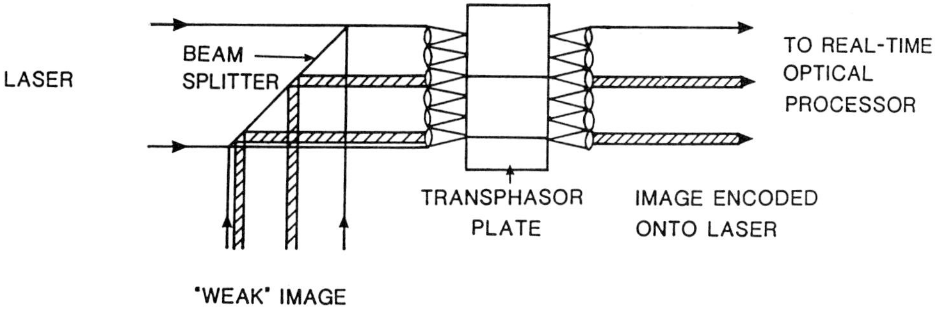

Fig. 23. Spatial light modulator schematic

but most likely from a laser diode; the bias laser switches through to high transmission on those pixels over which the write beam has passed. The written image is then held (displayed) for as long as the (cw) bias beam is maintained.

5.2 Digital optical computer architecture

Binary computation is based on the sets of logic operations that describe the possible outputs (0 or 1) for different combinations of multiple signal inputs. Table 3a shows the four possibilities for output combinations for a single input; table 3b shows eight possibilities for a pair of inputs. In the latter case there are a total of sixteen possible output combinations, the missing eight rely on one being able to distinguish between input signals (A = 0, B = 1 distinguishable from A = 1, B = 0).

All processes other than OFF and ON require a logic 'GATE' at which a decision is in effect made. For example, in the NOR-gate: IF signals A and B are both 0 (low-voltage electronically or low-power optically) THEN the output is 1 (high voltage or power), ELSE the output is 0. The names of the various gates are shown in

Table 3. Truth tables for binary logic a) One input, b) Two inputs

INPUT LEVEL	LOGIC GATE OUTPUTS OFF	NEGation	AFFirmation	ON
0	0	1	0	1
1	0	0	1	1

INPUT LEVELS A & B		LOGIC GATE PROCESSES OFF	AND	XOR	NOR	NAND	XNOR	OR	ON
0	0	0	0	0	1	1	1	0	1
0	1	0	0	1	0	1	0	1	1
1	1	0	1	0	0	0	1	1	1

the figure. Any operation or set of operations from which all others can be constructed by combinations in sequence may be used as the basic building blocks of a central processing unit. Example sets are (OR + NEG), (AND + NEG), or the single operations NOR or NAND. A computational logic processor might for example be constructed entirely of NOR-gates. Hence the emphasis on the inverter characteristic in the introduction. With a threshold at a signal level between 0 and 1 the inverter acts as a NOR logic gate.

Whilst the reflection-mode nonlinear Fabry-Perot does indeed provide a NOR-gate function, a combination of just two transphasor elements can in principle be constructed to produce any of the responses of table 3, subject to programming by adjustment of bias beam levels. The device is schemed in figure 24. The first plate (operating in reflection) is the logic processor, the second (in transmission) is used as a discriminator [49]. Programmed identical logic operations may be performed on complete 2-D arrays of image signals using such a device, enabling whole-image addition etc.

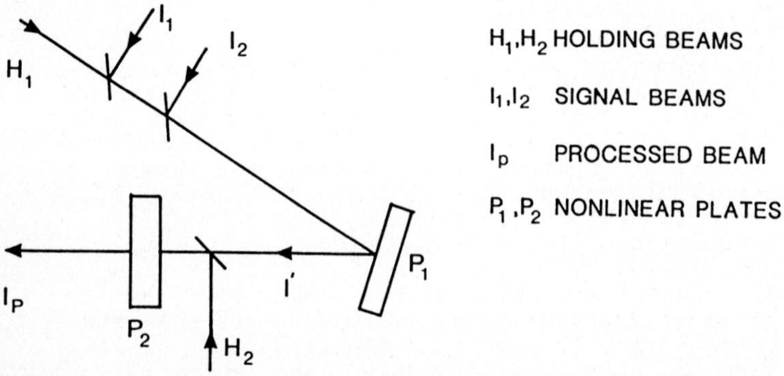

Fig. 24. Programmable optical logic processor

Sequential coupling of a few optical plates (as above) will be essential to any computational system. In the programmable processor, the high and low outputs of the first plate must have a signal difference large enough to drive the second in a noise-immune manner. A more precise requirement is placed on the tandem operation of bistable plates. Here it must be possible to operate a plate biased in the lower branch of an hysteretic loop, such that the high signal from a previous plate (attenuated or amplified if necessary) causes switch-on, whilst the low signal (identically modified) does not cause switch-on. The phenomenon is known as 'cascadability'. A one-plate to two-plate cascadability (or fan-out). is essential, and in principle sufficient for image-processing, because further fan-out may be achieved by repeated two-way cascading if necessary. Figure 25 shows for the two cw semiconductor systems that cascading is already possible [47]. The InSb contrast ratio is in excess of 3:1, which clearly allows one-to-two fan-out.

Cascading has been demonstrated experimentally in both systems. Further, in the InSb case two adjacent pixels on a single plate have been operated sequentially, figure 26. The reflection from one pixel was fed back as the switching signal for the reflection of the second. When the holding beam on the first pixel was

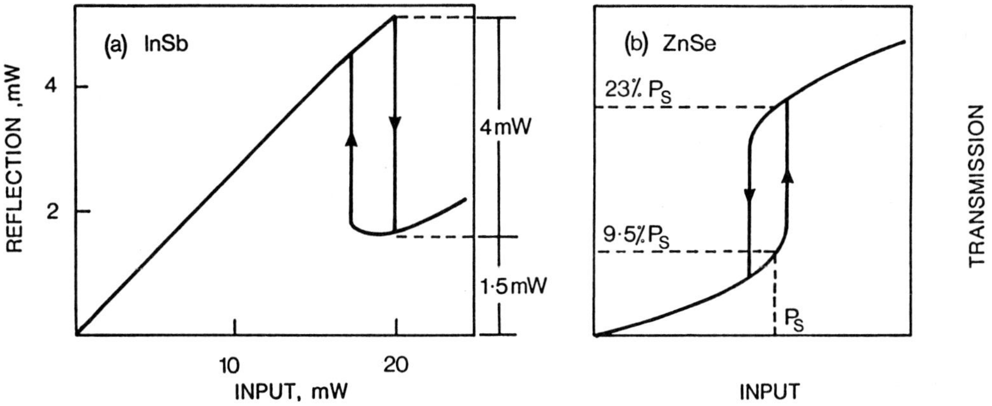

Fig. 25. Demonstration of cascadability potential in a) InSb, b) ZnSe-filters

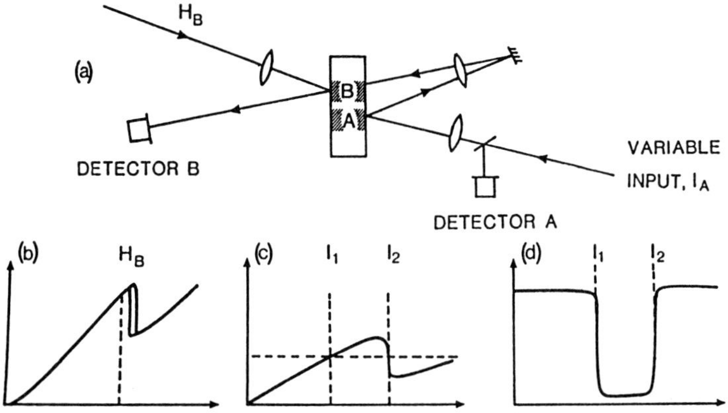

Fig. 26. a) Two coupled elements on a single optical chip.
b) Reflection characteristic of B, held at H_B for coupling.
c) Reflection of A as input I_A is increased. d) XNOR gate logic using the coupled circuit

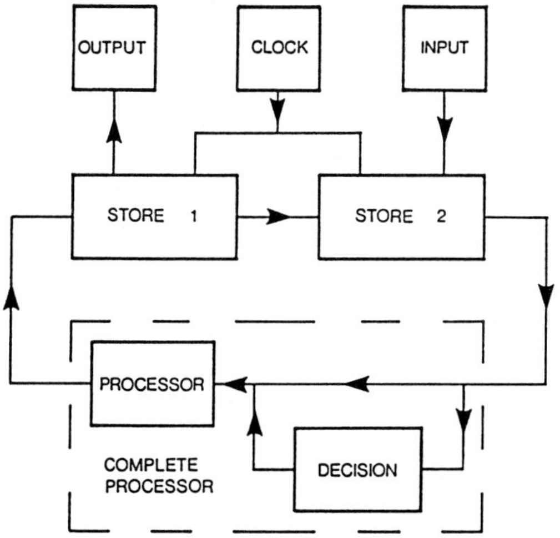

Fig. 27. Generic optical computation loop, a classic finite-state machine.

intensified to cause reflection switching of that pixel from high to low, then the reflection of the second pixel simultaneously switched from low to high [33].

Given even one-to-one cascadability, it is possible to construct a latching memory system such as would be required in a looped computational processor. Figure 27 shows a generic design for such a machine [49]. The input, which should be considered as an array image, is stored. At the appropriate time it is passed through to the central processor unit, which contains elements such as the programmable logic gate described above, elements that are not necessarily hysteretic. The processed image is therefore connected through to a second store, it must be held there and not allowed through into the first store before this is cleared, or else the two images (preprocessed and processed) will contaminate each other. The purpose of the clock is to control passage of images around the circuit.

A design for the store pair is shown in figure 28a. It consists of three bistable plates each addressed by holding beam arrays. The holding beam power-level is clocked to be either at the cente of the switch region (as shown for plates 1,3) or well below this region (perhaps totally off). An input image beam-split onto plate-3 will map through to the region S_3, producing a simultaneous associated spatial refractive index-variation across plate-3. Removing the input image, this index variation is held by H_3. The image is processed and passes through to plate-1 where it maps through directly to region S_1. Providing the holding beam on plate-2 is at a low level, then the output of this plate is uniformly dull and does not influence the operation of plate-3. The H_1 beam is now switched to low and the preprocessed image will be removed from region S_1. The processed image is simultaneously removed from S_p; however, it still remains in S_1, held by H_1. Continuing similar arguments, the processed image may be mapped through to S_2 etc., without image contamination, and the system is primed for a second cycle of processing. Such a loop, with many parallel information units, is known in computer jargon as a classic finite-state machine. Significantly, the ON-OFF image power levels are restored to standard values by plate-1 for each cycle. There is no error accumulation.

Optical clocking systems that work at bistability switching rates and therefore compatible with the store and processor time-scales have also been devised [9,49]. For example, figure 28b shows three plates operating in reflection such that a high

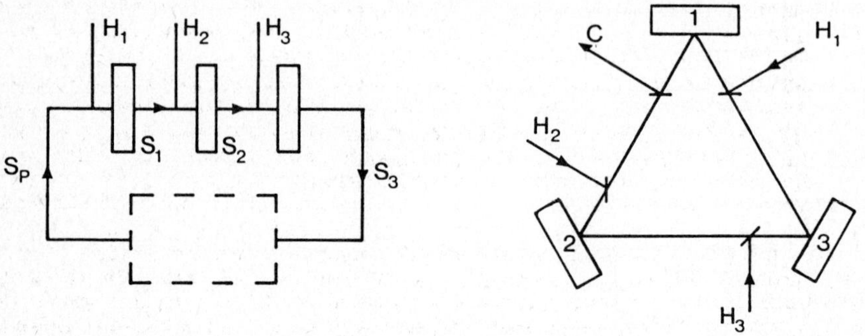

Fig. 28. a) Store-pair, or optical time delay. b) All-optical clock design

reflection from any one gate, fed onto the following gate, instigates switching and hence a low reflection from the second gate. Looping around this circuit, it can be seen that the looped signal incident at any particular gate on one cycle is always incompatible with the state of that gate as determined on the previous cycle. Therefore each gate reverses its state once per cycle and any signal (a single beam in this device), extracted either as a transmission or part of the reflection of a gate, will clock at the cycle rate.

The components of a parallel all-optical computational loop can thus in principle be constructed. A number of ideas, for example processing stages that accomplish particular physical tasks, have also been put forward - simultaneous (parallel) integration of sets of input binary numbers for example [50]. This device also demonstrates that one-element full-adder sum and carry operations are possible; by comparison, fourteen NOR gates are used in electronics. On the experimental side, a three-plate store and cycle system has been constructed using ZnSe-filters, and clocked manually [48]. Groups at the University of Southern California and at Ohio State University have demonstrated versions of the optical clock [51,52]. They have used liquid crystal light valves as very slow non-hysteretic NOR gates, and have mimicked electronic flip-flop clocks and master-slave clocks by interconnecting the outputs of specific valve pixels back around to different pixel positions. Computational loops with processing stages have not yet been set up, nor have multi-beam array loops. It must be stressed, however, that as far as the cw bistable devices are concerned, plates now exist which do have the essential optical response characteristics required of computational circuit elements. It is a matter of time, optical imaging, cavity optimization (as discussed in chapers 2, 3) Fabry-Perot/filter fabrication, and thermal engineering, as to when and whether machines that are viable complements to electronic computers will be constructed.

5.3 Summary

In view of the impressive achievements of electronic computers, it is perhaps difficult to imagine how improvements could be made, or indeed why one should wish to make them. A very brief history of computing may help explain why there is considerable interest and optimism at present concerning the development of digital optical computational machines.

The use of digital calculation aids goes back notably to the (undated) invention of the abacus, also to seventeenth century inventions such as Napiers 'bones' and Pascals decimal counter [53]. The nineteenth century saw the construction of various mechanical calculators, for example the Burrough's machines began to be retailed; and suggestions for complex mechanical computers (calculators with programmed instructions) such as Charles Babbage's difference engine began to appear. Despite this background, the invention of the electronic diode by Ambrose Fleming in 1904, and of de Forest's triode in 1907, did not by any means lead to an immediate recognition of the scope for electronics in computational devices. Early digital computers using programmed instructions were in fact produced from 1937, by IBM initially, using firstly magnetically controlled clutches and then electromagnetic relays. It was not until 1940 that using electronic valves, connected in relays based on the 1919

Eccles-Jordan trigger valve circuit, the first digital electronic machine, the "Electronic Numerical Integrator and Calculator" (ENIAC), was constructed. Designed in the USA by Eckert and Mauchley, the system contained eighteen thousand valves, and operated at a speed of five thousand additions per second. However, it required a room full of equipment and needed a twenty four hour replacement service for overheated components. ENIAC used a decimal digital system with rings of ten relay circuits. Binary electronic valve-computers did not arrive until almost a decade later with the post-war EDSAC, ACE and EDVAC machines. Also in the late 1940's Bardeen, Brittain and Schockly invented the transistor; once again, machine development was initially very slow. In 1958 Jack Kilby of Texas Instruments was the first person to fabricate two transistor components on a single silicon chip; thirty components per chip had been achieved by 1965. The next ten years saw the major breakthrough in electronic transistor computing, by 1975 large-scale integration (LSI) with thirty thousand components per chip was achieved, and VLSI can now boast one million gates on a five millimetre square chip. The computational power of ENIAC (and considerable more) has been reduced to microprocessor dimensions. The apparently humble home computer can achieve a million operations per second and at the present technological limits the Goodyear parallel array processor runs at 10^{11} operations per second. Despite this enormous ability there are even higher demands. For example, optical fibre communications links can carry far in excess of the processable information rate, and there is growing interest in putting this high bandwidth availability to use for mixed speech, data and video transmission. Also, NASA-launched Landsat satellites looking down at the earth's surface detect such a vast amount of information in optical images that 10^{14} operations per second would be required for real-time processing. Not surprisingly, military demands are growing for high-speed decision-making based on processing of huge data accumulation rates. In addition, the electronic interconnect bottleneck has already been described, individual electronic gate speeds are limited to 100 ns and 1 ns in Si and GaAs technology respectively, interfacing with optical images or fibre signals is in any event becoming necessary, chips are complex to fabricate and despite the enormous effort that has been invested in their manufacture, the wastage rate is still high.

In comparison, it has been suggested that optical switches might be constructed, based on materials other than those described in the bulk of this article, with switching times below 1 ps and with relatively low power consumption, figure 32 [54,55]. One has parallelism, free-space interconnects and already optical disc nonvolatile storage operating to the advantage of optics. It is possible to quote numbers that appear to give optical capacity far above that described for electronics; the following fairly conservative figures serve to demonstrate optical potentiality. Given operation at 0.5 μm radiation wavelength, a pixel separation of 10 μm appears quite feasible. Heat dissipation will be a problem, but so also is it for electronic machines such as the powerful Kray processors, cooled using Freon. A gate-number of 10^6 pixels on a one centimetre square plate (with the possible equivalence of each optical gate to several electronic NOR gates) is obtained for this separation. A switching time of 1 μs, marginally smaller than has already been demonstrated in cw devices, gives a capacity for 10^{12} operations per second per plate - that of the entire Goodyear processor.

We are currently at a similar stage to 1958 electronics. The equivalent of Jack Kilby's two elements on silicon has been achieved - we have coupled two bistable devices on a single 'optical chip'. We have the advantage of scientific hindsight in that electronic architectures and machines have been developed, but the disadvantage that the enormous investment in electronics will not be paralleled in optics unless superior specialist or perhaps even general purpose optical machines can be envisaged. However, given a fraction of the invested effort in microelectronic integrated circuitry over the past quarter of a century, it would be very surprising if digital optical devices of significant value were not constructed before the turn of the century.

REFERENCES

1 Papers in 'Proc. Symp. Optical & Electro-Optical Information Processing, Boston, 1964, Ed. J. Tippett, D. Berkowitz, L. Clapp, C. Koester and A. Vanderburgh, MIT Press (1968).

2 A. Szöke, V. Daneu, J. Goldhar and N.A. Kumit, Appl. Phys. Lett., 15, 376 (1969).

3 H.M. Gibbs, S.L. McCall and T.N.C. Venkatesan, Phys. Rev. Lett., 36, 1135 (1976).

4 T.N.C. Venkatesan and S.L. McCall, Appl. Phys. Lett., 30, 282 (1977).

5 T. Bischofberger and Y.R. Shen, Opt. Lett., 4, 40 (1979).

6 D.A.B. Miller, S.D. Smith and A.M. Johnston, Appl. Phys. Lett., 35, 658 (1979).

7 D. Weaire, B.S. Wherrett, D.A.B. Miller and S.D. Smith, Opt. Lett., 4, 331 (1979).

8 H. Kogelnik and T. Li, Appl. Opt., 15, 1550 (1966).

9 B.S. Wherrett, IEEE J. Quantum Electron., QE-20, 646 (1984).

10 D.A.B. Miller, IEEE J. Quantum Electron., QE-17, 306 (1981).

11 B.S. Wherrett, D. Hutchings and D. Russell, JOSA, to be published.

12 I. Janossy, M.R. Taghizadeh, J.G.H. Mathew and S.D. Smith, IEEE J. Quantum Electron., Special Issue on Optical Bistability, in press (1985).

13 P.A. Franken and J.F. Ward, Rev. Mod. Phys., 35, 23 (1963).

14 R.W. Terhune, P.D. Maker and C.M. Savage, Phys. Rev. Lett., 8, 401 (1962).

15 G. Mayer and F. Gires, C.r. Hebd. Seanc. Acad. Sci., Paris, 258, 2039 (1964).

16 P.D. Maker, R.W. Terhune and C.M. Savage, Phys. Rev. Lett., 12, 507 (1964).

17 P.D. Maker and R.W. Terhune, Phys. Rev. A, 137, 801 (1965).

18 C.K.N. Patel, R.W. Slusher and P.A. Fleury, Phys. Rev. Lett., 17, 1010 (1966).

19 e.g. O. Svelto, Prog. Optics, 12, 2 (1974).

20 A. Javan and P.L. Kelley, IEEE J. Quantum Electron., QE-2, 470 (1966).

21 J.P. Woerdman and B. Bolger, Phys. Lett. A, 30, 164 (1969).

22 K. Jarasuinas and J. Vaitkus, Physics Status Solidi A, 23, K19 (1974).

23 E.E. Bergmann, I.J. Bigio, B.J. Feldman and R.A. Fisher, Optics Lett., 3, 82 1978).

24 D.H. Auston, S. McAfee, C.V. Shank, E.P. Ippen and O. Teschke, Solid St. Electron., 21, 147 (1978).

25 H.J. Eichler, Ch. Hartig and J. Knof, Phys. Stat. Sol., a45, 433 (1978).

26 R.K. Jain and M.B. Klein, Appl. Phys. Lett., 37, 1 (1980).

27 R.K. Jain, Opt. Eng., 21, 199 (1982).

28 H.M. Gibbs, S.L. McCall, T.N.C. Venkatesan, A.C. Gossard, A Passner and W. Wiegmann, Appl. Phys. Lett., 35, 6 (1979).

29 D.A.B. Miller, C.T. Seaton and S.D. Smith, Phys. Rev. Lett., 47, 197 (1981).

30 B.S. Wherrett, Proc. Roy. Soc. (London), A390, 373 (1983).

31 e.g. B.G. Levich, "Theoretical Physics", vol. 2, N. Holland (Amsterdam) (1971).

32 Landolt-Börnstein Numerical Data & Functional Relationships in Science & Technology, Group III, vols. 17 a,b, Springer-Verlag, Berlin (1982).

33 A.C. Walker, F.A.P. Tooley, M.E. Prise, J.G.H. Mathew, A.K. Kar, M.R. Taghizadeh and S.D. Smith, Proc. Discussion Meeting "Optical Bistability, Dynamical Nonlinearity and Photonic Logic", London (1984), Ed. B.S. Wherrett and S.D. Smith, p. 59, The Roy. Soc. (1985).

34 F.A.P. Tooley, Ph.D. Thesis, Heriot-Watt University, unpublished.

35 F.A.P. Tooley, A.C. Walker and S.D. Smith, Proc. Discussion Meeting "Optical Bistability, Dynamical Nonlinearity and Photonic Logic", London (1984), Ed. B.S. Wherrett and S.D. Smith, p. 167, The Roy. Soc. (1985).

36 B.S. Wherrett, F.A.P. Tooley and S.D. Smith, Optics Commun., 52, 301 (1984).

37 A.K. Kar and B.S. Wherrett, JOSA, to be published.

38 S.D. Smith, J.G.H. Mathew, M.R. Taghizadeh, A.C. Walker, B.S. Wherrett and A. Hendry, Optics Commun., 51, 357 (1984).

39 J.A. Goldstone and E.M. Garmire, IEEE J. Quantum Electron., QE-17, 366 (1981).

40 W.J. Firth, private communication.

41 C.T. Seaton, S.D. Smith, F.A.P. Tooley, M.E. Prise and M.R. Taghizadeh, Appl. Phys. Lett., 42, 131 (1983).

42 J.G.H. Mathew, private communication.

43 F.A.P. Tooley, W.J. Firth, A.C. Walker, H.A. MacKenzie, J.J.E. Reid and S.D. Smith, IEEE J. Quantum Electron., to be published.

44 W.J. Firth, I. Galbraith and E.M. Wright, J. Opt. Soc. Am., in press.

45 D.J. Hagan, H.A. MacKenzie, H.A. Al-Attar and W.J. Firth, Opt. Lett., to be published.

46 J.G.H. Mathew and M.R. Taghizadeh, private communication.

47 S.D. Smith, Proc. Discourse at the Royal Institution, to be published.

48 S.D. Smith, A.C. Walker, B.S. Wherrett, F.A.P. Tooley, J.G.H. Mathew, M.R. Taghizadeh and I. Janossy, J. Opt. Soc. Am., to be published.

49 B.S. Wherrett, Appl. Opt., in press, Sept. (1985).

50 B.S. Wherrett, Optics Commun., in press (1985).

51 A.A. Sawchuk and T.C. Strand, Proc. IEEE, 72, 758 (1984).

52 S.A. Collins, S.F. Habiby and A.F. Zwilling, Proc. Agard Conf., Schliersee, Sept. (1984).

53 T.E. Ivall, "Electronic Computers", Iliffe & Sons Ltd., London (1960).

54 P.W. Smith, Bell Syst. Tech. J., 61, 1975 (1982).

55 D.A.B. Miller, D.S. Chemla, T.C. Damen, A.C. Gossard, W. Wiegmann, T.H. Wood and C.A. Burrus, Appl. Phys. Lett., 45, 13 (1984).

Optical Bistability and Nonlinearities of the Dielectric Function Due to Biexcitons

J.B. Grun

Laboratoire de Spectroscopie et d'Optique du Corps Solide,
Unité Associée au C.N.R.S. n° 232, Université Louis Pasteur,
5, rue de l'Université, F-67000 Strasbourg, France

The non-linear dielectric function due to biexcitonic optical transitions is being investigated. It is used to realize an optical bistable. Copper chloride is the material selected for these studies since it has simple and well-understood excitonic and biexcitonic properties [1].

I EXCITONS and BIEXCITONS in CuCl

I 1) Energy Levels and Symmetries

The dielectric properties of CuCl (point-group symmetry T_d) can be reasonably well described by the energy-level schema given in Fig. 1.

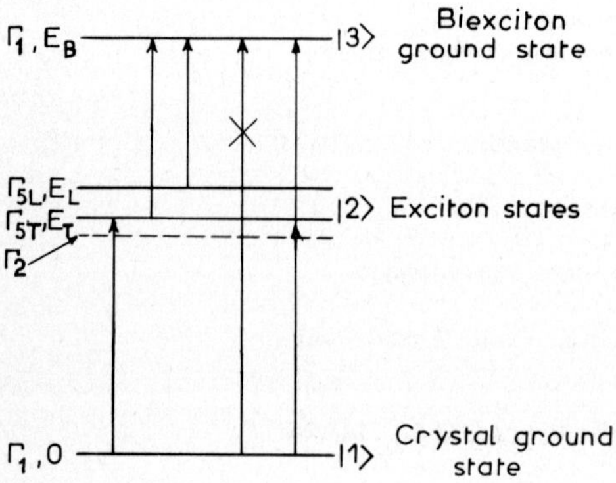

Fig. 1. Schema of exciton and biexciton energy-levels and of optical transitions in CuCl

In the crystal ground state $|1>$, all valence band states are occupied and all conduction band states empty. Its symmetry at the centre of the Brillouin Zone (BZ) is Γ_1.

Excitons $|2>$ are built from an electron in the lowest conduction band (Γ_6 symmetry) and a hole in the uppermost valence band (Γ_7 symmetry). Their ground state has the symmetry at the centre of the BZ : $\Gamma_6 \otimes \Gamma_7 \doteq \Gamma_2 \oplus \Gamma_5$.

The analytic part of the exchange interaction between electron and hole splits the Γ_2 and Γ_5 states. Further on, Γ_5 states are dipole-active. Therefore, the non-analytic exchange interaction leads to a splitting between the Γ_5 exciton states : Γ_{5L} represents the longitudinal exciton obtained when the dipole moment of the exciton is parallel to its wave-vector $\underset{\sim}{Q}$, and Γ_{5T} the two transverse excitons whose dipole moment is perpendicular to $\underset{\sim}{Q}$.

Two excitons can couple together to form a biexciton $|3>$. From Pauli's principle, the biexciton total wave-function has to be antisymmetric with respect to the exchange of the two electrons and the two holes. In analogy with the hydrogen molecule, we assume that the Bloch part of the biexciton ground state is antisymmetric with respect to the exchange of electrons and holes. Therefore, the envelope function is a symmetric function under the exchange of identical particles and transforms like Γ_1. The symmetry of the total wave-function is Γ_1.

$$\Gamma_{env} \otimes (\Gamma_6 \otimes \Gamma_6)^- \otimes (\Gamma_7 \otimes \Gamma_7)^- = \Gamma_1 \otimes \Gamma_1 \otimes \Gamma_1 = \Gamma_1$$

One-photon transitions between the crystal ground state Γ_1 and the Γ_{5T} transverse exciton state are dipole-allowed for both polarizations of the light field (symmetry Γ_5). Similarly, one-photon transitions are also allowed between Γ_5 exciton states and the Γ_1 biexciton ground state for both polarizations of the light field. One-photon transitions between the crystal ground state and the biexciton state are strictly forbidden. However, two-photon transitions are allowed if both photons have the same direction of polarization. This two-photon absorption is large, due to a resonance-enhancement by the exciton states and to a giant oscillator strength of the exciton-biexciton transition.

I 2) Dielectric Function and the Polariton Dispersion

The transverse excitons are strongly coupled to the electromagnetic field of the light. The eigenstates of the coupled exciton-photon system are called polaritons. Their dispersion $E(\underset{\sim}{Q})$ is related to the dielectric function $\varepsilon(\underset{\sim}{Q},E)$. In the case of the single oscillator model considered here, $E(\underset{\sim}{Q})$ is obtained from the solution of the equation :

$$\varepsilon(\underset{\sim}{Q},E) = \frac{\hbar^2 c^2 \underset{\sim}{Q}^2}{E^2(\underset{\sim}{Q})} = \varepsilon_b \left[1 + \frac{4\pi\beta\, E_T^2}{E_T^2(\underset{\sim}{Q}) - E^2(\underset{\sim}{Q})}\right] \qquad (1)$$

where the oscillator strength β is given by :

$$4\pi\beta = \frac{E_L^2 - E_T^2}{E_T^2} . \qquad (2)$$

$E_T(\underset{\sim}{Q}) = E_T + \hbar^2 Q^2/2m_x^*$ and $E_L(\underset{\sim}{Q}) = E_L + \hbar^2 Q^2/2m_x^*$ give the dispersions of the transverse and longitudinal exciton, respectively.

E_T and E_L are the energies of these excitons at $Q = 0$, and m_x^* is the effective exciton mass. ε_b is the background dielectric constant, which takes into account the different oscillators neglected in the one-oscillator model. These four parameters are necessary to define completely this polariton dispersion. They have been measured by hyper-Raman scattering [2] and resonant Brillouin scattering [3]. The values obtained by the first technique are the following : $E_T = 3.2025$ eV, $E_L = 3.208$ eV, $m_x^* = 2.5\ m_0$ and $\varepsilon_b = 5$.

I 3) Biexciton Dispersion

The biexciton energy $E_B(\underset{\sim}{K})$ can be expressed as a function of its wave-vector $\underset{\sim}{K}$ by the following relation :

$$E_B(\underset{\sim}{K}) = E_B + \frac{\hbar^2 K^2}{2m_B^*} . \qquad (3)$$

The energy E_B of the biexciton at the center of the Brillouin zone is equal to twice the energy of the Γ_2 exciton, minus its binding energy :

$$E_B = 2E_{\Gamma_2} - E_{binding} \qquad (4)$$

It has been measured by several methods : luminescence [4], hyper-Raman diffusion [2], four-wave mixing [5], exciton-biexciton-induced absorption [6]. The most precise technique is the absorption of two photons of a single laser beam, or of two different laser beams propagating in opposite directions. It corresponds to the transition between the crystal ground state and the biexciton. One finds [7] : $E_B = 6.372$ eV.

The biexciton dispersion has also been measured by two-photon absorption with two laser beams, using different configurations and photons (polaritons in the crystal) of different energies and wave-vectors. The value of the biexciton mass is the following [8] : $m_B^* = 5.4\ m_0$.

As can be seen, the spectroscopic parameters of excitons (polaritons) and biexcitons are well known in CuCl. It was therefore interesting to study in this compound the non-linear behaviour of the dielectric function in the polariton region, near half the biexciton energy, when a laser light creates a large density of polaritons. We shall first study theoretically the non-linear dielectric function. We shall then place this non-linear medium in a Fabry-Perot cavity and realize an optical bistable.

II NON-LINEAR DIELECTRIC FUNCTION

II 1) Matrix Density Formalism

Concerning the theoretical description of the dielectric function of CuCl, a three-level model is used [9] [10] [11] as indicated in Fig. 1. A light field A(t) can induce transitions between these states, as explained above.

The Hamiltonian of the system may be written as :

$$H = H_o - \mu A(t) \tag{5}$$

where H_o is the Hamiltonian of the non-interacting systems (light and solid) and μ the dipole operator.

The light field is given by :

$$A(t) = \tilde{A}(t)\cos(\omega t) , \tag{6}$$

where $\tilde{A}(t)$ is the envelope function which is assumed to be slowly-varying in time compared to $\cos(\omega t)$.

If ρ is the density matrix, the macroscopic polarization P(t) of the crystal is given in [12] by the following expression :

$$P(t) = N\mathrm{Tr}(\rho\mu) = N\,\overline{\langle\mu\rangle} . \tag{7}$$

$\overline{\langle\mu\rangle}$ denotes the ensemble average over the dipole matrix. $\rho(t)$ is calculated from Schroedinger's equation and P(t) derived from equation (7).

On the other hand, the polarization is related to the susceptibility χ and to the complex electric field A'(t) by the following convolution integral :

$$P(t) = \varepsilon_o \,\mathrm{Re}\left[\chi(t) * A'(t)\right] \tag{8}$$

Comparing (7) and (8), we can express the susceptibility tensor χ in terms of ρ.

II 2) Nonlinearities Due to Biexcitons

Using the parameters for CuCl [9], the dielectric function ε and the refractive index n have been calculated from the relation :

$$\varepsilon = \varepsilon_o(\varepsilon_b + \chi) \equiv n^2 \tag{9}$$

which defines the wave-vector $\underset{\sim}{Q}$ and the absorption coefficient α at a definite incident photon energy $\hbar\omega$ by :

$$Q = \frac{\omega}{\sqrt{2}\,c}\sqrt{\mathrm{Re}\,\varepsilon + \sqrt{(\mathrm{Re}\,\varepsilon)^2 + (\mathrm{Im}\,\varepsilon)^2}} \tag{10}$$

$$\alpha = \frac{\omega}{\sqrt{2}\,c}\sqrt{-\,\mathrm{Re}\,\varepsilon + \sqrt{(\mathrm{Re}\,\varepsilon)^2 + (\mathrm{Im}\,\varepsilon)^2}} \tag{11}$$

The real part n' of the refractive index n is equal to Qc/ω and the imaginary part n" to $C\alpha/\omega$.

The results are summarized by two figures [13]. Figure 2 represents the polariton dispersion near the biexciton resonance, for different photon densities n_p, and Fig. 3, the absorption coefficient.

An anomaly of the absorption coefficient α arises at $E_B/2$, due to the two-photon transition between the crystal ground state and the biexciton ground state. It increases with the light-field intensity. At the highest intensities, it is Stark-shifted, broadens and its amplitude diminishes. In the same spectral region, the dispersion of the polaritons is also drastically changed. The index of refraction becomes intensity-dependent. These effects have been observed in several experiments as : hyper-Raman scattering [14,15], four-wave mixing [16,17] and polarization spectroscopy [18].

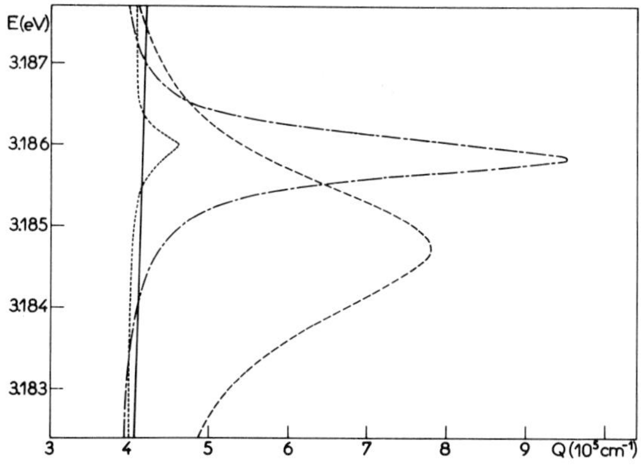

Fig. 2. Theoretical polariton dispersion E(Q), near half the biexciton energy, for different intensities of an incident light beam, in CuCl (full line : $n_p = 0$, dotted line : $n_p = 10^{15}$ cm^{-3}, dashed-dotted line : $n_p = 10^{16}$ cm^{-3}, dashed line : $n_p = 10^{17}$ cm^{-3})

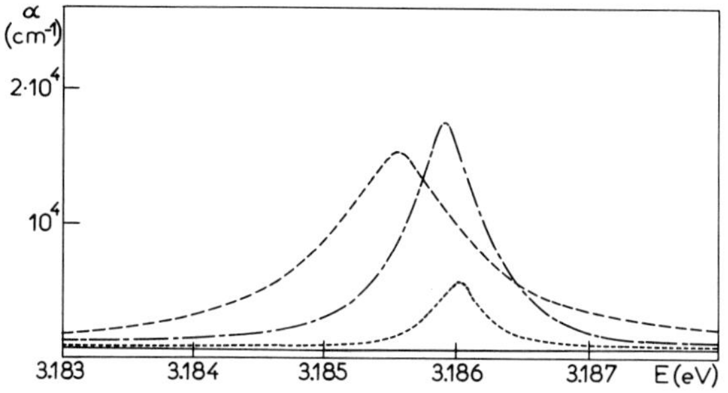

Fig. 3. Theoretical absorption coefficient $\alpha(E)$ of polaritons, near half the biexciton energy, for different intensities of an incident light beam, in CuCl (full line : $n_p = 0$, dotted line : $n_p = 10^{15}$ cm^{-3}, dashed-dotted line : $n_p = 10^{16}$ cm^{-3}, dashed line : $n_p = 4 \times 10^{16}$ cm^{-3})

This non-linear medium has been used to realize an optical bistable of biexcitonic type, the sample of CuCl being placed in a Fabry-Perot (F.P.) cavity which provides the necessary feedback. We have studied such a bistable both theoretically in a mean-field approximation [10,11] and experimentally. The main experimental characteristics of such a bistable will be reviewed here.

III STUDY of the BIEXCITON OPTICAL BISTABILITY in CuCl

III 1) Experimental Set-up

The samples of CuCl studied are platelets with a thickness of about 30 µm. They are grown by vapor phase transport in a reduced hydrogen atmosphere. They are placed inside a F.P. cavity which provides the feedback. The mirrors of the cavity are thin glass plates coated with platinium films which do not react chemically with CuCl. The reflection coefficient of the mirrors is 90 % but the intensity variation of the light transmitted through the device is only 40 %. This is due to the residual absorption of the sample and also to the lack of parallelism of the crystal surfaces. The device is cooled down to pumped liquid helium temperature (∿ 2 K) in a double pyrex dewar.

The experimental set-up is drawn in Fig. 4. We use a XeCl excimer laser (EMG 101 from Lambda Physik) to pump a dye laser working with BiBuQ in toluene in a grazing

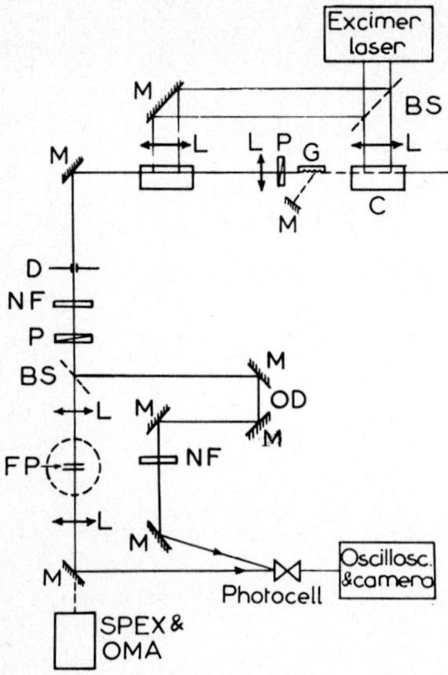

Fig. 4. Experimental set-up.
M : mirror, BS : beam splitter, L : lens, C : dye cell, P : polarizer, D : diaphragm, NF : neutral filter, FP : Fabry-Perot, G : grating, OD : optical delay

incidence configuration. This laser emission is then amplified in a second dye cell pumped by the same excimer laser. Great care is taken to keep the superradiant emission small (< 1 %) compared to the laser emission. The temporal shape of the pulses is well defined with a half-width of about 3 ns. The laser emission has a spectral half-width of 0.05 meV. After passing through a diaphragm, neutral density filters, a variable NRC neutral filter and a glan polarizer, the beam is split into two parts by a glass plate. One part of the beam is focused onto the Fabry-Perot etalon containing the CuCl crystal within a spot of 100 µm diameter. The power density can be varied up to 60 MW/cm^2. The transmission of the laser pulses through the F.P. is detected by a fast photocell (UVHC 20 from RTC-Philips) and is analysed in time by a fast oscilloscope (7104 from Tektronix with a 7A 29 amplifier). The time resolution is better than 500 ps. The transmission of the beam can also be spectrally analysed through a 3/4 Spex spectrograph and an OMA system (1205 D from PAR). The other part of the beam passes through neutral density filters and a variable NRC neutral filter, and is, after a delay of 5.9 ns, detected by the same photocell. This beam is used as a reference for the temporal shape of the pulse. The signal, visualized on the oscilloscope, is then photographed with a camera so that single shots can be analysed.

III) 2) Optical Bistability [19,20]

Figure 5 shows the transmitted and incident laser pulses measured at a photon energy close to half the biexciton energy and with a maximum intensity of 30 MW/cm^2. The transmitted pulse is clearly deformed when compared to the incident pulse.

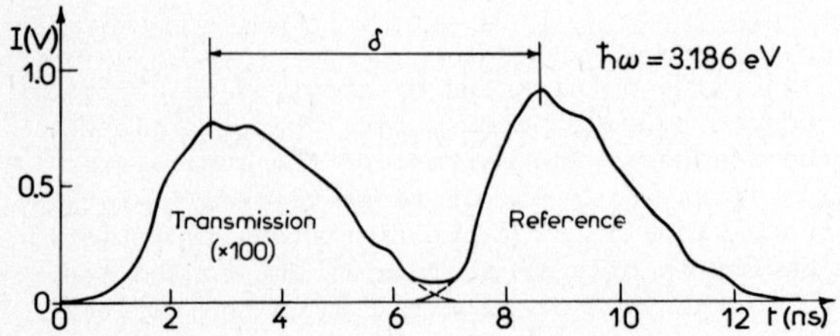

Fig. 5. Transmitted and incident laser pulses measured by a fast photocell and oscilloscope in CuCl at a photon energy of 3.2 eV. δ is the optical delay between the two pulses

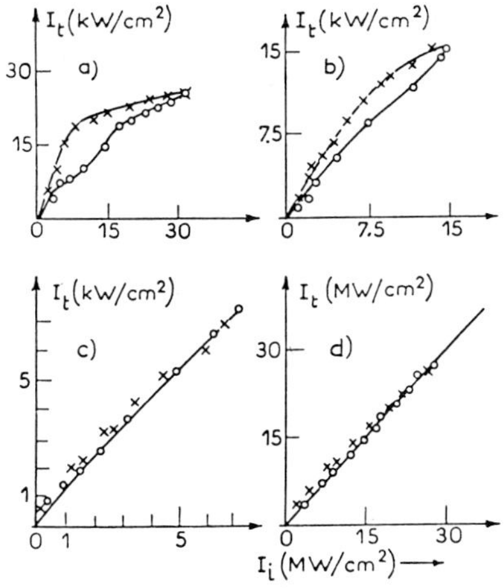

Fig. 6. Transmitted versus incident intensity curves for different maximum intensities of the laser pulse a,b,c, in CuCl ; d : there is no sample inside the cavity ; o : increasing laser intensity I_i ; x : decreasing I_i

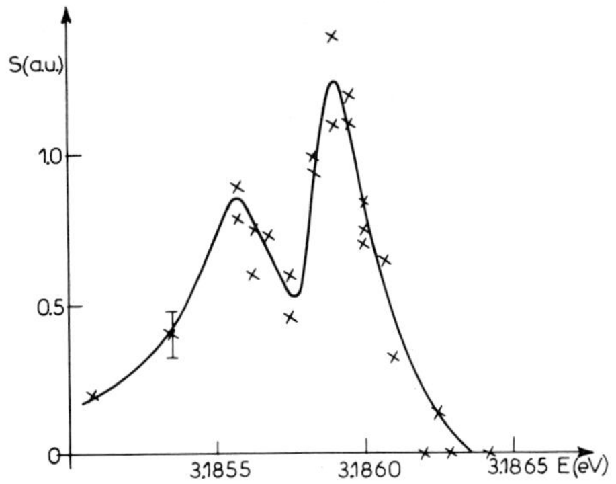

Fig. 7. Areas of the bistability loop as a function of the photon energy of the incident laser beam

The intensity of the transmitted pulse is now analysed in detail by plotting in Fig. 6 this intensity as a function of the incident intensity. In Fig. 6a, the bistability is clearly observed with a switch-up intensity of 15 MW/cm^2 and a switch-down of 5 MW/cm^2. The switching point remains at a fixed intensity when the maximum intensity of the pulse is decreased. Below 15 MW/cm^2, no switching is observed, as can be seen in Fig. 6c. In the last figure, there is no sample inside the cavity.

We have also studied the dependence of the hysteresis loop on the photon energy of the laser beam. In Fig. 7, we have plotted the area of the hysteresis loop as a function of the photon energy of the laser. This quantity is proportional to the energy stocked inside the cavity. We have obtained here a curve with two maxima : one corresponds to a maximum of transmission of the F.P. at 3.1855 eV ; the second, to the biexciton resonance $E_B/2$. The structure observed depends on the position of the maximum of transmission compared to $E_B/2$. Excitation spectra with one maximum were also observed. This proves that the switching behaviour is related to biexciton resonance.

III 3) Time Response of the Bistable Device

Analysing the pulse deformation, we could show that the switch-up and -down times are quite short, less than the resolution time of the detection system (< 500 ps), when working slightly away from the biexciton resonance. The switch-down time increases when we work closer to the resonance ($\sim$ 3 ns).

These results can be explained as follows : when slightly-off resonance, biexcitons are created only virtually ; at resonance, however, the biexciton are really created and the index changes are related to the presence of excitons and biexcitons, as we shall see. Therefore, their lifetime affects the switch-down time of the device.

To measure more precisely time responses, we have improved our experimental set-up [21]. First, we have replaced our detection system (photocell and oscilloscope) by a streak camera (Hadland) and an OSA system (WP3 from BM Spektronik) for data processing. We have also improved the dye laser in order to have highly reproducible laser pulses with a smooth envelope. We use a Hänsch configuration with a grating as wavelength selector and a F.P. outside the cavity. The laser pulse is again sent into an amplifier. For the measurement, the reference signal is first obtained by accumulating about 15 laser shots. The transmitted signal is recorded immediately after by accumulating the same number of shots. The time-resolution of the measurement depends on the time-scale of the streak, on the width of its entrance slit and on the reproducibility of the laser pulses. When going to the fastest sweeps of the streak camera, only parts of the signals, selected by changing the time-delays of the trigger of the camera, can be visualized.

We first plot the intensity of the transmitted signal as a function of the intensity of the incident one. Figure 8a gives the curve obtained when no sample is placed inside the F.P. cavity. Its linear characteristics give an estimate of the laser fluctuations. In Fig. 8b, the hysteresis loop plotted is obtained for a photon energy $\hbar\omega$ of the laser below the biexciton resonance $E_B/2$. When the photon energy is varied, the hysteresis keeps the same characteristics and shows a strong resonance at $E_B/2$.

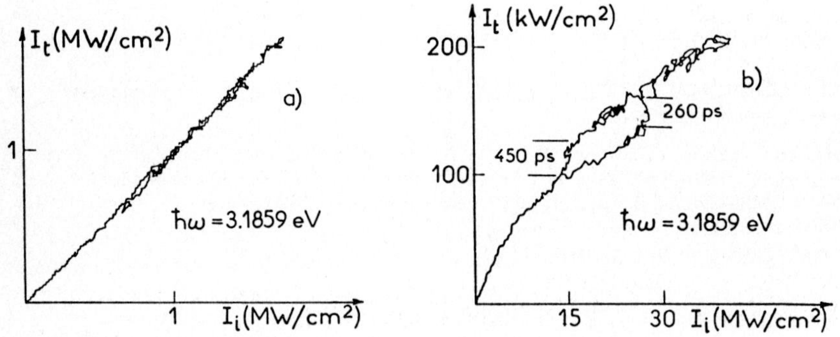

Fig. 8. Transmitted versus incident intensity curves for a cavity a) without sample inside, b) a CuCl platelet inside

We are observing in this case real optical bistability and no simple transient effects. It can be clearly seen as follows : our incident pulses have secondary intensity maxima and, for each of them, the same branch of the hysteresis loop is described several times before the switching-up of the device occurs. The same effect is observed before the switching-down of the device. If we had observed transient effects, the transmission would not have stayed on the high or low level of the device but would have shown open structures.

The switching times of the device, obtained at this photon energy, are indicated in Fig. 8b : 260 ps for the switch-up and 450 ps for the switch-down. They are the fastest we have measured. With other samples, they could be longer. In that case, the absorption was found to be very important. However, even the fastest times measured are much longer than the ones expected theoretically by HANAMURA (∿ ps). As it has been pointed out in [22], the main parameter for the switching time is probably the polariton escaping-time from the cavity, which can be rather long when the group velocity is taken into account. Furthermore, near the biexciton resonance, the absorption plays an important role even if the device is of dispersive type. When it starts to switch, the absorption changes due to the variation of the exciton and the biexciton populations. The transmission of the light through the device will be influenced by the relaxation of these populations. This will be particularly crucial for the way the system is driven back to its equilibrium.

III 4) Transient Non-linear Absorption

These absorption effects may even lead to disfunctionings of the device : hysteresis loop displayed in the reverse direction, no definite up-to-down commutation and no stable hysteresis [21]. See for instance the loop obtained with a thicker sample in Fig. 9.

For strong absorption, we simply observe a single path transmission through the device. This transmission is governed not only by linear or two-photon absorption but also by higher-order processes. We have observed these effects when studying the transmission of a laser pulse of several nanosecond duration through a 30 μm thick sample of CuCl, which was not placed in a F.P. cavity.

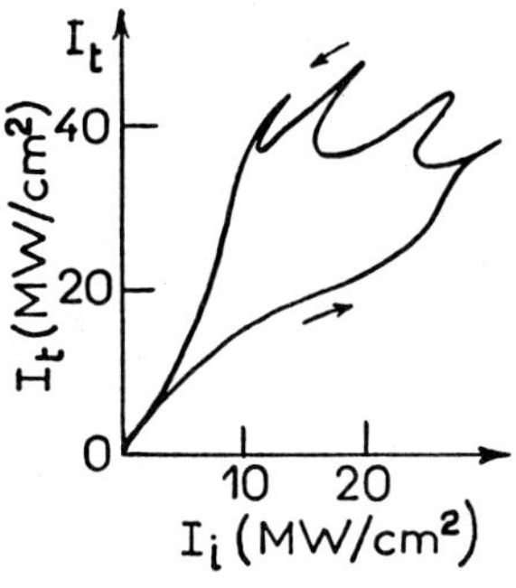

Fig. 9. Transmitted versus incident intensity loop for a thicker sample of CuCl. A disfunctioning of the bistable can be seen

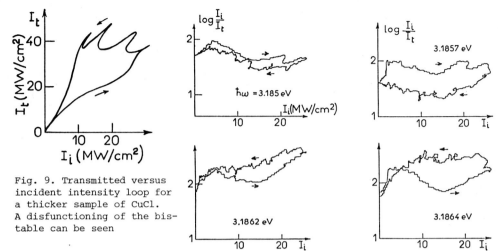

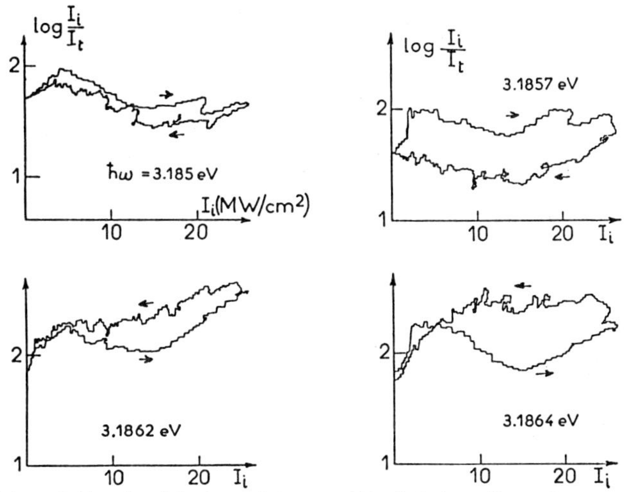

Fig. 10. Logarithm of the ratio of the incident and transmitted intensity of a laser pulse as a function of its incident intensity at different photon energies, for a CuCl sample (not placed inside a F.P. cavity)

In Fig. 10, we plot the logarithm of the ratio of the incident and transmitted intensity as a function of the incident laser intensity for different photon energies. The arrows indicate increasing and decreasing intensities.

Three main features emerge :

First, the absorption follows a different path in the increasing than in the decreasing part of the laser pulse. We have a hysteresis loop or to express it differently a memory effect.

Second, this memory effect depends on the photon energy of the laser : below half the biexciton energy, the overall absorption measured is smaller in the decreasing than in the increasing part of the pulse. Above the biexciton resonance, it is the reverse.

Third, no real switching is observed.

These effects can be similarly observed with samples inside a F.P. cavity when the feedback is no more playing its role. They cannot be explained by simple one- or two-photon absorption, but can be related to the dynamics of the exciton and biexciton population as we have shown [23] and may be also due to propagation effects and intensity-dependent damping constants.

IV CONCLUSION

Optical bistability due to biexcitons has been observed at low temperatures with CuCl platelets placed in Fabry-Perot cavities. Switching times (switch-up and switch-down) are subnanosecond. Commutation energies are of several microjoules, corresponding to pulse intensities of several megawatts/cm^2.

Disfunctionings of these optical bistables have been shown to be due to a strong transient absorption partly related to exciton and biexciton population dynamics.

ACKNOWLEDGEMENTS

Most of the experimental and theoretical results presented here have been obtained in collaboration with J.Y. BIGOT, B. HÖNERLAGE, M. FRINDI, R. LEVY and F. TOMASINI. I would like to thank B. HÖNERLAGE for several discussions and a careful reading of the manuscript, and F. FIDORRA and Prof. C. KLINGSHIRN for their help in performing some of these experiments.

Part of this work has been carried out in the framework of an operation launched by the Commission of the European Communities under the experimental phase of the European Community Stimulation Action (1983-1985).

REFERENCES

1 J.B. Grun, B. Hönerlage and R. Levy : in Excitons, eds E.I. Rashba, M.D. Sturge, "Modern Problems in Condensed Matter Science", Vol.2, North Holland, Amsterdam (1982)

D.S. Chemla and A. Maruani : Prog. in Quantum Electronics, 8, 1 (1982)

2 Vu Duy Phach, A. Bivas, B. Hönerlage and J.B. Grun : Phys. Stat. Sol.b), 86, 159 (1978)

B. Hönerlage, A. Bivas and Vu Duy Phach : Phys. Rev. Lett. 41, 49 (1978)

3 J.C. Merle, F. Meseguer and M. Cardona : Int. Conf. on the Physics of Semiconductors, San Francisco (1984), to be published

J.C. Merle : Festkörperprobleme XXV (1985), to be published

4 R. Levy, C. Klingshirn, E. Ostertag, Vu Duy Phach and J.B. Grun : Phys. Stat. Sol.b), 77, 381 (1976)

5 A. Maruani, J.L. Oudar, E. Batifol and D.S. Chemla : Phys. Rev. Lett. 41, 1372 (1978)

6 A. Bivas, Vu Duy Phach, B. Hönerlage and J.B. Grun : Phys. Stat. Sol.b), 84, 235 (1977)

7 Vu Duy Phach, A. Bivas, B. Hönerlage and J.B. Grun : Phys. Stat. Sol.b), 84, 731 (1977)

G.M. Gale and A. Mysyrowicz : Phys. Rev. Lett. 32, 727 (1974)

8 T. Mita, K. Satome and M. Ueta : J. Phys. Soc. Japan, 48, 496 (1980)

9 J.Y. Bigot and B. Hönerlage : Phys. Stat. Sol.b), 121, 649 (1984)

B. Hönerlage and J.Y. Bigot : Phys. Stat. Sol.b), 123, 201 (1984)

10 J.Y. Bigot, Thesis, Strasbourg

11 J.B. Grun, B. Hönerlage and R. Levy : J. Lum. 30, 217 (1985)

12 A. Yariv : Quantum Electronics, 2nd ed., Wiley & Sons, New York (1975)

13 B. Hönerlage, J.Y. Bigot : Phys. Stat. Sol.b), 124, 221 (1984)

14 B. Hönerlage, J.Y. Bigot, R. Levy, F. Tomasini and J.B. Grun : Sol. Stat. Comm. 48, 803 (1983)

15 R. Levy, B. Hönerlage and J.B. Grun : Helv. Phys. Acta, 58, 252 (1985)

16 R. Levy, F. Tomasini and J.B. Grun : J. Lum. 31/32, 870 (1984)

17 R. Levy, F. Tomasini, J.Y. Bigot and J.B. Grun : to be published

18 M. Kuwata, T. Mita and N. Nagasawa : Sol. Stat. Com. 40, 911 (1981)

M. Kuwata and N. Nagasawa : J. Phys. Soc. Japan, 51, 2591 (1982)

T. Itoh and T. Katohno : J. Phys. Soc. Japan, 51, 707 (1982)

19 R. Levy, J.Y. Bigot, B. Hönerlage, F. Tomasini and J.B. Grun : Sol. Stat. Comm. 48, 705 (1983)

N. Peyghambarian, H.M. Gibbs, M.C. Rashford, D.A. Weinberger : Phys. Rev. Lett. 51, 1692 (1983)

20 R. Levy, B. Hönerlage and J.B. Grun : Phil. Trans. R. Soc., London, A313, 229 (1984)

21 J.Y. Bigot, F. Fidorra, C. Klingshirn and J.B. Grun : J. Quant. Electr. to be published

22 J.W. Haus, C.M. Bowden, C.C. Sung : Phys. Rev. A 31, 1936 (1985)

23 J.Y. Bigot, J. Miletic, B. Hönerlage : to be published

Instabilities and Chaos in Nonlinear Optical Beam Interactions

C. Flytzanis

Laboratoire d'Optique Quantique*, Ecole Polytechnique,
F-91128 Palaiseau, Cedex, France
and
Max-Planck-Institut für Quantenoptik, D-8046 Garching, F.R.G.

1. Introduction

It is a well-established fact that intense optical beams interacting nonlinearly suffer instabilities when their intensities exceed certain critical values. Then a stationary regime becomes unstable with respect to small fluctuations and may even go over to a chaotic state. Such a behavior has been predicted and observed in many cases both in passive as well as active media and is presently a subject of intensive study as it is related to important fundamental and practical problems in nonlinear optics.

The occurrence of such effects in nonlinear optics provides an additional and conclusive demonstration of the universality of certain features previously found in other nonlinear dynamical systems in mechanics and hydrodynamics. Furthermore the precise understanding and description of these effects is of considerable practical interest as it allows one to establish the limits and range of reliability of the nonlinear optical devices, a field of central importance in the future technology of optical telecommunications and optical treatment of information and imaging; by the same token it provides crucial clues to the physicist and optical engineer to improve the performances within these limits.

Instabilities in nonlinear optical cavities were observed in the early days of laser physics but were most frequently suppressed as erratic or undesirable features. This attitude drastically changed in recent years once it was established that the occurrence of these instabilities follows certain patterns and obeys quite general laws that are shared in common by a large class of nonlinear dynamical systems irrespective of their physical content. This universality allowed the transfer and use in nonlinear optics of new concepts and approaches from other areas; at the same time it unraveled a whole new class of phenomena of fundamental importance.

In the following we will discuss some cases in nonlinear optics that illustrate these new features and at the same time are relevant for the technology of nonlinear optical devices. The study of optical bistable devices gave the main impetus to this area and particular emphasis below will be paid to the topic of optical bistability. The next section will serve as a general introduction to the most important concepts and methods of approach illustrated by the simple quadratic map. In the subsequent four sections the following cases will be presented: instabilities and transition to chaos in a passive nonlinear Fabry-Perot cavity, instabilities and transition to chaos in nonlinear optical systems. The presentation will be kept at a simple and rather informative level.

2. Instabilities and Transition to Chaos in Nonlinear Dynamical Systems

Many nonlinear dynamical systems in general and most nonlinear optical ones in particular can be described [1] by flows, that is a set of first-order differential

---

* Laboratoire Propre du Centre National de la Recherche Scientifique (C.N.R.S.)

equations

$$\dot{\bar{x}} = F(\bar{x},\lambda) , \qquad (2.1)$$

where $\bar{x} = (x_1, x_2,...x_d)$ is a d-dimensional vector quantity and F is a nonlinear function of the x_i, the dot stands for derivation with respect to time t and λ is an external control parameter. We may always assume that the system is autonomous, that is the set of equations (1) does not explicitly contain the time since such dependence can be eliminated by increasing by one the dimension d. Furthermore we will assume that the system is dissipative as is always the case in practice in nonlinear optics, noise and absorption losses being a general feature there. We are interested in the long time behavior of such systems.

If $\bar{x}$ is one-dimensional the solution may asymptotically evolve to a fixed point and for $\bar{x}$ two-dimensional, a limit cycle is in addition a possible asymptotic behavior. For higher dimensions, three or more, the asymptotic behavior in addition may evolve to a chaotic one or to the so-called deterministic chaos; we present a simple description of the latter below.

It is often convenient to study the flow (1) by introducing the Poincaré map which amounts to cutting the trajectory by a (d-1)-dimensional hyperplane and denoting the points that are generated with increasing time by $\bar{x}(1)$, $\bar{x}(2)$,... so that the long time behavior of (1) can be replaced by that of the return map

$$\bar{x}(n+1) = G(\bar{x}(n),\lambda) . \qquad (2.2)$$

For simplicity in the following we shall admit that flows can be replaced by maps and we shall concentrate our attention on the latter since they correspond to a situation that naturally arises in optical cavities and some other nonlinear optical devices of finite extension.

The one-dimensional quadratic map [1,2]

$$x_{n+1} = 4\lambda x_n(1-x_n) = f(x_n) \qquad (2.3)$$

will serve to introduce and substantiate the main concepts that will be needed in the subsequent sections where specific nonlinear optical interactions are treated. The values of the external control parameter λ will always be restricted in the range (0,1) so that any iterate of x in (0,1) always remains in (0,1); with appropriate variable changes all quadratic maps can be cast in the form (2.3). This is the simplest nonlinear map and any sequence of iterates of (2.3) can also be visualized by the usual iterative graphical contruction used to find the intersection of the two curves

$$y = 4\lambda x(1-x) = f(x), \qquad (2.4)$$

$$y = x, \qquad (2.5)$$

as depicted in Fig. 1.

From the outset the sequences

$$(x_0, x_1, x_2, ...) \qquad (2.6)$$

obtained through such iterative procedure are expected to be quite complicated after many iterations for arbitrary values of λ in the range (0,1). A more careful analysis however reveals that this complexity, whenever it occurs, is reached in successive stages as λ is monotonically varied in (0,1) and follows well-defined patterns which are shared in common by a large class of nonreversible maps, (2.3) being the simplest one. We proceed to characterize the patterns that the system (2.3) exhibits after many iterations.

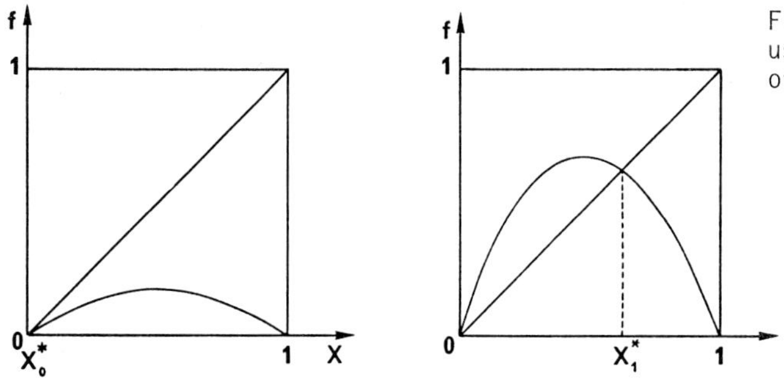

Fig.1. Graphical construction used to visualize a sequence of iterates of (2.3)

2.1 <u>Stationary or periodic state:</u> If the generator x_0 of a sequence (2.6) is chosen to be a fixed point of f(x) or $x_0 = x^*$ where

$$x_0^* = f(x^*), \tag{2.7}$$

then (2.6) reduces to a sequence of self-producing points

$$(x^*, x^*, x^*, \ldots\ldots), \tag{2.8}$$

which can be viewed as <u>stationary</u> or <u>periodic</u> with period 1. One easily finds that the map (2.3) has two fixed points for $\lambda > 1/4$, namely

$$x_0^* = 0 \ ,$$

$$x_1^* = 1 - 1/4\lambda \ ,$$

which coincide for $\lambda = 1/4$; for $\lambda < 1/4$ there is only one fixed point $x_0^* = 0$ since x is bound in (0,1).

If the generator x_0 is not a fixed point of f(x) the generated sequence (2.6) will be shaped by the fixed points that are stable. Since the fixed points are intersections of y = x, a straight line of slope 1, and a curve f(x), a fixed point is termed stable or an attractor if $|f'(x)| < 1$ where $f'(x)$ is the derivative (slope) of f(x) at the point x; if $|f'(x^*)| > 1$, the fixed point x^* is unstable: only attractors eventually determine the long time behavior of the generated sequence.

Now the slope of (2.4) at a point x is $f'(x) = 4\lambda(1-2x)$. If $0 < \lambda < 1/4$ the single point $x_0^* = 0$ is an attractor and if one disregards the transient behavior, namely the first iterates of the sequence generated by any x, after many iterations the sequence (1.6) will asymptotically evolve to the stationary one

$$(x_0^*, x_0^*, x_0^*, \ldots\ldots\ldots), \tag{2.9}$$

irrespective of the initial x (see also Fig. 1a).

For $\lambda = 1/4$, y = x is tangent to the curve (2.4) at the fixed point $x_0^* = 0$ which is now a doubly degenerate fixed point; for $\lambda > 1/4$ an additional fixed point $x_1^* = 1 - 1/4\lambda$ appears, which is an attractor as long as $1/4 < \lambda < 3/4$ while $x_0^* = 0$ altogether ceases to be so. For $1/4 < \lambda < 3/4$ then, if we disregard its first iterates any sequence of iterates will asymptotically evolve to the stationary one

$$(x_1^*, x_1^*, x_1^*, \ldots\ldots\ldots) \tag{2.10}$$

again irrespective of the initial x_0 (see also fig. 1b).

The above result is an universal property of maps namely the eventual behavior of the sequence generated by the map is determined by an attractor irrespective of the starting point x_0 provided the latter lies within the <u>bassin of attraction</u> of the attractor

2.2 <u>Period doubling:</u> If λ is increased beyond 3/4, both fixed points x_0^* and x_1^* become unstable and the question then arises what determines the eventual behavior of a sequence (2.6). To answer this question we first note that any sequence (2.6) can also be viewed as the intercalation of the two sequences

$$(x_0, x_2, x_4, \ldots\ldots) \tag{2.11}$$

$$(x_1, x_3, x_5, \ldots\ldots) \tag{2.12}$$

each one generated by the map

$$x_{n+2} = f^2(x_n) = f(f(x_n)) \tag{2.13}$$

and the condition $x_1 = f(x_0)$; these can also be visualized as the sequences of iterates obtained in the graphical determination of the intersection of the curves

$$y = f^2(x) = f(f(x)), \tag{2.14}$$

$$y = x. \tag{2.15}$$

Since the eventual behavior of each sequence (2.11) and (2.12) is determined by the attractors of $f^2(x)$, namely

$$x^* = f^2(x^*) = f(f(x^*)) \tag{2.16}$$

so will be the case of their "composite" sequence (2.6). We proceed to find this behavior by repeating and extending the previous approach.

It is clear from the definitions (2.7) and (2.16) that a fixed point of $f(x)$ is also a fixed point of $f^2(x)$ as well, but not vice versa. Furthermore since

$$\{f^2(x_0)\}' = f'(x_0)f'(x_1) \tag{2.17}$$

if $x_1 = f(x_0)$, one also has $\{f^2(x^*)\}' = (f'(x^*))^2$ for a fixed point of $f(x)$, so that if $|f'(x^*)|<1$ a fortiori $\{f^2(x^*)\}'<1$ and accordingly a stable fixed point of $f(x)$ is also a stable fixed point of $f^2(x)$. But in addition $f^2(x)$ may have fixed points, intersections of (2.14) and (2.15) that are not fixed points of $f(x)$; as we will shortly see, the attractors among these additional points will determine the eventual behavior of the sequences (2.11) and (2.12) and consequently that of (2.6). How do these additional fixed points make their appearance in the map (2.13) when $f(x)$ is the quadratic map (2.3)? The behavior is depicted in Fig. 2.

First we notice from (2.4) that $f(x) = 4\lambda x(1-x)$ has a single extremum (maximum) at $f(x) = 0$ or $\tilde{x} = 1/2$ while $f^2(x)$ has a minimum there but also two maxima at the points $\tilde{x}'$ and $\tilde{x}''$ such that $\tilde{x} = f(\tilde{x}') = 1/2$ since then $f(\tilde{x}) = 0$ and similarly for $\tilde{x}''$. The fixed point $x_1^* = 1-1/4\lambda$, which is an attractor for $1/4<\lambda<3/4$, is flanked by the minimum at $\tilde{x}=1/2$ and the maximum at $x = \tilde{x}'$ of $f^2(x)$ and furthermore there is only one intersection point between $y=x$ and $y = f^2(x)$ as long as $1/4<\lambda<3/4$.

At $\lambda = \Lambda_1 = 3/4$, $f'(x_1^*) = -1$ but $\{f^2(x_1^*)\}' = 1$ so that $y = x$ becomes the tangent of $f^2(x)$ at x_1^* and x_1^* becomes a triply degenerate fixed point of $f^2(x)$. As λ is increased beyond $\Lambda_1 = 3/4$, this triple degeneracy is lifted and two new fixed points of $f^2(x)$ appear on either side of the initial one; these two new fixed points, x_2^* and x_3^*, are attractors in contrast to the initial one x_1^* which becomes unstable once $\lambda > 3/4$. These two fixed points x_2^* and x_3^* of $f^2(x)$ are not fixed points of $f(x)$ but are related to each other by the relations

$$\begin{aligned} x_3^* &= f(x_2^*) \\ x_2^* &= f(x_3^*) \end{aligned} \tag{2.18}$$

so they are mirror images to each other with respect to $f(x)$; from (2.17) and (2.18)

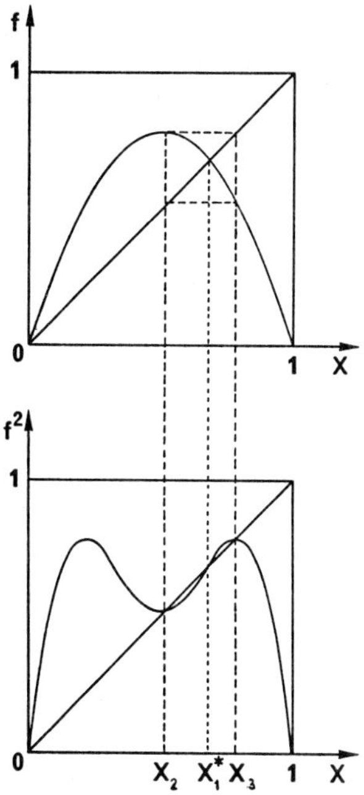

Fig.2. The appearence of additional fixed points in the map (2.13) when f(x) is the quadratic map (2.3)

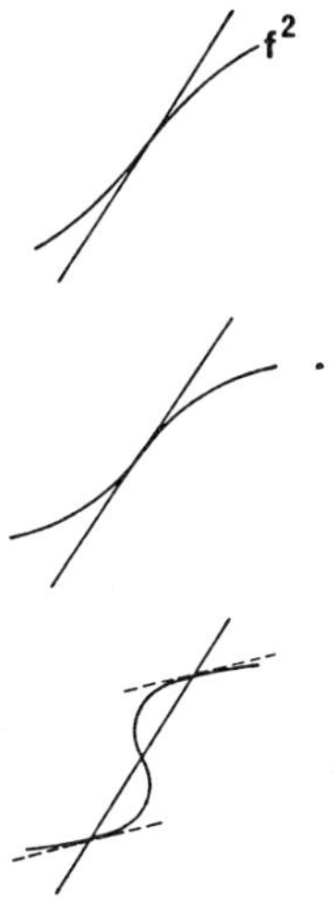

one also sees that

$$\{f^2(x^*)\}' = f'(x_2^*)f'(x_3^*) = f'(x_3^*)f'(x_2^*) = \{f^2(x_3^*)\}'$$

so that $f^2(x)$ has the same slope at these two points and this slope is positive and smaller than unity as long as $\lambda>\Lambda_1$ but below Λ_2 (to be defined below). As λ is increased beyond Λ_1 there will be first a value λ_1 were the slope becomes zero and one of the points (2.18) will coincide with the minimum of $f^2(x)$ and the other with its maximum. Beyond this value λ_1 the slope changes sign, becomes negative and increases in absolute value; at a value $\lambda = \Lambda_2$ it becomes -1 and beyond this value the two fixed points x_2^* and x_3^* cease to be stable simultaneously.

For each of these fixed points of $f^2(x)$ we reached a situation similar to the one we had for the fixed point x_1^* of f(x) from which these points were issued. It is easy to convince oneself that the subsequent behavior will be determined by the pair $f^2(x)$ and $f^4(x)$ in a way similar to that of the pair f(x) and $f^2(x)$ and f(x) may be altogether abandoned. Actually we only need to concentrate our attention at one of the fixed points of $f^2(x)$, the one closest to $\tilde{x} = 1/2$, and repeat the same procedure as before. The very structure of f insures that this pattern will be self-repeating and self-reproducing itself ad infinitum at smaller and smaller (fig.3) but also more and more numerous regions of the space; as λ is increased and crosses successively the values λ_1, λ_2, ...the period will be doubled each time. This is a consequence of the general chain rule

$$\{f^n(x_0)\}' = f'(x_0)f'(x_1)\dots f'(x_{n-1}).$$

One also says that the system exhibits a sequence of period-doubling bifurcations; the dimension of the space where these bifurcations occur can be described with fractals [1].

The sequence Λ_n has a limit Λ_∞ beyond which the system is no longer periodic: beyond this point the system is said to exhibit deterministic chaos. It has been shown [2] that the series Λ_n converges geometrically at the rate

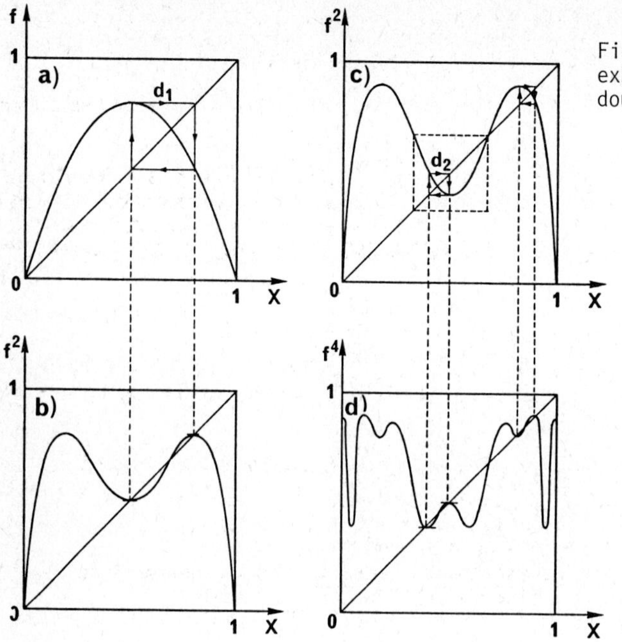

Fig.3. Representation of a system exhibiting a sequence of period-doubling bifurcations

$$\lim_{n\to\infty} \frac{\Lambda_n - \Lambda_{n-1}}{\Lambda_{n+1} - \Lambda_n} = 4.66......$$

and furthermore

$$\lim_{n\to\infty} \frac{d_n}{d_{n+1}} = -2.5029 ,$$

where d_n is the algebraic distance from $\tilde{x} = 1/2$ to the attractor of order n.

The previous discussion concerned the so-called bifurcation route to chaos (Feigenbaum [2]), which shows a close analogy with the renormalization group theory for the second-order phase transitions. Many systems were found to exhibit such a transition to chaos and as will be shown below theoretically we may expect that a certain class of nonlinear optical systems will qualitatively exhibit similar behavior (Ikeda et al. [3-6]). At present there is no firm experimental investigation to validate or confirm this expectation; some indications were however obtained concerning the very first steps of transition to chaos with a period-doubling scenario (Matsuoka et al [7] , Harrison et al [8,9]).

There are other routes to chaos like the intermittency one (Manneville and Pomeau [10]) where a signal shows long regular phases randomly interrupted by relatively short irregular bursts; the frequency of these bursts increases continuously with an external parameter. It is not clear yet whether such a route to chaos may occur in a nonlinear optical system and the same may be said for any other conceivable route to chaos. At present the commonest definition of chaos is the one introduced by Ruelle and Takens [11] which states that, in a chaotic system, two points that start close together in a bassin of an strange attractor in the phase space, diverge exponentially. This definition of the chaos has been extensively used in the modern theory of turbulence in hydrodynamics [12].

3. Instabilities and chaos in a passive nonlinear Fabry-Perot cavity

The nonlinear Fabry-Perot cavity has been extensively studied both in the active (laser) and in the passive (optical bistability) regimes. In line with the scope of these lectures we summarize below the main results concerning instabilities in the passive regime only.

In its simplest realization the Fabry-Perot cavity consists of two plane parallel mirrors a distance L apart with a nonlinear medium of refractive index n_0 and optical Kerr coefficient n_2 inserted between them. We wish to study the way instabilities are manifested in the transmission of this device when a monochromatic beam of electric field

$$E(z,t) = \mathrm{Re}\{\mathcal{E}\, e^{-i\omega t}\} \tag{3.1}$$

is incident perpendicular to the mirrors in the direction z; the mirror reflectivity and the absorption losses in the nonlinear medium at the frequency ω are R and α_0 respectively.

This electric field induces inside the cavity a polarization at the same frequency

$$P(z,t) = P^{(1)} + P^{(3)} = \mathrm{Re}\{\mathcal{P} e^{ikz-i\omega t}\}, \tag{3.2}$$

where we neglected nonlinearities higher than the third order. We write

$$\mathcal{P} = (\chi^{(1)} + \chi^{(3)}|\mathcal{E}|^2)\mathcal{E} = \chi(|\mathcal{E}|^2)\,\mathcal{E}, \tag{3.3}$$

where $\chi^{(3)}$ is the third-order susceptibility related to the optical Kerr effect coefficient through

$$n_2 = 3\chi^{(3)}/n_0^2\varepsilon_0 c \quad ; \tag{3.4}$$

furthermore we assume that the dynamics of the effective susceptibility $\tilde{\chi}$ are described by a Debye relaxation equation

$$\tau\dot{\tilde{\chi}} + \tilde{\chi} = \chi^3|\mathcal{E}(t)|^2, \tag{3.5}$$

which can also be cast in the integral form

$$\chi = \chi^{(1)} + \frac{\chi^{(3)}}{\tau}\int_{-\infty}^{t}|\mathcal{E}(s)|^2 e^{-(t-s)/\tau} ds \tag{3.6}$$

and τ is a phenomenological relaxation time which can range from ms for electrostriction-induced nonlinearities up to ps or fs for those of electronic origin .

For transverse waves the Maxwell equations give

$$\frac{\partial^2 E}{\partial z^2} = \frac{1}{c^2}\frac{\partial^2}{\partial t^2}\left(E + \frac{P}{\varepsilon_0}\right) \tag{3.7}$$

and assuming the envelopes $\mathcal{E}$ and $\mathcal{P}$ to be slowly varying functions of the time and space variables, one obtains in slow varying envelope approximation [13,14]

$$\frac{\partial\mathcal{E}}{\partial z} + \frac{n_0}{c}\frac{\partial\mathcal{E}}{\partial t} = \frac{ik}{2\varepsilon_0}\mathcal{P}, \tag{3.8}$$

which has to be solved together with (3.5) and the boundary conditions

$$\begin{aligned}\mathcal{E}(0,t) &= (1-R)^{1/2}\mathcal{E}_i + Re^{ikL}\,\mathcal{E}(L,t-L/c)\\ \mathcal{E}_t(t) &= (1-R)^{1/2}\mathcal{E}(L,t) \quad .\end{aligned} \tag{3.9}$$

We introduce the phase change $\phi = 4\pi nL/\lambda$ after a round trip in the cavity, and after some transformations [15,3,6] one obtains the set of equations

$$\tilde{\mathcal{E}}(t) = A + B\,\tilde{\mathcal{E}}(t-t_R)e^{i(\phi(t)-\phi_o)} \tag{3.10}$$

$$\tau\dot{\phi}(t) = -\phi(t) + \mathrm{sgn}(n_2)|\tilde{\mathcal{E}}(t-t_R)|^2 , \tag{3.11}$$

where $\tilde{\mathcal{E}} = \mathcal{E}/\mathcal{E}_o$, $\mathcal{E}_o = \{k|n_2|(1-\exp(i\alpha_o L))/\alpha_o\}^{12}$, $B=R\exp(-\alpha L)$, $A = (1-R)^{12}|\mathcal{E}\ |/\mathcal{E}_o$, ϕ_o is a mistuning parameter of the cavity and $t_R = L/c$ is the delay from propagation inside the cavity (round trip time).

This set of equations completely describes the transmission properties of the nonlinear Fabry-Perot cavity when the transverse effects are neglected. As it stands this is a combination of a flow and a map. For instantaneous relaxation of the nonlinearity, $\tau = 0$, one may assume stationary regime and this set of equations reduces to the relation

$$E(t) = A + BE(t-t_R)e^{i(|E(t-t_R)|^2-\phi_o)} = U(\,E(t-t_R)) , \tag{3.12}$$

which is a mapping if we concentrate on the value of the electric field after n round trips or $t = nt_R$ or

$$E_n = A + BE_{n-1}e^{i(|E_{n-1}|^2-\phi_o)} . \tag{3.13}$$

However it should be stressed that this is not a Poincaré map in the sense introduced in the previous section and any conclusion drawn from (3.13) concerning instabilities and chaos should not necessarily be valid for the initial set of equations.

The stationary solutions of (3.13) are given by

$$\{1 + B^2 - 2B\cos(|\mathcal{E}_s|^2-\phi_o)\}|\mathcal{E}_s|^2 = A^2 \tag{3.14}$$

and are multivalued functions of A; relation (3.14) is identical to the one given in Werrett's lectures in this volume. For $2A^2B \ll 1$, relation (3.14) possesses stable solutions for the low transmittivity branch of the hysteresis loop; the high transmittivity branch was found piecewise unstable towards self-pulsation [16,17].

In more careful analysis, along the lines of the previous section it was found that for arbitrary values of $|A|^2$ the behavior was more complex. It was found that the stationary solutions of (3.14) are unstable for $A^2B \gg 1$ and that the iterates $E_n = E(nt_R)$ of the mapping (3.13) undergo successive bifurcations and period doubling and finally behave in a chaotic way associated with the existence of a strange attractor whose rough shape is the spiral [3,4]

$$|\mathcal{E}-A|^2 = B^2\{\arg(\mathcal{E}-A) + \phi_o\} \tag{3.15}$$

for small B. This transition to chaos was termed delay-induced chaos and to substantiate it we turn back to the equations (3.10) and (3.11) which describe the dynamic evolution of the system.

We assume $\tau \ll t_R$ and $B \ll 1$ with $A^2 B$ kept fixed. Then this set of equations can be approximated by a one-dimensional flow

$$\tau\dot{\phi}(t) = -\phi(t) + f(A^2;\ \phi(t-t_R)) , \tag{3.16}$$

where

$$f(A^2;x) = A^2(1+2B\cos(x-\phi_o)) . \tag{3.17}$$

It was found that as $\mu = A^2$ is increased beyond a certain value μ_A, the stationary

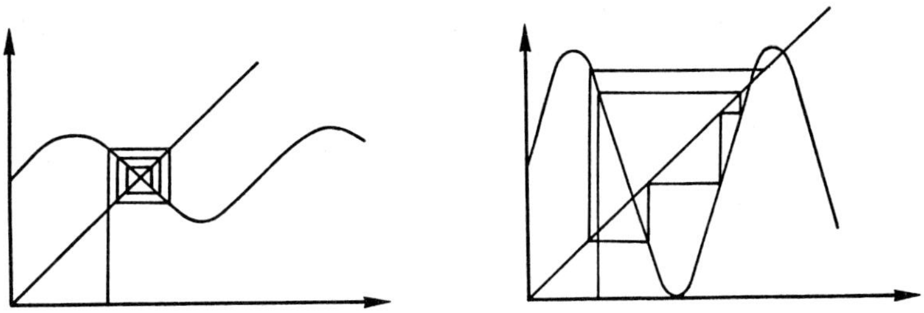

Fig. 4 Graphical representation of a stable and an unstable behavior in the nonlinear passive Fabry-Perot cavity

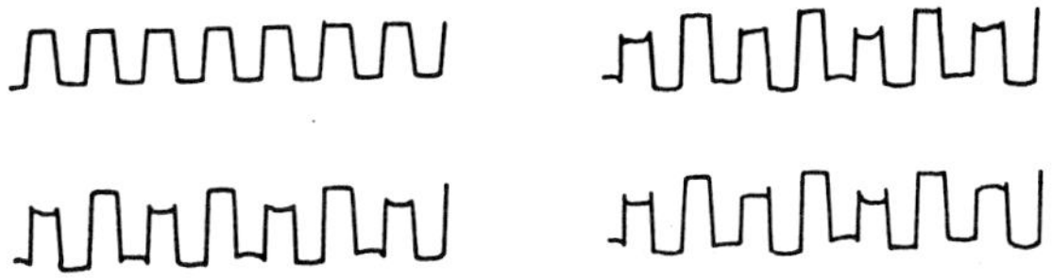

Fig. 5 Period doubling route to chaos in the nonlinear passive Fabry-Perot cavity (from ref. 6)

solution (3.14) becomes unstable and a square wave of period $T_0=2t_R$ appears and as μ is further increased the pattern outlined in the previous section sets in: the period of the square wave successively doubles $T_0 \to 2T_0 \to 2^2T_0 \to$ .... and a certain value μ_F the period becomes infinite so that the wave cannot recover at any later instant its initial form. [3-6] (see Figs 4 and 5).

For $\mu > \mu_F$ the form of this chaotic wave still preserves coarse features of a square wave of period T_0 (square wave chaos). As μ is increased well above μ_F then a new chaotic behavior sets in, termed developed chaos, where now the period changes discontinuously in the sequence $T_0 \to T_0/3 \to T_0/5 \ldots T_0/n_{max}$ where n_{max} is an odd integer which increases with t_R/τ. Thus higher harmonics now appear in the power spectrum. As μ increases beyond a value μ_B, the fully developed chaos established itself and the power spectrum becomes a smooth function of frequency. At the same time the phase $\phi(t)$ becomes a random function (see Figs 5-7).

In the state of fully developed chaos for $t \gg \tau/2\pi\mu B$ the motion of $\phi(t)$ becomes almost a gaussian process which can be explained easily in terms of the equation

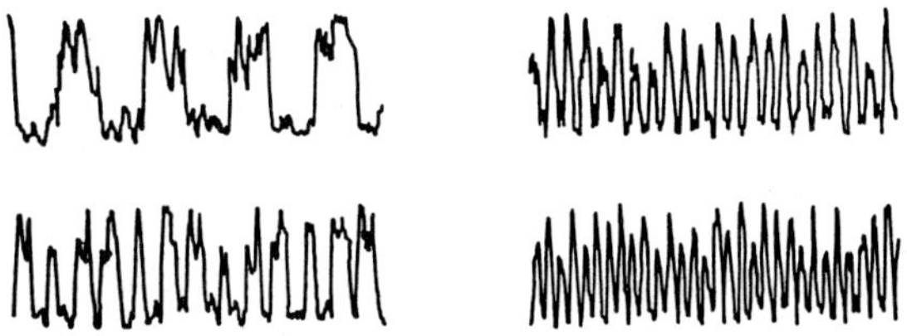

Fig. 6 Appearance of higher harmonics in developed chaos (from ref. 6)

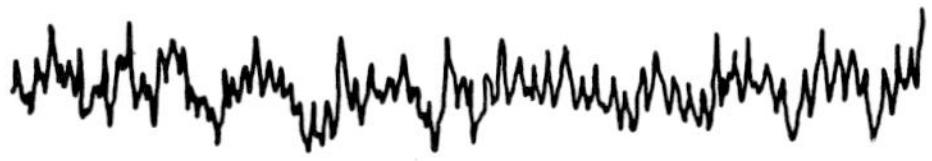

Fig. 7 Evolution in time of the phase in the fully developed chaos (ref. 6)

$$\phi(t) = \frac{2\pi\mu B}{\tau}\int_{-\infty}^{t} e^{-(t-s)/\tau}\cos(\phi(t-t_R)-\phi_o)ds + \text{const} , \qquad (3.18)$$

which is the integral transform of (3.11); for large μ the fluctuation of ϕ is much enhanced which in turn induces rapid changes in the integrand over very short and uncorrelated time intervals and by the central limit theorem, ϕ being a superposition of such short times, its motion becomes gaussian. For times $t \ll \tau/2\pi\mu B$, however, the motion is not gaussian as was shown by calculating the moments of ϕ. [4,6]

The delay induced chaos occurs when $t_R \gg 1$ and $A^2B \gg 1$ which, according to (3.18) implies large changes in the phase ϕ ($\phi \gg 1$). In the other extreme $t_R/\tau \ll 1$ and $\phi \ll 1$ it has been predicted that a quite different chaotic behavior occurs dominated by the relaxation time τ and the longer coherence in the polarization and has been termed nutation chaos as the relevant equations in this case reduce to Bloch type equations in contrast to the previous case which was dominated by the propagation process inside the cavity and the delay that the beam suffers [5,6].

A very interesting situation connected with the above nonlinear interaction configuration is the case where the Fabry-Perot cavity is replaced by a monomode optical fiber of length L pumped by a periodic train of pulses when the group velocity dispersion is taken into account. The nonlinear propagation of a pulse due to the combined effect of the Kerr effect and the group velocity dispersion is described by the nonlinear Schrödinger equation

$$i\frac{\partial\phi}{\partial z} + \alpha\frac{\partial^2\phi}{\partial t^2} + |\phi|^2\phi + i\Gamma\phi = 0 ,$$

where Γ is the loss and α the sign of the group velocity dispersion. When an input field [18]

$$\phi(t,z=0) = A\ \Sigma\text{sech}(t-nt_R) ,$$

where t_R is the round trip time synchronized to the cavity length, is introduced into such a cavity with $\alpha > 0$ and A is monotonically increased then for $t_R \gg 1$ (no pulse overlap in cavity) the output field after a transient regime shows a succession of period doubling bifurcations which starts at $A \simeq 1.526$ and accumulates in $\simeq 1.97$. Beyond this limit it shows a chaotic behavior.

Experimentally, the situation is not as clear and at presently the few indications concerning the occurrence of chaos in nonlinear optical systems do not follow exactly the above pattern. For instance the first steps of period-doubling have been observed in a number of recent experiments [7-9,19,20] following the predictions of ref. [3] but the subsequent pattern does not completely fit into this theory and more investigations are needed to clarify the situation in the passive nonlinear Fabry-Perot cavity. Recently the occurrence of instabilities and transition to chaos in active nonlinear Fabry-Perot cavities (lasers) have received much attention too [21,22].

4. Intrinsic instabilities and chaos

The occurrence of the delay-induced instabilities and chaos discussed in the previous section was conditioned by the cavity but this is not a prerequisite for such a behavior. Such a situation can also arise from the intrinsic nonlinearities in the equation of motion of the charges in a dielectric. It has been predicted that when the anharmonicity in the equation of motion of the bound charges or collective modes is fully taken into account, bistable operation will be possible [23,24] and a hysteresis will show up for relatively low intensities of the electric field; for still higher intensities, instabilities and transition to chaos sets in along the lines discussed before [23,25]. This is because the polarization state of the medium can be a multivalued function of the electric field with some of these values becoming unstable and leading thus to instabilities and chaos.

To fix the ideas the effect of the anharmoncity in the classical equation of motion will be introduced with the help of the Duffing oscillator equation of a damped anharmonic oscillator which in the presence of an external sinusoidal driving force reads

$$\ddot{Q} + \frac{1}{T}\dot{Q} + \omega_0^2 Q + \gamma Q^3 = \frac{e}{m}(\mathcal{E}(z)e^{-i\omega t}). \quad (4.1)$$

The response of diverse systems can be modelled on the basis of equation (4.1): bound electrons in dielectrics, molecular vibrations, phonons and excitons in crystals etc. In (4.1) T is a phenomenological relaxation time, e /m is the coupling strength of the mode amplitude Q with the external field. The amplitude at the fundamental frequency we are interested in, is of the form

$$Q = \frac{e}{m}\,\mathrm{Re}\{R(z)e^{-i\omega t}\} \quad (4.2)$$

and if we neglect the higher harmonics the amplitude R of the fundamental component is determined from the equation

$$R(z) = \mathcal{E}(z)/\{(\Delta_2 + \alpha_1/2) - i\omega/T\}, \quad (4.3)$$

where $\Delta_2 = \omega^2 - \omega_0^2$ and $\alpha_1 = 3\gamma|R|^2/2$ (note that both Δ_2 and γ can be positive or negative).

For fixed frequency ω it is clear from (4.3) that the system exhibits hysteresis and bistability. The critical fields where the jumps occur can be easily obtained [23] . Quite revealing is also the trajectory of the maximum oscillator amplitude as a function of ω for various field strengths. From (4.3) we see that for $\gamma = 0$, one simply obtains

$$R_0(z) = \mathcal{E}(z)/(\Delta_2 - i\omega T), \quad (4.4)$$

which is the solution of an harmonic oscillator of constant frequency ω_0 while for arbitrary γ , from (4.3) one has

$$R(z) = \mathcal{E}(z)/\{(\Delta_2 + \alpha_1/2) - i\omega T\}, \quad (4.5)$$

which is similar in form to (4.4) but with an apparent frequency that changes with the amplitude. This is a consequence of the anharmonicity in (4.1) and leads to bistable behavior and instabilities.

For weak fields and amplitudes R the motion can be analysed in terms of simple limit cycles and fixed points. As the electric field becomes large the motion becomes much more complicated and certain aperiodic solutions appear that are very sensitive to the initial conditions [25]. This behavior is particular easy to visualize when $|R|^2$ is depicted as a function of ω/ω_0 for given intensities of the electric field that exceed certain critical values. For $\omega/\omega_0 > 1$ the solutions correspond to limit cycles but as ω/ω_0 decreases below 1 then a set of cascading

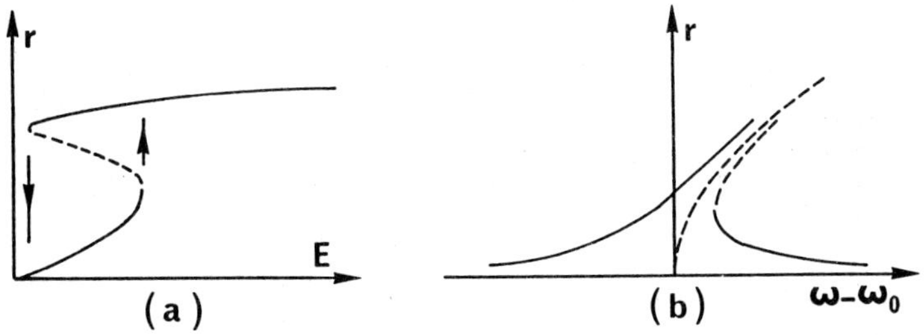

Fig. 8 Bistability and hysteresis in the intrinsic case (from ref. 23)

Fig. 9

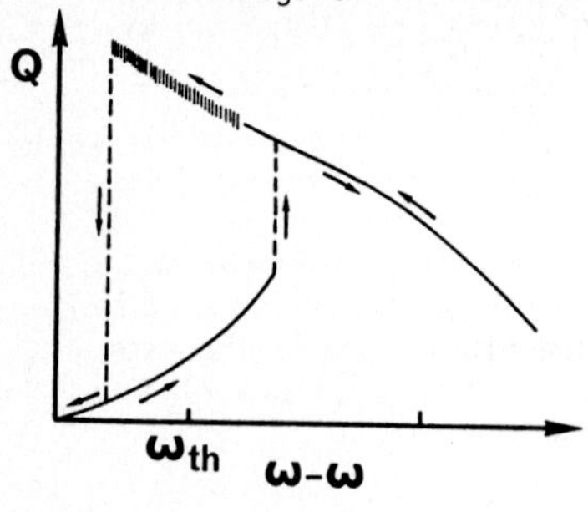

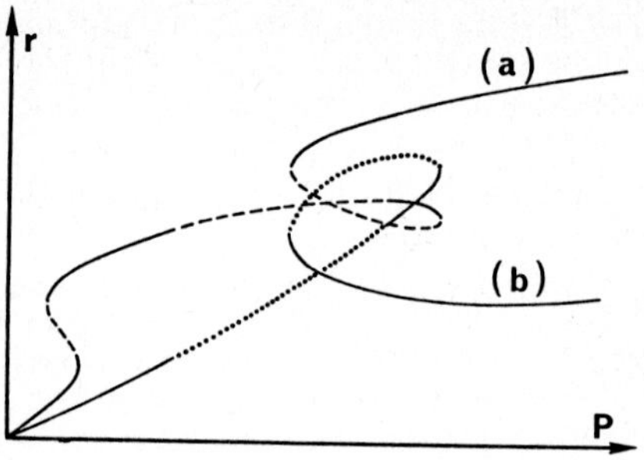

Fig. 10

Fig. 9 Period doubling transition to chaos in the intrinsic case Duffing nonlinear oscillator (from ref. 25)

Fig. 10 Hysteresis and instabilities in (a) the fundamental and (b) the third harmonic of the forced Duffing oscillator

bifurcation at frequencies ω_n starts which ends at a frequency ω_{th} where a chaotic state is established characterized by the appearance of strange attractor in phase space [25]. Quite revealing is also the power spectrum which shows sharp peaks for $\omega > \omega_{th}$ but becomes smooth as $\omega < \omega_{th}$ (see Figs 8 and 9).

Quite interesting is also the behavior of the harmonics of ω depicted in Fig. 10, and as can be seen there hysteresis and instabilities occur there as well; this occurs at precisely the values of $\mathcal{E}$ where R, the amplitude of the fundamental, becomes unstable.

The above behavior cannot be observed directly since the propagation of the field generated by the nonlinear polarization inside the medium must also be obtained. This is a very complex problem and only approximate solutions have been obtained in the case of the bistable stationary regime, namely the regime where (4.3) is valid [24,26] . For higher intensities the situation drastically changes leading to huge numerical difficulties.

Of course, the question that arises is to what extent the previous behavior, based on the classical description of the motion, also persists when the quantum mechanical description is used. It has been shown that the quantum mechanical Duffing oscillator also exhibits bistable behavior [27,28] but no indications have been obtained yet concerning the occurrence of instabilities there. The occurence of chaos in quantum mechanical systems is a problem of considerable fundamental interest and is actively studied presently [29].

5. Instabilities in intracavity four wave nonlinear interactions

The occurrence ob bistable operation, instabilities and transition to chaos in the beam transmitted through a nonlinear Fabry-Perot cavity results from the mutual interaction of the counterpropagating beams and the enhancement of this interaction through the feedback provided by the cavity. It is of outmost importance therefore to single out just this interaction and probe its dynamics in a more direct way than through the transmitted beam. This can be done by the optical phase conjugation technique [30, 31] or its variations.

The principle and the configuration scheme used in the optical phase conjugation have been analysed elsewhere in this volume. As represented in Fig. 11, two intense pump beams of same frequency ω and amplitudes $\mathcal{E}_p$ and $\boldsymbol{\mathcal{E}}_p$ counterpropagate along the z-axis and constitute the total pump field

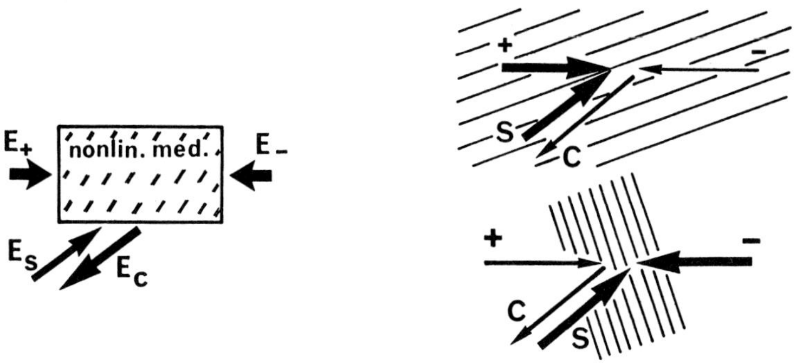

Fig. 11 Beam configuration in optical phase conjugation (right) and the two spatial gratings (left)

$$E = \mathrm{Re}\{\mathcal{E}_p e^{ikz-i\omega t} + \mathcal{E}_p\, e^{-ikz-i\omega t}\} \tag{5.1}$$

inside a medium which for convenience is assumed optically isotropic and characterized by scalar linear and third-order susceptibilities, $\chi^{(1)}$ and $\chi^{(3)}$ respectively. The mutual interaction of these two beams is interrogated by a third beam of same frequency ω propagating along a ζ -axis different from the z-axis.

$$E_s = \mathrm{Re}\{\mathcal{E}_s e^{ik\zeta-i\omega t}\}\,. \tag{5.2}$$

For moderately intense beams, the polarization induced at frequency ω jointly by the three beams is

$$P_c^{(3)}(\omega) = \mathrm{Re}\{\varepsilon_0\chi^{(3)}(\omega,-\omega,\omega)\mathcal{E}_p\mathcal{E}_p\,\mathcal{E}_s^{*}e^{-ik\zeta-i\omega t}\} = \mathrm{Re}\{\mathcal{P}^{(3)}e^{-ik\zeta-i\omega t}\}\,, \tag{5.3}$$

which will radiate a beam of electric field amplitude $\mathcal{E}_c$ propagating counter to $\mathcal{E}_s$ along the ζ-axis

$$E_c = \mathrm{Re}\{\mathcal{E}_c e^{-ik\zeta-i\omega t}\}\,. \tag{5.4}$$

This is the direction of perfect phase matching and therefore the process is very efficient. The amplitudes $\mathcal{E}_s$ and $\mathcal{E}_c$ in the slow varying envelope approximation with the pumps $\mathcal{E}_p$ and $\mathcal{E}_{p'}$ undepleted, satisfy the equations[30]

$$\frac{\partial\mathcal{E}_c}{\partial\zeta} = i\kappa\,\mathcal{E}_s^{*}\,, \qquad \frac{\partial\mathcal{E}_s}{\partial\zeta} = i\kappa\,\mathcal{E}_c\,, \tag{5.5}$$

with $\kappa = k\chi^{(3)}\mathcal{E}_p\mathcal{E}_p/2$ which can be easily solved but the main physical results can be simply interpreted in terms of the optical gratings which are induced in the medium: each beam is scattered by the grating formed by the other two and contributes to $\mathcal{E}_c$. These contributions correspond to the three terms in

$$\mathcal{P}_c^{(3)} = \chi_{ps}^{(3)}(\mathcal{E}_p\cdot\mathcal{E}_s^{*})\mathcal{E}_{p'} + \chi_{p's}^{(3)}(\mathcal{E}_{p'}\cdot\mathcal{E}_s^{*})\,\mathcal{E}_p + \chi_{pp'}^{(3)}(\mathcal{E}_p\cdot\mathcal{E}_{p'})\mathcal{E}_s^{*}\,, \tag{5.6}$$

which can be derived from (5.3) by appropriate vector operations. The two first terms correspond to spatial optical gratings of small and large spacings respectively and the third term in (5.6) is a temporal grating and is related to the so-called self-diffraction [30] (see also fig. 11).

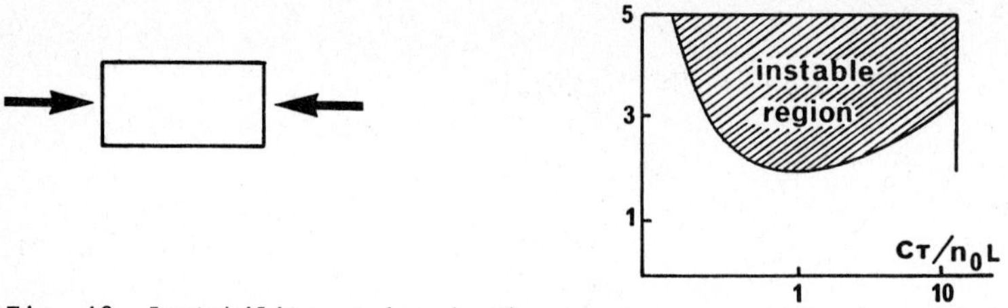

Fig. 12 Instability region in the counterpropagating beam interaction (ref. 33)

As it was shown first by Kukhtarev et al.[32] the stationary solutions of (5.5) are unstable when the temporal variations of the amplitudes and of the Kerr susceptibility as described by Debye type equations (compare equ (3.5)) are included in (5.5). Such a behavior was also found in more recent calculations by Silberberg and Bar Joseph [33] in the extreme case of degenerate four wave interaction where the pump $\mathcal{E}_p$ and the signal $\mathcal{E}_s$ coincide in a single field and similarly for $\mathcal{E}_p$ and $\mathcal{E}_c$ in which case only the self-diffraction (temporal) grating in (5.6) is effective.

In these two calculations and similar ones, the beams were propagating in a nonlinear medium of length L not inserted in a cavity; in particular no bistable operation or related behavior can take place as long as only the third-order polarisation (5.6) is assumed to be induced in the nonlinear medium. It was suggested[34] that the situation will drastically change when the nonlinear medium is inserted in a Fabry-Perot cavity; in such a configuration it was shown that the intracavity phase conjugated reflectiviy may show bistable operation, instabilities or transition to chaos whenever such effects occur in transmission. More recent extensions and experimental demonstration of the nearly degenerate four wave interaction by Nakajima and Frey[35] indicate that such configurations provide powerful techniques to directly probe the beam interaction inside the nonlinear Fabry-Perot cavity and selectively study the transition to chaos in frequency domain. We analyse below only the intracavity optical phase conjugation scheme [34].

In this scheme the mutual interaction of the forward and backward traveling beams, $\mathcal{E}_+$ and $\mathcal{E}_-$ respectively, that are set up inside the nonlinear Fabry-Perot cavity by an external incident pump beam (compare section 3) is interrogated with a weak external beam propagating in the direction ζ and its phase conjugated beam is studied as a function of E_0. The interaction configuration is shown in Fig.13 The total field inside the cavity is $E = E_p + E'$, where

$$E_p = \mathrm{Re}\{\mathcal{E}_+ e^{ikz} + \mathcal{E}_- e^{ikz}\} e^{-i\omega t}\} = \mathrm{Re}\{\mathcal{E}_p e^{-i\omega t}\} \tag{5.7}$$

$$E' = \mathrm{Re}\{\mathcal{E}_s e^{ik\zeta} + \mathcal{E}_c e^{ik\zeta}\} e^{-i\omega t}\} = \mathrm{Re}\{\mathcal{E}' e^{-i\omega t}\}. \tag{5.8}$$

We only consider the stationary regime so that the total polarization $P = \mathrm{Re}\{\mathcal{P} e^{-i\omega t}\}$ can be written

$$\mathcal{P} = \varepsilon_0 \chi(|\mathcal{E}|^2)\mathcal{E}. \tag{5.9}$$

We further assume that $|\mathcal{E}_+|, |\mathcal{E}_-| >> |\mathcal{E}_c|, |\mathcal{E}_s|$ so that we may expand $\chi(|\mathcal{E}|^2)$ to first order in $\mathcal{E}'$ around $\mathcal{E}_p$ and retain terms linear in $\mathcal{E}'$,

$$\chi(|\mathcal{E}|^2) = \chi(|\mathcal{E}_p|^2) + \mathcal{E}_p \frac{\partial \chi}{\partial \mathcal{E}_p} \mathcal{E}' + \mathcal{E}_p \frac{\partial \chi}{\partial \mathcal{E}_p^*} \mathcal{E}. \tag{5.10}$$

The propagation equation gives

$$(\nabla^2 + k^2)\mathcal{E} = -k^2 \chi(|\mathcal{E}|^2)\mathcal{E} \tag{5.11}$$

and indentifying terms of same order in $\mathcal{E}'$ we obtain

$$(\nabla^2 + k^2)\mathcal{E}_p = -k^2\chi(|\mathcal{E}_p|^2)\mathcal{E}_p \tag{5.12}$$

$$(\nabla^2 + k^2)\mathcal{E}' = -k^2\{\{\chi + |\mathcal{E}_p|^2\chi'\}\mathcal{E}' + \mathcal{E}_p^2\chi'^*\mathcal{E}'^*\}, \tag{5.13}$$

where $\chi' = \partial\chi/\partial|\mathcal{E}|^2$. Expanding χ and χ' in Fourier series and introducing in (5.12) and (5.13) the slow varying envelope approximation, one obtains [34,35]

$$\frac{\partial\mathcal{E}_+}{\partial z} = \frac{ik}{2}\{\chi_0 + \left|\frac{\mathcal{E}_-}{\mathcal{E}_+}\right|\chi_1\}\mathcal{E}_+ \tag{5.14}$$

$$\frac{\partial\mathcal{E}_-}{\partial z} = -\frac{ik}{2}\{\chi_0 + \left|\frac{\mathcal{E}_+}{\mathcal{E}_-}\right|\chi_1\}\mathcal{E}_- \tag{5.15}$$

$$\frac{\partial\mathcal{E}_s}{\partial\zeta} = -\alpha\mathcal{E}_s + i\kappa\mathcal{E}_c^* \tag{5.16}$$

$$\frac{\partial\mathcal{E}_c^*}{\partial\zeta} = \alpha^*\mathcal{E}_c^* + i\kappa\mathcal{E}_s \tag{5.17}$$

with $\alpha = -ik(\chi_o + \chi_o)/2$, $\kappa = k\chi/2$ and $\chi_0 = \langle\chi\rangle_z$, $\chi_0 = \langle\chi\cos 2kz\rangle_z$, $\chi_0 = \langle|\mathcal{E}_p|^2\chi\rangle_z$ where $\langle\ldots\rangle_z$ denotes spatial averaging. These four equations must be solved with the boundary conditions

$$\mathcal{E}_+(0) = \sqrt{R}\,\mathcal{E}_-(0) + \sqrt{1-R}\,\mathcal{E}_0 \tag{5.18}$$

$$\mathcal{E}_-(L) = \sqrt{R}\,\mathcal{E}_+(L)e^{2ikL} \tag{5.19}$$

$$\mathcal{E}_s(0) = \mathcal{E}_s \tag{5.20}$$

$$\mathcal{E}_c(\ell) = 0\ , \tag{5.21}$$

where L is the cavity length, R is the mirror reflectivity and ℓ is the length of the nonlinear medium along the ζ-axis; $\mathcal{E}_0$ is the incident field outside the cavity. Equations (5.14,5.15) together with (5.18,5.19) give a relation between $|\mathcal{E}_0|^2$ and $|\mathcal{E}_+|^2$ (or $|\mathcal{E}|^2$) which under suitable conditions can have stable solutions; these are the conditions for the occurrence of the optical bistability discussed in section 3. With these solutions one can then proceed to solve (5.16) and (5.17) with the boundary conditions (5.20) and (5.21). This can be done numerically but under certain conditions analytical expressions can be obtained as well: the objective is to obtain the phase conjugated reflectivity

$$R_c = \left|\frac{\mathcal{E}_c(0)}{\mathcal{E}_s(0)}\right|^2 .$$

The calculations were done for the absorptive one- and two-photon resonant cases respectively [34] ; as shown in Fig. 13 one obtains bistable reflectivity. The two cases drastically differ because of the dynamic Stark shift effect in the case of two-photon resonanc configuration. One can also compute the behavior of the phase conjugated signal when the solutions of (5.14) and (5.15) are not stable and the system undergoes the period-doubling route to chaos as discussed in sections 2 and 3; this calculation is quite involved.

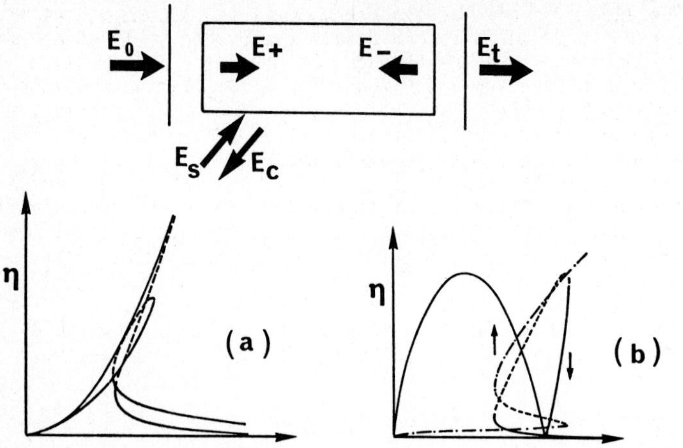

Fig. 13 Intracavity optical phase conjugation, hysteresis loops in (a) one-photon resonant case and (b) two-photon resonant case (ref. 34)

The above behavior can occur in a number of different situations and in particular in the degenerate four interaction in a nonlinear medium with a distributed feedback (Winful and Marburger [36]) which has the same effect as the cavity in the previous case: indeed when the interaction inside the Fabry-Perot cavity is unfolded, the successive round trips there play the same role as the spatial periodicity in the ditributed feedback configuration and a complete analogy between the two can be drawn.

Actually the presence of the cavity or any other spatial periodic conditions is not a prerequisite for the occurrence of bistability or instabilities in the degenerate four wave interaction. It has been shown [23] that the intrinsic nonlinear equations of motion of the charges or the collective modes in a dielectric (see section 4) lead to intrinsic instabilities in the degenerate four wave interaction. In particular the instabilities and transition to chaos discussed in sect. 4 in the transmission regime have their analogue in optical phase conjugated reflectivity [23]; their origin is a Bragg type reflection from a time grating (see Fig. 14).

A most elegant and far-reaching extension of the intracavity optical phase conjugation was proposed and experimentally demonstrated in a very beautiful experiment by Nakajima and Frey [35]; a semiconductor laser cavity was used in this experiment which consisted in using nearly degenerate four wave interaction in an amplifying medium. This sidesteps some very serious experimental problems connected with the off-axis geometry that is necessary in the exact degenerate case |34| and

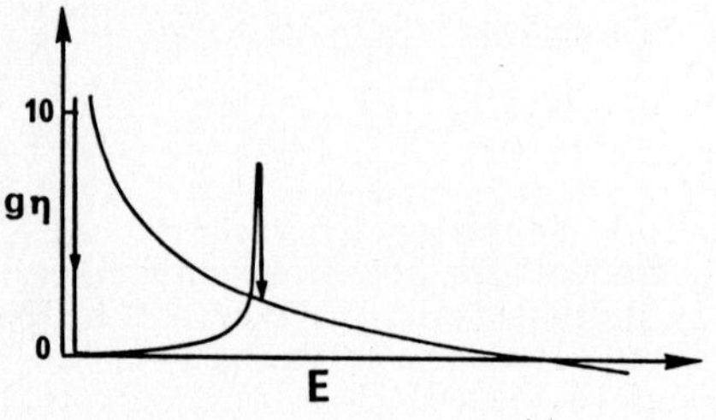

Fig. 14 Intrinsic instabilities and hysteresis in optical phase conjugation (from ref. 23)

introduces a new degree of freedom, namely the frequency. In their scheme a high intensity pump field of frequency ω and a low intensity probe beam of frequency $\omega - \delta\omega$ ($\delta\omega \ll \omega$) are sent propagating parallel through the nonlinear medium which exhibits saturable gain; because of the nonlinear four wave interaction inside the cavity, a phase conjugated beam at frequency at frequency $\omega + \delta\omega$ is generated and all beams can be analyzed independently because of their different frequencies. The theoretical analysis can be done as above with some additional complications due to the larger number of fields inside the cavity and the dynamic Stark effect which now plays a very important role. This technique opens new and far reaching possibilities for the study of the instabilities and transition to chaos in optical nonlinear systems; these can now be studied selectively in the frequency domain instead of in the time domain as has been the case up till now. Thus in the case of period-doubling route to chaos, by varying the mismatch $\delta\omega$, one can follow the appearance of the different frequencies $\Omega_n = 2^{-2}/T_0$ where T_0 is the fundamental period, which is $2t_R$ (the round trip time) in the case of the delay-induced chaos in a nonlinear cavity.

6. Conclusion

We have limited the previous discussion to some few cases which were chosen out of the immense and highly sophisticated literature on the subject because they can be treated without very extensive mathematical and numerical developments and also allow one to grasp the working of some quite general principles. A thorough discussion of the optical instabilities in general is beyond the scope of these lectures and besides loses much of its impact when presented separately from a discussion of the instabillities and chaos in nonlinear dynamical systems. But even with this limited and partial material at hand it is quite clear that the occurrence of instabilities and chaos is a very common feature in nonlinear optical interactions. It is a very fundamental one too which sets essential limitations on the performances of nonlinear devices and thus justifies the immense effort that is presently concentrated there. Despite this effort our understanding is still partial and only qualitative at most. Out of this situation only careful experimental work can allow us to establish unambiguously the actual ways that instabilities and chaos occur in nonlinear optical systems and exploit them.

Before closing the subject of these lectures let us summarize a few aspects common to the cases we considered previously. Throughout this discussion, the instabilities and chaos had their origin in the time evolution of the relative phase of the electric field with itself as it emerges from the nonlinear interaction and propagation inside the medium. Since only plane waves were used, all these effects occur along the propagation direction and thus can be termed longitudinal instabilities and chaos. The introduction of transverse effects, which are unavoidable because of the finite cross section of the real optical beams, substantially complicates the problem but also introduces some fundamentally new aspects related to spatial structures and chaos in contrast to the temporal ones considered above.

7. References

1. See for instance, H.G. Schuster, Deterministic Chaos, an Introduction (Physical Verlag, Weisheim 1984)
2. M.J: Feigenbaum, J. Stat. Phys. 19, 25 (1978); Physica. 7D, 16 (1983)
3. K. Ikeda, H. Daido & O. Akimoto, Phys. Rev. Lett. 45, 709 (1980)
4. K. Ikeda, K. Kondo & O. Akimoto, Phys. Rev. Lett. 49, 1467 (1982)
5. K. Ikeda,& O. Akimoto, Phys. Rev. Lett. 48, 617 (1982)
6. K. Ikeda, Journal de Physique, 44, C2-183 (1983)
7. H. Nakatsuka, S. Asaka, H. Itoh, K. Ikeda & M. Matsúoka, Phys. Rev. Lett. 50, 109 (1983)
8. R.G. Harrison, W.J. Firth, C.A. Emshary & I.A. Al-Saidi, Phys. Re . Lett. 51, 562 (1983)
9. R.G. Harrison, W.J. Firth & I.A: Al-Saidi, Phys. Rev. Lett. 53, 258 (1984)

!0. P. Manneville & Y. Pomeau, Phys. Lett. 75A 1 (1979); Physica, 1D, 219 (1979)
11. D. Ruelle & F. Takens, Comm. Math. Phys. 20, 167 (1971)
12. R.May, Nature(London) 261, 459 (1976)
13. F.S. Felber & J.H. Marburger, Phy. Rev. A17, 335 (1978)
14. R. Bonifacio & L.A. Lugiato, Opt. Comm. 19, 172 (1976)
15. K. Ikeda, Opt. Comm. 30, 257 (1979)
16. S.L. McCall, Appl. Phys. Lett. 32, 284 (1978)
17. R. Bonifacio, M. Gronchi & L.A. Lugiato, Opt. Comm. 30, 129 (1979)
18. K.T. Blow & N. J: Doran, Phys. Rev. Lett. 52, 526 (1984)
19. J.V. Moloney, F.A. Hopf & H.M. Gibbs, Phys. Rev. A25, 3442 (1982)
20. H.M. Gibbs, F.A. Hopf, D.L. Kaplan,& R.L. Shoemaker, Phys. Rev. Lett. 46, 474 (1981)
21. see for instance the review articles in Laser Physics, eds J.D. Harvey & D.F. Walls (Springer-Verlag, Berlin 1983)
22. F. T. Arecchi in Order and Fluctuations in Equilibrium and Nonequilibrium Statistical Mechanics eds G. Nicolis, G. Dewel & J.W. Turner (John Wiley, New York 1981)
23 C. Flytzanis & C.L. Tang, Phys. Rev. Lett. 45, 441 (1980); see also C. Flytzanis G.P. Agrawal & C.L. Tang in Lasers and Applications eds W.O.N. Guimares, C.T.Lin & A. Mooradian (Springer-Verlag, Berlin 1981) p. 317
24. J.A. Goldstone & E. Garmire, Phys. Rev. Lett. 53, 910 (1984)
25. B.A. Huberman & J.P. Crutchfield, Phys. Rev. Lett. 43, 1743 (1979)
26. L.A. Ostrowski, Zh. Eksp. Teor. Fis. 51, 1189 (1987) (Sov. Phys. JETP, 24, 797 (1967))
27. N.N. Rozanov & V.A. Smirnov, Pris'ma Zh. Eksp. Teor. Fiz. 33, 504 (1981) (JETP Letters, 33, 488 (1981))
28. B: Ritchie & C.M. Bowden, Phys. Rev. to be published
29. see for instance G. Casati, Chaotic Behavior in Quantum Systems, Theory and Applications (Plenum Press, New York 1984)
30. see for instance Optical Phase Conjugation, ed. R.A. Fisher (Academic Press 1983)
31. Proceedings of the International Workshop on Optical Phase Conjugation and Instabilities, Cargèse (Corsica) 1982, Journal de Physique, C2-1983
32. N.V. Kukhtarev & T.I. Semenets, Sov. J. Quant. El. 11. 1216 (1981); see also A. Borshch, M. Brodin, V. Volkov & N. Kukhtarev, Opt. Comm. 41, 213 (1982)
33. Y. Silberberg & I. Bar Joseph, Phys. Rev. Lett. 48, 1541 (1982); J. Opt. Soc. Am. B1,662 (1984)
34. G. P. Agrawal, C. Flytzanis, R. Frey & F Pradère, Appl. Phys. Lett. 38, 492 (1981)
35. H. Nakajima & R. Frey, Phys. Rev. Lett. 54, 1798 (1985)
36. H.G. Winful & J.H. Marburger, Appl. Phys. Lett. 36, 613 (1980)

Index of Contributors